Smart Nanomaterials Technology

Nanotechnology is a rapidly growing scientific field and has attracted a great interest over the last few years because of its abundant applications in different fields like biology, physics and chemistry. This science deals with the production of minute particles called nanomaterials having dimensions between 1 and 100 nm which may serve as building blocks for various physical and biological systems. On the other hand, there is the class of smart materials where the material that can stimuli by external factors and results a new kind of functional properties. The combination of these two classes forms a new class of smart nanomaterials, which produces unique functional material properties and a great opportunity to larger span of application. Smart nanomaterials have been employed by researchers to use it effectively in agricultural production, soil improvement, disease management, energy and environment, medical science, pharmaceuticals, engineering, food, animal husbandry and forestry sectors.

This book series in Smart Nanomaterials Technology (SNT) publishes cutting edge research monographs and edited volumes by top researchers and professionals across the globe. This book series aims to comprehensively cover topics in the fabrication, synthesis and application of these materials for applications in the following fields:

- Energy Systems—Renewable energy, energy storage (supercapacitors and electrochemical cells), hydrogen storage, photocatalytic water splitting for hydrogen production
- Biomedical—controlled release of drugs, treatment of various diseases, biosensors
- Agricultural—agricultural production, soil improvement, disease management, animal feed, egg, milk and meat production/processing
- Forestry—wood preservation, protection, disease management
- Environment—wastewater treatment, separation of hazardous contaminants from wastewater, indoor air filters

Review Process

The proposal for each volume is reviewed by the main editor and/or the advisory board. The chapters in each volume are individually reviewed single blind by expert reviewers (at least two reviews per chapter) and the main editor.

Ethics Statement for this series can be found in the Springer standard guidelines here—https://www.springer.com/us/authors-editors/journal-author/journal-author-helpdesk/before-you-start/before-you-start/1330#c14214.

Jen-Tsung Chen
Editor

AI-Driven Nanotechnology in Precision Medicine and Healthcare

Springer

Editor
Jen-Tsung Chen
Department of Life Sciences
National University of Kaohsiung
Kaohsiung, Taiwan

ISSN 3004-8273 ISSN 3004-8281 (electronic)
Smart Nanomaterials Technology
ISBN 978-981-92-1689-5 ISBN 978-981-92-1690-1 (eBook)
https://doi.org/10.1007/978-981-92-1690-1

This Springer imprint is published by the registered company Springer Nature Singapore Pte Ltd.
The registered company address is: 152 Beach Road, #21-01/04 Gateway East, Singapore 189721, Singapore

Preface

This book offers emerging knowledge of the current integration and coordination of two frontier approaches, i.e., artificial intelligence (AI) and nanotechnology, for a range of medical applications, which can be highly expected to advance precision medicine and personalized healthcare. It covers burning subtopics highlighting nanomedicines, nanorobotics, nanocarriers, nanosensors, and nanomotors, in their advanced versions driven by AI models, which drive the design and engineering of smart nanomaterials.

Nanomaterials in AI-driven smart forms can be applied as the system of "AI-enabled nanotechnology" to organize next-generation healthcare systems for advancing global public health. Undoubtedly, AI-enabled nanotechnology holds promise to be applied in medical fields, involving combat against various diseases, such as cancers, infectious diseases, inflammatory diseases, and autoimmune disorders, and was comprehensively discussed in this book for the emerging AI-assisted approaches of disease prediction, medical diagnostics, tissue engineering, cell therapy, gene therapy, and vaccine development. Additionally, this book shares opinions on safety and environmental risks in medical applications of nanotechnology and the integration of AI technology.

The readers include students, teachers, professors, researchers, experts, in the medical field and materials science, and those who are interested in or learning ways to develop precision medicine and next-generation healthcare.

Kaohsiung, Taiwan — Jen-Tsung Chen

Contents

Contributors

Oumayma Abbassi Laboratory for Agricultural Production Improvement, Biotechnology and Environment, Mohammed First University, Oujda, Morocco

Kazi Asraf Ali Department of Pharmaceutical Technology, Maulana Abul Kalam Azad University of Technology, Haringhata, West Bengal, India

Vijaya Anand Arumugam Department of Human Genetics and Molecular Biology, Bharathiar University, Coimbatore, Tamil Nadu, India

S. Aruna Sharmili Department of Biotechnology, Stella Maris College (Autonomous), Chennai, Tamil Nadu, India

Alwin Robert Asirvatham Department of Endocrinology and Diabetes, Prince Sultan Military Medical City, Ministry of Defense, Riyadh, Kingdom of Saudi Arabia

Salah-eddine Azizi Higher Institute of Nursing Professions and Health Techniques, Oujda, Morocco

Chahrazad Belkhiri Laboratory for Agricultural Production Improvement, Biotechnology and Environment, Mohammed First University, Oujda, Morocco

Bichitra Biswas Department of Pharmaceutical Technology, Maulana Abul Kalam Azad University of Technology, Haringhata, West Bengal, India

Mohammed Dalli Higher Institute of Nursing Professions and Health Techniques, Oujda, Morocco;
Laboratory of Bioresources, Biotechnology, Ethnopharmacology and Health, Faculty of Sciences, University Mohammed First, Oujda, Morocco

Shweta Dang Department of Biotechnology, Jaypee Institute of Information Technology, Noida, Uttar Pradesh, India

Debnath Das Department of Pharmaceutics, ISF College of Pharmacy, Moga, Punjab, India

Bandita Datta Department of Chemistry, Amity Institute of Applied Sciences, Amity University, Kolkata, India

Brijnandan Singh Dehiya Department of Chemical Engineering, Deenbandhu Chhotu Ram University of Science and Technology, Murthal, Haryana, India

Balaji Esakkimuthu Sudha Hospital, Hyderabad, Telangana, India

Reema Gabrani Department of Biotechnology, Jaypee Institute of Information Technology, Noida, Uttar Pradesh, India

Thilagavathi Gangadharan Department of Information Technology, Coimbatore Institute of Technology, Coimbatore, Tamil Nadu, India

Preeti Garg Hindu College of Pharmacy, Sonipat, Haryana, India

Rupesh K. Gautam Centre for Pharmacology, Amity Institute of Pharmacy, Amity University Uttar Pradesh, Noida, India

Ekambaram Gayathiri Department of Plant Biology and Plant Biotechnology, Guru Nanak College (Autonomous), Chennai, India

Joyeta Ghosh Department of Dietetics and Applied Nutrition, Amity Institute of Applied Sciences, Amity University, Kolkata, India

Mohini Goyal Department of Botany, Kurukshetra University, Kurukshetra, India

Gratiela Gradisteanu-Pircalabioru eBio-Hub Research Center, University Politehnica of Bucharest—CAMPUS, Bucharest, Romania;
Research Institute of the University of Bucharest (ICUB), University of Bucharest, Bucharest, Romania;
Academy of Romanian Scientists, Bucharest, Romania

Hinde Hami Laboratory of Biology and Health, Faculty of Science, Ibn Tofail University, Kenitra, Morocco

Shama Anzum Ismail Department of Biochemistry, PSG College of Arts and Science, Coimbatore, Tamil Nadu, India

S. Jayashree Department of Biotechnology, Stella Maris College (Autonomous), Chennai, Tamil Nadu, India

Antony Prabhu Jeyabal Philomenathan Department of Human Genetics and Molecular Biology, Bharathiar University, Coimbatore, Tamil Nadu, India

Jyoti Hindu College of Pharmacy, Sonipat, Haryana, India

Sonia Kesavan Department of Biochemistry, Ethiraj College for Women (Autonomous), Chennai, Tamil Nadu, India

Manish Kumar KIET School of Pharmacy, Krishna Institute of Engineering & Technology (KIET), Ghaziabad Delhi NCR, Uttar Pradesh, India

Pawan Kumar Department of Biotechnology, Kurukshetra University, Kurukshetra, Haryana, India

Mohammed Amine Lafraxo Higher Institute of Nursing Professions and Health Techniques, Oujda, Morocco;
Laboratory of Biology and Health, Faculty of Science, Ibn Tofail University, Kenitra, Morocco

Bouchra Legssyer Laboratory for Agricultural Production Improvement, Biotechnology and Environment, Mohammed First University, Oujda, Morocco

Chirag Madan Hindu College of Pharmacy, Sonipat, Haryana, India

Hitesh Malhotra Faculty of Pharmaceutical Sciences, PCTE Group of Institute, Ludhiana, Punjab, India

Mahendrakumar Mani Department of Biotechnology, Guru Nanak College (Autonomous), Chennai, Tamil Nadu, India

Manoj Mani Department of Human Genetics and Molecular Biology, Bharathiar University, Coimbatore, Tamil Nadu, India

Uma Shankar Maurya Goel Institute of Pharmacy and Sciences, Lucknow, India

Irina Negut Plasma and Radiation Physics, National Institute for Lasers, Magurele, Romania

Palanisamy Prakash Department of Biotechnology, Vinayaka Missions Kirupananda Variyar Engineering College, Vinayaka Missions Research Foundation (Deemed to be University), Salem, Tamil Nadu, India

Ritu Raj Department of Biotechnology, Jaypee Institute of Information Technology, Noida, Uttar Pradesh, India

D. Rajesh Department of Microbiology, Chennai National Arts and Science College, Chennai, Tamil Nadu, India

Sanjit Kr. Roy Department of Pharmaceutical Technology, Maulana Abul Kalam Azad University of Technology, Haringhata, West Bengal, India

Meenu Saini Department of Physics, Capital University, Koderma, Jharkhand, India

Shobana Sampath Department of Biotechnology, Vel Tech Rangarajan Dr. Sagunthala R&D Institute of Science and Technology, Avadi, Chennai, Tamil Nadu, India

Nagarajan Sankaranarayanan Department of Chemistry, National Institute of Technology Manipur, Imphal, India

Jayashree Shanmugam Department of Biotechnology, Stella Maris College (Autonomous), Chennai, Tamil Nadu, India

Velayuthaprabhu Shanmugam Department of Biotechnology, Bharathiar University, Coimbatore, Tamil Nadu, India

Jitender Sharma Department of Biotechnology, Kurukshetra University, Kurukshetra, Haryana, India

Anil Sindhu Department of Biotechnology, Deenbandhu Chhotu Ram University of Science and Technology, Murthal, Haryana, India

Keerthana Sivakumar Department of Biotechnology, Vel Tech Rangarajan Dr. Sagunthala R&D Institute of Science and Technology, Avadi, Chennai, Tamil Nadu, India

Shivani Sivaskandhamohan Department of Biotechnology, Vel Tech Rangarajan Dr. Sagunthala R&D Institute of Science and Technology, Avadi, Chennai, Tamil Nadu, India

K. Sonia Department of Biochemistry, Ethiraj College for Women (Autonomous), Chennai, Tamil Nadu, India

Shruti Srivastava Directorate of Research, Innovation & Consultancy, Chandigarh University, Uttar Pradesh, India

Aruna Sharmili Sundararaj Department of Biotechnology, Stella Maris College (Autonomous), Chennai, Tamil Nadu, India

Sneha Thakur Department of Dietetics and Applied Nutrition, Amity Institute of Applied Sciences, Amity University, Kolkata, India

Sekar Vijayakumar Engineering Research Center of Environment-Friendly Functional Materials, College of Material Science and Engineering, Huaqiao University, Ministry of Education, Xiamen, People's Republic of China

Anita Visan National Institute for Lasers, Plasma and Radiation Physics, Măgurele, Romania

Anita Ioana Visan Plasma and Radiation Physics, National Institute for Lasers, Magurele, Romania

Bancha Yingngam Department of Pharmaceutical Chemistry and Technology, Faculty of Pharmaceutical Sciences, Ubon Ratchathani University, Ubon Ratchathani, Thailand

Integrating AI Models and Nanotechnology for Precision and Personalized Medicine: Current Achievements and Prospects

K. Sonia, S. Jayashree, S. Aruna Sharmili, D. Rajesh, and Sekar Vijayakumar

Abstract The synergistic combination of artificial intelligence (AI) and nanotechnology is reshaping the landscape of personalized medicine by enabling predictive, data-driven, and patient-specific therapeutic strategies. Nanotechnology offers unique opportunities to design nanoscale systems for sensitive diagnostics, targeted drug delivery, and controlled therapeutic release, while AI provides advanced computational tools capable of analyzing complex and heterogeneous biomedical datasets. Together, these technologies address the limitations of conventional treatment approaches by accounting for individual variability in genetics, disease biology, and therapeutic response. This chapter explores the role of AI-enabled nanotechnology in precision healthcare, with emphasis on disease detection, treatment planning, and personalized drug delivery. The use of machine learning and deep learning techniques in molecular imaging, biosensing, and digital pathology is discussed, highlighting their contribution to early diagnosis, accurate disease characterization, and real-time monitoring. The chapter further examines the use of AI-driven models to support pharmacogenomics and personalized drug development by integrating multi-omics data, clinical records, and real-world evidence to predict patient-specific drug responses and optimize drug strategies. In addition, AI-assisted modeling

K. Sonia
Department of Biochemistry, Ethiraj College for Women (Autonomous), Chennai, Tamil Nadu, India

S. Jayashree (✉) · S. Aruna Sharmili
Department of Biotechnology, Stella Maris College (Autonomous), Chennai, Tamil Nadu, India
e-mail: jayashreesanmugam2@gmail.com

D. Rajesh
Department of Microbiology, Chennai National Arts and Science College, Chennai, Tamil Nadu, India

S. Vijayakumar
Engineering Research Center of Environment-Friendly Functional Materials, College of Material Science and Engineering, Huaqiao University, Ministry of Education, Xiamen, People's Republic of China
e-mail: vijaysekar05@gmail.com

J. Chen (ed.), *AI-Driven Nanotechnology in Precision Medicine and Healthcare*, Smart Nanomaterials Technology, https://doi.org/10.1007/978-981-92-1690-1_1

aids in the rational design of nanoparticles with tailored physicochemical properties to improve therapeutic efficiency and minimize adverse effects. Emerging concepts such as intelligent nanodevices, adaptive delivery platforms, and AI-guided nanomedicine optimization are also addressed. The integration of AI with advanced therapeutic modalities, including gene therapy and precision oncology, is presented as a promising direction for future research.

Keywords Nanotechnology · Artificial intelligence · Personalized medicine · Gene therapy · and Precision diagnostics

1 Introduction

Nanotechnology and Artificial Intelligence (AI) represent two of the most transformative technological advancements of the modern era, with profound implications for healthcare innovation. Individually, both fields have demonstrated remarkable potential; however, their integration has opened new avenues for the development of precision and personalized medicine. The convergence of nanoscale materials with AI-driven analytical and predictive models enables the extraction of meaningful insights from complex biological data, facilitating more accurate diagnosis, targeted therapy, and patient-specific treatment strategies.

AI, a rapidly evolving branch of computer science, focuses on the creation of intelligent systems capable of learning, reasoning, and decision-making, thereby mimicking key aspects of human cognition [1]. When applied to nanotechnology-based medical platforms, AI algorithms can analyze large-scale biological and clinical datasets to guide the rational design of nanomaterials, predict their biological behavior, and optimize their interactions within physiological environments [2, 3]. This synergy is particularly valuable in precision and personalized medicine, where inter-individual variability in genetics, metabolism, immune response, and disease progression poses significant challenges to conventional therapeutic approaches [4].

Recent achievements in AI-generated nanomedicine include the development of smart diagnostic nanomaterials capable of generating patient-specific disease profiles, AI-assisted optimization of nanoparticle size, surface chemistry, and drug-loading efficiency, as well as predictive modeling of nanomaterial interactions with blood components, immune cells, and target tissues [5]. Despite these advances, the translation of AI-guided nanotechnologies into clinical practice remains complex due to biological heterogeneity, data integration challenges, and regulatory constraints.

This chapter explores the current achievements and prospects of integrating AI models with nanotechnology for precision and personalized medicine, highlighting emerging trends, technological challenges, and the potential of AI-enabled nanomedicine to transform next-generation healthcare.

2 AI and Nanotechnology in Precision Diagnostics

The convergence of nanotechnology and AI is reshaping precision diagnostics by enabling highly sensitive and patient-specific disease detection. Nano-enabled diagnostic platforms provide molecular-level information, while AI-driven analytical methods interpret complex biological data with high accuracy [6]. Together, these technologies support advanced diagnostic systems capable of early disease identification, real-time monitoring, and personalized clinical decision-making.

AI further enhances clinical diagnostics by improving the accurate identification and evaluation of complex diseases, including cancer, cardiovascular disorders, and neurodegenerative conditions. When integrated with nano-enabled diagnostic tools such as biosensors and advanced imaging agents, AI enables objective interpretation of intricate biological signals. Although conventional diagnostic approaches remain valuable, their reliance on subjective assessment can introduce variability and delay clinical decision-making limitations that AI-assisted nano-diagnostic platforms are well positioned to address [7].

Advances in AI have significantly improved the interpretation of medical images, with deep learning frameworks demonstrating high accuracy in the analysis of radiological datasets. Convolutional neural network-based models are particularly well suited for processing MRI, CT, and radiographic images, enabling the recognition of subtle disease-related patterns that may not be readily detected by human observers [8]. This enhanced analytical capability facilitates earlier diagnosis and improves the precision of clinical decision-making.

Nanotechnology facilitates the selective delivery of therapeutic agents to cancerous tissues, while AI enhances diagnostic accuracy and enables patient-specific treatment optimization [9]. Advanced AI algorithms are increasingly employed to predict optimal nanoparticle dimensions, surface characteristics, and morphology, thereby improving tumor penetration, cellular uptake, and overall therapeutic performance.

Yasaka et al. [10] reported that convolutional neural network-based deep learning models achieved superior accuracy in identifying liver tumors from CT images when compared with conventional diagnostic approaches.

Lai et al. [11] have shown that surface-charged gold nanoparticles can function as effective CT contrast enhancers for improving image resolution in small hepatocellular carcinoma. When combined with AI-based segmentation frameworks such as 3D U-Net and Trans U-Net, these nano-contrast agents enable precise tumor delineation and support accurate image analysis. This integrated approach has the potential to address major limitations in adaptive radiotherapy by improving target definition and treatment planning.

Conventional evaluation of neurodegenerative disorders using radiological images has largely depended on specialist interpretation of MRI, CT, and PET scans, a process constrained by observer-dependent variability and the difficulty in identifying subtle alterations during early disease stages. The introduction of AI, particularly machine learning and deep learning approaches, has markedly enhanced the accuracy and reliability of imaging-based diagnostics. These computational models enable automated identification of neuroanatomical changes, including brain volume loss, cortical thinning, white matter abnormalities, and characteristic metabolic patterns, often preceding the onset of clinical manifestations [12].

In Alzheimer's disease, deep learning-based MRI analysis has demonstrated superior performance in detecting hippocampal degeneration compared to traditional volumetric assessments. Similarly, in Parkinson's disease, AI-assisted imaging improves the recognition of nigrostriatal impairment, even during preclinical phases. Despite these advances, clinical translation remains limited by factors such as variability in imaging data across centers, the absence of standard acquisition protocols, and regulatory hurdles associated with approving AI-based diagnostic systems for routine clinical use [13].

Histopathology, which involves microscopic evaluation of tissue specimens, is an area where AI is demonstrating significant potential. Conventional examination of biopsy samples relies on manual assessment by pathologists, a process that is often labor-intensive and subject to inter-observer variability. In contrast, AI-enabled digital pathology platforms can rapidly and consistently assess histological images, facilitating accurate detection of pathological features such as malignant cells and inflammatory changes [14].

Flach et al. [15] investigate the incorporation of the AI-enabled system Paige prostate detects into routine clinical workflows for prostate cancer diagnosis. The potential of AI to enhance both diagnostic precision and efficiency in the evaluation of prostate biopsy samples has been assessed. Cone beam computed tomography (CBCT) guided online adaptive radiotherapy for postoperative esophageal cancer integrates deep learning and AI to enhance the accuracy and adaptability of radiation treatment delivery [16].

The integration of AI into remote patient monitoring (RPM) platforms substantially improves their diagnostic and predictive capabilities. AI-driven algorithms can process continuous data streams obtained from wearable devices to detect patterns, deviations, and early warning signals associated with disease progression. For instance, in patients with heart failure, AI models can analyze longitudinal data on body weight and fluid balance to predict clinical exacerbations, enabling timely medical intervention that may avert the need for hospitalization [17].

Nanosensor technologies enable the ultra-sensitive detection of disease-related biomarkers present at extremely low concentrations, supporting early and accurate diagnosis. In oncology, nanosensors have been applied to identify biomarkers such as CA 15–3, HER2, BRCA1/2, CEA, and CYFRA 21–1, while in neurological disorders they facilitate the monitoring of indicators including β-amyloid, tau protein levels, electroencephalographic changes, and thyroid hormone levels [18]. The incorporation of advanced nanomaterials like gold nanoparticles and graphene significantly enhances detection sensitivity, allowing identification of biomarkers that were previously below detectable limits, thereby improving early disease recognition.

3 AI and Nanotechnology to Predict Personalized Drugs

Variability in therapeutic outcomes represents a persistent challenge in modern pharmacology, as individuals frequently display different levels of efficacy, toxicity, or resistance when treated with the same drug. Such differences arise from complex interactions among genetic variation, molecular signaling pathways, disease heterogeneity, and physiological conditions. Conventional drug development and treatment strategies, which are largely guided by population-level averages, often fail to accommodate this biological diversity. Consequently, there is a growing emphasis on predictive approaches that can anticipate individual drug responses and support more precise, patient-aligned therapeutic decisions.

AI has emerged as a key enabler of predictive personalized drug strategies due to its capacity to analyze complex, high-dimensional biomedical data. Advanced machine learning (ML) and deep learning (DL) algorithms can identify nonlinear relationships and subtle patterns within large datasets that are difficult to capture using traditional analytical methods. In the context of personalized pharmacotherapy, AI-based models are increasingly used to predict drug efficacy, adverse reactions, and treatment resistance by integrating molecular, clinical, and phenotypic information. These predictive capabilities allow for more informed drug selection and optimization of therapeutic regimens. Nanotechnology provides complementary tools that enhance predictive drug research by enabling precise investigation at biologically relevant length scales. Nanoscale materials and systems facilitate improved characterization of drug properties, interactions with biomolecules, and disease-associated molecular features. In addition, nano-enabled and diagnostic platforms generate high-resolution biological data that can improve the accuracy and robustness of predictive models. By offering detailed insight into drug-biological interactions, nanotechnology strengthens the mechanistic foundation of personalized drug prediction.

The integration of AI and nanotechnology establishes a synergistic framework that advances predictive personalized drug development beyond empirical trial-and-error approaches. AI-driven computational models can interpret complex datasets generated through nanoscale technologies, while nanotechnology enhances the quality, specificity, and relevance of the biological inputs used for prediction. This interdisciplinary integration supports rational evaluation of therapeutic candidates and enables more accurate anticipation of patient-specific responses.

Importantly, predictive personalized drug strategies supported by AI and nanotechnology extend beyond therapeutic effectiveness to include safety and long-term treatment outcomes. Early identification of individuals at risk of adverse drug reactions or therapeutic failure allows for timely intervention and treatment modification. Such predictive insight is particularly valuable in complex and heterogeneous diseases, where effective therapy can lead to significant clinical consequences. By enabling risk stratification and response prediction, these technologies contribute to safer and more effective pharmacotherapy.

Within this context, personalized medicine aims to improve therapeutic precision by recognizing the genetic and biological differences among patients that influence their responses to treatment [19]. Unlike traditional "one-size-fits-all" approaches, personalized medicine tailors interventions based on an individual's genetic makeup, lifestyle, and other biological characteristics [20]. This approach goes beyond clinical treatment to encompass diagnostics, drug development, and drug delivery systems, offering opportunities to predict and prevent adverse drug reactions and optimize therapeutic efficacy [21]. Innovations may include modified drug formulations or the integration of companion diagnostic tools, allowing health care providers to deliver patient-specific therapies more effectively and efficiently [22].

AI has emerged as a transformative technology in personalized medicine, enabling the analysis of complex, high-dimensional datasets that traditional statistical methods cannot handle [23]. The growing importance of AI in this field is largely driven by the rapid expansion of genomics and other omics datasets, including proteomics and metabolomics [24]. ML and DL techniques, including support vector machines, random forests, and neural networks, have become central tools for predicting patient-specific drug responses. By integrating genomic, transcriptomic, proteomic, metabolomic, and clinical data, these models can identify biomarkers associated with drug efficacy and safety guide clinical decision-making, and optimize drug selection and dosing [25]. AI can also combine data from diverse sources—such as electronic health records, clinical trials, and real-world evidence to design individualized treatment strategies.

One of the most significant applications of AI in personalized medicine is pharmacogenomics, which examines how genetic variation influences individual drug responses. AI-based analysis of patient genetic information enables the prediction of therapeutic outcomes, assisting clinicians in selecting the most effective pharmacological agents and determining optimal dosing regimens. Predictive modeling has shown particular promise in oncology. Dercle et al. [26] demonstrated that ML models trained on treatment-specific quantitative features derived from longitudinal CT imaging could predict patient sensitivity to chemotherapy, targeted therapy, and immunotherapy in non-small cell lung cancer.

Similarly, Johannet et al. [27] integrated convolutional neural networks trained on histopathological images with clinical data to predict responses to checkpoint immunotherapy in advanced melanoma. In addition, Liu et al. [28] developed a logistic regression-based framework that incorporated genomic, transcriptomic, and clinical datasets to anticipate resistance to PD-1 inhibitor therapy in treatment-naïve melanoma patients. Together, these studies highlight the potential of AI to generate actionable insights for tailoring therapies to individual patients.

Nanotechnology further enhances personalized medicine by enabling precise and targeted delivery of therapeutic agents. Nanocarriers, including polymeric nanoparticles, liposomes, and lipid-based systems, can be engineered to deliver controlled drug doses to specific anatomical sites. When combined with AI-driven predictive modeling, nanotechnology enables the rational design of customized drug delivery platforms optimized according to patient-specific genetic profiles, disease characteristics, and treatment requirements. This integrated approach improves therapeutic outcomes, minimizes adverse effects, and represents a significant step toward patient-centric health care.

Beyond predicting patient responses, AI and ML are also transforming drug discovery and development. By analyzing large datasets of chemical structures, biological targets, and molecular interaction profiles, these approaches support virtual screening, target identification, and lead optimization, thereby accelerating and refining the drug development process [29]. The integration of AI-driven drug discovery with nanotechnology-based delivery systems facilitates the development of individualized nanocarrier-based therapies, establishing a new paradigm for personalized and precision medicine.

4 AI and Nanotechnology to Predict Personalized Drug Delivery

The integration of AI with nanotechnology has significantly advanced patient-specific drug delivery systems by enabling precise and targeted therapeutic interventions. Nanoparticles can be rationally engineered with adjustable size, surface characteristics, and biological compatibility to achieve controlled drug release and enhanced bioavailability. Such tailored delivery platforms improve treatment efficiency while minimizing non-specific distribution, thereby reducing systemic toxicity and associated side effects [30].

Nanotechnology has transformed advanced drug delivery through the development of diverse nanocarrier platforms that synergize effectively with AI-based predictive tools. Various nanocarriers, including liposomes, gold nanoparticles, and quantum dots, have shown considerable promise in enhancing the therapeutic performance of anticancer drugs while minimizing unintended toxicity to healthy tissues [31]. More recently, AI-enabled nanocarrier platforms have leveraged predictive modeling techniques to refine nanoparticle architecture, allowing precise control over parameters such as size, morphology, and surface properties to achieve targeted drug delivery. Collectively, these advances contribute to improved treatment efficacy and more efficient delivery of therapeutics to tumor sites [32].

Controlled drug delivery platforms are engineered to provide sustained release of therapeutics over prolonged durations, thereby decreasing dosing frequency and enhancing patient adherence. Targeted delivery strategies are designed to concentrate drugs at disease-specific sites while limiting unintended effects on healthy tissues. AI contributes significantly to the optimization of both sustained and targeted delivery systems by predicting how material selection and formulation parameters affect release behavior and targeting precision [33]. For instance, AI integrates in vitro release data, imaging outputs, and clinical evidence to optimize hydrogel, polymer, and nanoparticle formulations for controlled and targeted drug delivery.

The combination of AI with nanotechnology has enabled the development of intelligent nanocarriers that support real-time monitoring of adaptive therapeutic strategies and theranostic applications. Smart nanoparticles incorporating suitable imaging agents allow clinicians to track dynamic drug distribution and tumor responses, enabling AI-guided modifications of therapeutic regimens [34, 35].

Deep learning and machine learning, the two major branches of AI, play pivotal roles in this process. Deep learning enables the identification of complex data patterns and the prediction of interactions between nanocarriers and biological systems, while machine learning supports the estimation of nanocarrier physicochemical characteristics and biological behavior through analysis of structural attributes. Additionally, AI-assisted platforms can integrate multiple therapeutic modalities into a single intelligent system, enhancing treatment efficiency and enabling personalized management of diverse clinical conditions [36].

A range of ML algorithms, including support vector machines, random forest models, gradient boosting techniques (including XG Boost), and ensemble-based frameworks, have been effectively employed to predict key physicochemical and biological characteristics of nanocarriers [37]. Leveraging comprehensive experimental datasets, machine learning approaches can effectively model intricate nonlinear dependencies between synthesis variables and nanomaterial properties [38]. These predictive modeling approaches minimize the need for expensive and time-intensive experimental cycles while enabling prioritization of optimal formulations before preclinical and clinical evaluation [39].

Recent progress in deep learning has significantly enhanced predictive modeling by enabling the characterization of complex, high-dimensional datasets describing nanomaterial interactions within biological systems. In this context, convolutional neural networks (CNNs) and recurrent neural networks (RNNs) have employed to enhance predictive accuracy, linking key design variables, including polymer composition, ligand density, and surface charge, with biodistribution, cellular uptake, and therapeutic performance [40].

The growing application of AI in generative design enables the de novo generation of nanocarrier design. In this content, variational autoencoders (VAEs) and generative adversarial networks (GANs) capture latent representations of nanocarrier attributes and generate designs with optimized size, surface charge, and controlled release profiles [41]. In parallel, reinforcement learning frameworks conceptualize therapeutic optimization as a dynamic decision-making process, enabling iterative refinement of delivery strategies to maximize therapeutic efficacy while minimizing adverse off-target interactions.

Computational approaches that originated in materials discovery are now finding growing relevance in nanomedicine applications. GAN-based models have been employed to predict nanoparticle distribution within tumors, while variational VAE and diffusion-based models are increasingly used to navigate complex lipid and polymer design spaces. Additionally, protein design platforms such as AlphaFold and RFdiffusion support the development of receptor-specific ligands that can be functionalized onto nanocarriers to enhance targeting precision [42].

The integration of AI with Feedback System Control (FSC) frameworks further facilitates the rational optimization of combinatorial cancer chemotherapy. By continuously analyzing therapeutic response data, FSC-based AI systems guide nanocarrier-mediated drug delivery to improve therapeutic precision, efficacy, and adaptability [43]. Overall, AI-driven nanotechnology platforms enable molecular-level control over drug delivery, allowing precise spatial and temporal distribution of therapeutics and establishing an unprecedented level of accuracy in personalized drug delivery systems.

5 AI in Personalized Gene Therapy

The convergence of AI and gene therapy represents a significant step forward in precision healthcare, uniting the complementary capabilities of both technologies to improve clinical outcomes. AI has the potential to support gene therapy at multiple levels, including optimized patient identification, informed treatment planning, and continuous monitoring of therapeutic responses. By processing large and complex datasets such as genomic profiles, clinical histories, and imaging data, AI-driven models can accurately identify individuals most likely to respond to specific gene-based treatments. This targeted approach enhances therapeutic efficiency and contributes to a higher success rate in gene therapy interventions.

AI further enables individual optimization of gene therapy by continuously evaluating patient-specific responses over time. Through real-time data integration and advanced analytical models, AI systems can detect emerging response patterns and predict when treatment modifications or supplementary interventions may be required. This adaptive, patient-focused strategy facilitates highly customized care and enhances overall therapeutic outcomes. In addition to efficacy optimization, AI plays a critical role in improving the safety profile of gene therapies. By analyzing historical clinical data and prior treatment outcomes, AI-driven models can predict potential adverse reactions and identify individuals at increased risk of complications. These predictive risk assessment tools enable early warning signals and timely clinical interventions, thereby minimizing adverse events and improving the overall reliability and safety of gene therapy approaches [44].

Beyond its direct therapeutic use, AI-based approaches in genomics, pharmacogenomics, and gene editing serve as key facilitators in the development, refinement, and personalization of gene therapy strategies across a wide range of diseases. AI-based algorithms have also been applied in the clinical section of patients with Leber congenital amaurosis (LCA) carrying pathogenic RPE65 mutations who are suitable candidates for Luxturna gene therapy. Post treatment, AI-assisted analysis of optical coherence tomography (OCT) images facilitates the evaluation of retinal structural changes and therapeutic efficacy. Similarly, AI tools assist in identifying patients with retinitis pigmentosa harboring RPGR mutations for targeted gene-based therapies. Automated retinal image analysis further enables continuous monitoring of photoreceptor integrity and disease progression [45].

The integration of AI with Pharmacogenomics and Clustered Regularly Interspaced Short Palindromic Repeats (CRISPR) technologies has strengthened precision medicine by enhancing the prediction of drug-gene interactions, improving gene-editing accuracy and enabling advanced modeling of therapeutic outcomes. AI-driven algorithms support the rational design of CRISPR guide RNAs, minimizing off-target modifications while increasing editing efficiency. At the same time, pharmacogenomic data guide the selection for CRISPR-based strategies tailored to individual genetic profiles. This multidisciplinary convergence is accelerating the development of AI-assisted gene therapies and pharmacogenomics-informed drug discovery, thereby reshaping the future of personalized treatment approaches [46].

Pharmacogenomics focuses on understanding how genetic variation influences drug efficacy, metabolism, and safety, thereby supporting personalized treatment strategies. CRISPR technology offers a highly precise method for genome editing, enabling the correction of disease-causing mutations and functional validation of genetic variants. AI acts as an integrative platform by analyzing complex multi-omics datasets, predicting gene-editing outcomes, and optimizing drug-target interventions that enhance therapeutic effectiveness while minimizing off-target effects and adverse outcomes [19]. Genome-wide CRISPR-based knockout and activation screening studies have facilitated the identification of synthetic lethal interactions in cancer, supporting the discovery of novel therapeutic targets for precision oncology. When combined with AI-driven predictive modeling, data generated from a CRISPR screening platform can accelerate drug discovery pipelines and support the development of targeted therapies for genetically stratified patient populations [47].

The synergistic application of AI and CRISPR technologies is also redefining cancer immunotherapy and personalized oncology by enhancing immune checkpoint targeting, optimizing CAR-T cell engineering, and regulating the tumor microenvironment. AI-based patient stratification models can predict responsiveness to immune checkpoint inhibitors, including PD-1 and CTLA-4 therapies, thereby guiding treatment selection and improving clinical outcomes. Additionally, AI-driven optimization of CAR-T cell engineering and modulation of the tumor microenvironment further enhances therapeutic efficacy in cancer treatment [46].

AI-powered single-cell transcriptomic analysis and T-cell receptor sequencing enable the identification of optimal tumor antigens for CAR-T cell therapies, thereby enhancing both specificity and therapeutic effectiveness. Furthermore, AI-integrated CRISPR knockout screening approaches have uncovered key mechanisms underlying immune evasion in tumors, informing the development of next-generation cancer immunotherapies [48].

Integrating multi-omics datasets is crucial for elucidating disease mechanisms, understanding drug responses, and mapping gene–environment interactions. Advanced AI-based computational frameworks facilitate the combined analysis of genomic, transcriptomic, proteomic, and metabolomic data, offering a comprehensive review of disease progression and therapeutic outcomes. When paired with CRISPR-mediated gene perturbation studies, AI-driven multi-omics analyses can identify novel drug targets and validate pharmacogenomic biomarkers. Additionally,

AI-enabled single-cell multi-omics approaches reveal cell-type-specific therapeutic responses, supporting the development of highly personalized treatment strategies [49].

Machine learning techniques, encompassing both traditional algorithms and deep learning models, have been widely applied to predict drug responses in cancer therapy. These approaches effectively manage high-dimensional datasets, capture complex biological interactions, and provide valuable insights to support personalized medicine in oncology [50]. Deep learning approaches have emerged as a powerful tool in genomic research, capable of learning complex patterns from large-scale genomic datasets. Its applications in genomics include functional genome annotation, identification of regulatory sequence elements, and the design of synthetic genomic constructs. Deep learning models have also been employed to predict viral integration sites and DNA-protein binding regions, underscoring their versatility in genomic analysis [51]. The ability of these models to handle multimodal and highly heterogeneous data makes genomics an ideal field for leveraging deep learning techniques.

Natural language processing (NLP) has been increasingly used to identify novel drug–gene interactions. NLP methods have also been applied for biomarker discovery and gene–disease association mapping, both of which play a crucial role in identifying potential drug–gene relationships [52]. AI platform PARAMESAN leverages NLP-based text mining to predict drug–gene interactions in genetic disorders, highlighting the growing role of language-based AI tools in precision medicine and gene therapy research [53].

6 AI and Nanotechnology-Driven Therapeutics

The convergence of nanoarchitectonics with AI represents a sophisticated approach to designing drug delivery platforms that can be programmable structures responsive to external stimuli, and continuously refined through automated optimization. While nanoarchitectonics focuses on assembling multilevel nanostructures with fine control over spatial organization and functional timing, AI contributes powerful tools for simulation, predictive analysis, and systematic refinement during system development. Within this context, an AI-enabled nanoarchitectonics paradigm has been proposed to enhance precision in targeted drug delivery [54].

AI has demonstrated strong potential in tailoring drug delivery strategies to patient-specific characteristics, improving formulation efficiency, and forecasting key physicochemical and pharmacokinetic properties of therapeutic agents. The integration of ML algorithms with large-scale datasets can significantly shorten development timelines for safe and effective nanotechnology-based drug delivery systems (DDS), thereby contributing to better therapeutic outcomes and progress in pharmaceutical research. Moreover, the application of AI in DDS is expected to transform treatment personalization and elevate standards of patient care [55].

Looking ahead, emerging directions include the combined use of AI with quantum computing technologies, as well as the expansion of digital healthcare frameworks such as telemedicine and the Internet of Medical Things (IoMT). These advancements could enable real-time, data-driven clinical decisions, supporting more personalized and timely health care delivery [56].

Hamid et al. [35] examined the influential role of AI in the design and functional optimization of smart multifunctional nanocarriers (SMNs), highlighting its ability to address key limitations in targeted delivery, regulated release, and theranostic applications. The review emphasizes the importance of interdisciplinary approaches to fully realize AI-driven innovations in nanomedicine, supporting the translation of SMNs into clinically effective and safer therapeutic solutions.

Lang et al. [57] reviewed recent advancements and existing limitations in the emerging area of machine learning–assisted nanotheranostics, highlighting opportunities for developing next-generation platforms through the use of robust datasets and sophisticated ML models to enhance clinical outcomes. AI-enabled quality assurance in manufacturing is expected to improve production efficiency while minimizing operational errors. In parallel, the integration of robotic automation into laboratory workflows can enhance throughput and reproducibility, thereby accelerating and reducing the cost of drug discovery. Cloud-based collaborative infrastructures further support data sharing and collective innovation, expediting research progress and strengthening cooperation in nanomedicine [58]. With continued advancements in AI methodologies, their integration with nanotechnology is expected to drive significant innovations across a wide range of sectors, including healthcare, electronics, diagnostics, and robotics. This cross-disciplinary convergence is likely to expand research horizons and catalyze transformative progress in bioengineering, neuroscience, and related fields [59].

The convergence of AI and nanotechnology represents a transformative shift with significant implications across scientific research and industrial innovation. In the healthcare domain, AI-enabled nanotechnological approaches are reshaping disease diagnosis and treatment by enabling faster development of patient-specific drugs and highly targeted therapeutic interventions. Beyond biomedical applications, AI-driven advances in nanotechnology support the rational design of materials with customized functional properties, facilitating progress in fields such as electronics, aerospace, and advanced manufacturing. Collectively, these developments are driving new directions in materials science and engineering while expanding the scope of nanotechnology-enabled solutions [4].

Owing to their unique physicochemical characteristics, nanomaterials exhibit exceptional potential in a wide range of biomedical applications. They are extensively employed in areas such as antimicrobial therapies, diagnostics, cell culture systems, scaffold fabrication, regenerative medicine, targeted drug delivery, and tissue engineering [60].

Advances in characterization methods, together with their distinctive physicochemical properties, have enabled a wide range of nanomaterials, including MXenes, polymers, graphene, and carbon nanotubes, to be extensively applied in biomedical fields, particularly in biosensing and bioimaging applications [61, 62].

Building on recent technological advances and illustrative case studies, this work demonstrates how AI-based tools ranging from generative design frameworks to predictive pharmacokinetic modeling are reshaping the field by shifting formulation strategies from empirical approaches to mechanism-informed, AI-driven intelligent design. Furthermore, by addressing existing challenges and proposing future directions for the integration of AI with nanoarchitectonics, the discussion emphasizes pathways toward the development of clinically translatable nanomedicine platforms [54].

Enhanced therapeutic outcomes can be achieved through the optimized use of existing drugs by leveraging precision diagnostic tools and personalized drug-tailoring strategies to design patient-specific combination therapies that simultaneously target multiple disease pathways. Such approaches offer a means to address drug resistance and improve overall treatment effectiveness. AI and related computational modeling techniques are expected to play a crucial role in the development, optimization, and clinical implementation of these advanced nanotechnological solutions [5]. AI has been applied to the rational design of nanoparticles for targeted drug delivery, enhancing both their therapeutic efficiency and safety profiles.

Artificial intelligence has been applied to the rational design of nanoparticles for targeted drug delivery, enhancing both their therapeutic efficiency and safety profiles. AI-enabled nanotechnology can enhance molecular profiling and early diagnosis, refine the design of nanomedicines, and improve their efficacy. By refining nanomedicine characteristics, enhancing therapeutic synergy, and minimizing nanotoxic effects, AI improves targeting precision and expedites the development of personalized therapeutic strategies [63].

7 Conclusion and Future Perspectives

Artificial intelligence is transforming nanotechnology and pharmaceutical research by enhancing drug discovery workflows, refining formulation design, and enabling personalized therapeutic strategies through advanced predictive modeling and data-driven analysis. In the absence of AI, pharmaceutical development is often slower and more resource-intensive, as limitations in discovery efficiency, clinical trial optimization, and regulatory assessment contribute to increased costs and delayed patient access to new treatments.

AI has emerged as a transformative force in intelligent drug delivery systems and precision medicine, significantly enhancing drug profiling, patient stratification, and therapeutic interventions. By enabling continuous patient monitoring, pharmacovigilance, and integrated diagnostic platforms, AI-driven approaches support early detection of adverse responses and facilitate the rational optimization of nanoparticle-based treatment strategies. The development of real-time, individualized point-of-care diagnostic tools and predictive algorithms has further enabled precise dose customization, reduced off-target effects, and improved patient adherence, collectively advancing the safety and efficacy of drug delivery systems.

The integration of AI with genomic, proteomic, and biomarker data has established a robust foundation for personalized treatment protocols. Nevertheless, translating the potential of digital health into practice will depend on coordinated global collaboration, supportive policy frameworks, and sustained strategic investment. Guided by insights gained from previous technological revolutions, prioritizing digital health is essential to achieve fair access to precision health care and support proactive disease management strategies. Looking ahead, emerging AI-enabled healthcare innovations are expected to increasingly integrate sensor-based health monitoring with AI, the internet of things (IoT), robotics, and next-generation medical devices. This technological convergence has the potential to enable early disease detection, personalized digital therapeutics, and remote patient monitoring, thereby reducing healthcare costs and alleviating the societal burden of chronic diseases. Despite existing challenges related to data standardization, clinical validation, and regulatory approval, future research should prioritize translational and clinically deployable solutions over purely theoretical developments. Emphasis on real-time diagnostics, patient-centric self-management platforms, and adaptive treatment strategies to improve clinical outcomes and transform healthcare delivery. Overall, AI is driving health care toward a proactive, individualized, and technologically integrated paradigm, marking the advent of a digital health revolution aimed at advancing global health and well-being.

References

1. Ching T, Himmelstein DS, Beaulieu-Jones BK, Kalinin AA, Do BT, Way GP, Ferrero E, Agapow PM, Zietz M, Hoffman MM, Xie W, Rosen GL, Lengerich BJ, Israeli J, Lanchantin J, Woloszynek S, Carpenter AE, Shrikumar A, Xu J, Cofer EM et al (2018) Opportunities and obstacles for deep learning in biology and medicine. J R Soc Interface 15(141):20170387
2. Rosenblum D, Joshi N, Tao W, Karp J, Peer D (2018) Progress and challenges towards targeted delivery of cancer therapeutics. Nat Commun 9:1410
3. Cascini F, Beccia F, Causio FA, Melnyk A, Zaino A, Ricciardi W (2022) Scoping review of the current landscape of AI-based applications in clinical trials. Front Public Health 10:949377
4. Chakroborty S, Nath N, Sahoo S, Singh BP, Bal T, Tiwari K, Hailu YK, Singh S, Kumar P, Chakraborty C (2025) A review of emerging trends in nanomaterial-driven AI for biomedical applications. Nanoscale Adv 7(12):3619–3630
5. Adir O, Poley M, Chen G, Froim S, Krinsky N, Shklover J, Shainsky-Roitman J, Lammers T, Schroeder A (2020) Integrating artificial intelligence and nanotechnology for precision cancer medicine. Adv Mater (Deerfield Beach, Fla.) 32(13):e1901989
6. Hassan YM, Wanas A, Ali AA, El-Sayed WM (2025) Integrating artificial intelligence with nanodiagnostics for early detection and precision management of neurodegenerative diseases. J Nanobiotechnol 23(1):668
7. Parvin N, Joo SW, Jung JH, Mandal TK (2025) Multimodal AI in biomedicine: pioneering the future of biomaterials, diagnostics, and personalized healthcare. Nanomaterials (Basel, Switzerland) 15(12):895
8. Nair A, Ong W, Lee A, Leow NW, Makmur A, Ting YH, Lee YJ, Ong SJ, Tan JJH, Kumar N, Hallinan JTPD (2025) Enhancing radiologist productivity with artificial intelligence in magnetic resonance imaging (MRI): a narrative review. Diagnostics (Basel, Switzerland) 15(9):1146

9. Bhange M, Telange D (2025) Convergence of nanotechnology and artificial intelligence in the fight against liver cancer: a comprehensive review. Discov Oncol 16(1):77
10. Yasaka K, Akai H, Abe O, Kiryu S (2018) Deep learning with convolutional neural network for differentiation of liver masses at dynamic contrast-enhanced CT: a preliminary study. Radiology 286(3):887–896
11. Lai J, Luo Z, Liu J, Hu H, Jiang H, Liu P, He L, Cheng W, Ren W, Wu Y, Piao JG, Wu Z (2024) Charged gold nanoparticles for target identification-alignment and automatic segmentation of CT image-guided adaptive radiotherapy in small hepatocellular carcinoma. Nano Lett 24(34):10614–10623
12. Mohsen S (2025) Alzheimer's disease detection using deep learning and machine learning: a review. Artif Intell Rev 58:262
13. Feng X, Provenzano FA, Small SA, Alzheimer's Disease Neuroimaging Initiative (2022) A deep learning MRI approach outperforms other biomarkers of prodromal Alzheimer's disease. Alzheimer's Res Ther 14(1):45
14. Kang JS (2023) Application of artificial intelligence-based digital pathology in biomedical research. Biomed Sci Lett 29(2):53–57
15. Flach RN, van Dooijeweert C, Nguyen TQ, Lynch M, Jonges TN, Meijer RP, Suelmann BBM, Willemse PM, Stathonikos N, van Diest PJ (2025) Prospective clinical implementation of Paige prostate detect artificial intelligence assistance in the detection of prostate cancer in prostate biopsies: CONFIDENT P trial implementation of artificial intelligence assistance in prostate cancer detection. JCO Clin Cancer Inform 9:e2400193
16. Ward A, Graham S, Almond E, Withers D (2024) CBCT (cone-beam CT)-based online adaptive radiotherapy in the post-operative oesophageal cancer patient. Cureus 6(10):e70971
17. Barrett M, Boyne J, Brandts J, Brunner-La Rocca HP, De Maesschalck L, De Wit K, Dixon L, Eurlings C, Fitzsimons D, Golubnitschaja O, Hageman A, Heemskerk F, Hintzen A, Helms TM, Hill L, Hoedemakers T, Marx N, McDonald K, Mertens M, Müller-Wieland D et al (2019) Artificial intelligence supported patient self-care in chronic heart failure: a paradigm shift from reactive to predictive, preventive and personalised care. EPMA J 10(4):445–464
18. Ghorbian M, Ghobaei-Arani M, Babaei MR, Ghorbian S (2025) Nanotechnology and nanosensors in personalized healthcare: a comprehensive review. Sens Bio-Sens Res, 47, 100740
19. Ho D, Quake SR, McCabe ERB, Chng WJ, Chow EK, Ding X, Gelb BD, Ginsburg GS, Hassenstab J, Ho CM, Mobley WC, Nolan GP, Rosen ST, Tan P, Yen Y, Zarrinpar A (2020) Enabling technologies for personalized and precision medicine. Trends Biotechnol 38(5):497–518
20. Kaushik A, Jayant RD, Bhardwaj V, Nair M (2018) Personalized nanomedicine for CNS diseases. Drug Discov Today 23(5):1007–1015
21. Ciardiello F, Ciardiello D, Martini G, Napolitano S, Tabernero J, Cervantes A (2022) Clinical management of metastatic colorectal cancer in the era of precision medicine. CA Cancer J Clin 72(4):372–401
22. Prendergast ME, Burdick JA (2020) Recent advances in enabling technologies in 3D printing for precision medicine. Adv Mater (Deerfield Beach, Fla.) 32(13):e1902516
23. Ahmed Z, Mohamed K, Zeeshan S, Dong X (2020) Artificial intelligence with multi-functional machine learning platform development for better healthcare and precision medicine. Database: J Biol Databases Curation 2020:baaa010
24. Hamamoto R, Suvarna K, Yamada M, Kobayashi K, Shinkai N, Miyake M, Takahashi M, Jinnai S, Shimoyama R, Sakai A, Takasawa K, Bolatkan A, Shozu K, Dozen A, Machino H, Takahashi S, Asada K, Komatsu M, Sese J, Kaneko S (2020) Application of artificial intelligence technology in oncology: towards the establishment of precision medicine. Cancer 12(12):3532
25. Le NQK, Tran TX, Nguyen PA, Ho TT, Nguyen VN (2024) Recent progress in machine learning approaches for predicting carcinogenicity in drug development. Expert Opin Drug Metab Toxicol 20(7):621–628

26. Dercle L, Fronheiser M, Lu L, Du S, Hayes W, Leung DK, Roy A, Wilkerson J, Guo P, Fojo AT, Schwartz LH, Zhao B (2020) Identification of non-small cell lung cancer sensitive to systemic cancer therapies using Radiomics. Clin Cancer Res 26(9):2151–2162
27. Johannet P, Coudray N, Donnelly DM, Jour G, Illa-Bochaca I, Xia Y, Johnson DB, Wheless L, Patrinely JR, Nomikou S, Rimm DL, Pavlick AC, Weber JS, Zhong J, Tsirigos A, Osman I (2021) Using machine learning algorithms to predict immunotherapy response in patients with advanced melanoma. Clin Cancer Res 27(1):131–140
28. Liu D, Schilling B, Liu D, Sucker A, Livingstone E, Jerby-Arnon L, Zimmer L, Gutzmer R, Satzger I, Loquai C, Grabbe S, Vokes N, Margolis CA, Conway J, He MX, Elmarakeby H, Dietlein F, Miao D, Tracy A, Gogas H et al (2019) Integrative molecular and clinical modeling of clinical outcomes to PD1 blockade in patients with metastatic melanoma. Nat Med 25(12):1916–1927
29. Parvatikar PP, Patil S, Khaparkhuntikar K, Patil S, Singh PK, Sahana R, Kulkarni RV, Raghu AV (2023) Artificial intelligence: machine learning approach for screening large database and drug discovery. Antivir Res 220:105740
30. Sengar A (2024) Precision in practice: nanotechnology and targeted therapies for personalized care. Int J Adv Nano Comput Anal 3(2)
31. Shirzad M, Shaban M, Mohammadzadeh V, Rahdar A, Fathi-karkan S, Sadat Hoseini Z, Najafi M, Kharaba Z, Ali Aboudzadeh M (2025) Artificial intelligence-assisted design of nanomedicines for breast cancer diagnosis and therapy: advances, challenges, and future directions. BioNanoScience 15:354
32. Curtis C, McKenna M, Pontes H, Toghani D, Choe A, Nance E (2019) Predicting in situ nanoparticle behavior using multiple particle tracking and artificial neural networks. Nanoscale 11:22515–22530
33. Deori C, Hujuri L (2024) Artificial intelligence (AI): It's role in drug discovery and novel drug delivery system, Int J Sci Res, 13:1426–1430
34. Patel P, Kumar K, Jain VK, Popli H, Yadav AK, Jain K (2023) Nanotheranostics for diagnosis and treatment of breast cancer. Curr Pharm Des 29(10):732–747
35. Noury H, Rahdar A, Romanholo Ferreira LF, Jamalpoor Z (2025) AI-driven innovations in smart multifunctional nanocarriers for drug and gene delivery: a mini-review. Crit Rev Oncol Hematol 210:104701
36. Chaudhary A, Mukhopadhyay S (2020) Artificial intelligence in health sector. Int J Sci Eng Res 11(8):1141–1145
37. Chou WC, Canchola A, Zhang F, Lin Z (2025) Machine learning and artificial intelligence in nanomedicine. Wiley Interdiscip Rev Nanomed Nanobiotechnol 17(4):e70027
38. Almansour K, Alqahtani AS (2025) Utilization of machine learning approach for production of optimized PLGA nanoparticles for drug delivery applications. Sci Rep 15:8840
39. Jones DE, Ghandehari H, Facelli JC (2016) A review of the applications of data mining and machine learning for the prediction of biomedical properties of nanoparticles. Comput Methods Prog Biomed 132:93–103
40. Bhosekar S, Singh P, Garg D, Ravi V, Diwakar MA (2025) Review of deep learning-based multi-modal medical image fusion. Open Bioinform J 18:e1875036237069
41. Elton DC, Boukouvalas Z, Fuge MD, Chung PW (2019) Deep learning for molecular design—a review of the state of the art. Mol Syst Des Eng 4:828–849
42. Han Y, Kim DH, Pack SP (2025) Nanomaterials in drug delivery: leveraging artificial intelligence and big data for predictive design. Int J Mol Sci 26(22):11121
43. Zhou Z, Kearnes S, Li L, Zare RN, Riley P (2019) Optimization of molecules via deep reinforcement learning. Sci Rep 9:10752
44. Esteva A, Robicquet A, Ramsundar B, Kuleshov V, DePristo M, Chou K, Cui C, Corrado G, Thrun S, Dean J (2019) A guide to deep learning in healthcare. Nat Med 25(1):24–29
45. Chen TC, Lim WS, Wang VY, Ko ML, Chiu SI, Huang YS, Lai F, Yang CM, Hu FR, Jang JR, Yang CH (2021) Artificial intelligence-assisted early detection of retinitis Pigmentosa—the most common inherited retinal degeneration. J Digit Imaging 34(4):948–958

46. Bhat AA, Nisar S, Mukherjee S, Saha N, Yarravarapu N, Lone SN, Masoodi T, Chauhan R, Maacha S, Bagga P, Dhawan P, Akil AA, El-Rifai W, Uddin S, Reddy R, Singh M, Macha MA, Haris M (2022) Integration of CRISPR/Cas9 with artificial intelligence for improved cancer therapeutics. J Transl Med 20(1):534
47. Bock C, Datlinger P, Chardon F, Coelho MA, Dong MB, Lawson KA, Lu T, Maroc L, Norman TM, Song B, Stanley G, Chen S, Garnett M, Li W, Moffat J, Qi LS, Shapiro RS, Shendure J, Weissman JS, Zhuang X (2022) High-content CRISPR screening. Nat Rev Methods Prim 2(1):9
48. Tao R, Han X, Bai X, Yu J, Ma Y, Chen W, Zhang D, Li Z (2024) Revolutionizing cancer treatment: enhancing CAR-T cell therapy with CRISPR/Cas9 gene editing technology. Front Immunol 15:1354825
49. Subramanian I, Verma S, Kumar S, Jere A, Anamika K (2020) Multi-omics data integration, interpretation, and its application. Bioinform Biol Insights 14:1177932219899051
50. Taherdoost H, Ghofrani A (2024) AI's role in revolutionizing personalized medicine by reshaping pharmacogenomics and drug therapy. Intell Pharm 2(5):643–650
51. Eraslan G, Avsec Ž, Gagneur J, Theis F (2019) Deep learning: new computational modelling techniques for genomics. Nat Rev Genet 7:389–403
52. Bhatnagar R, Sardar S, Beheshti M, Podichetty JT (2022) How can natural language processing help model informed drug development?: a review. JAMIA Open 5(2):ooac043
53. Wu HY, Chiang CW, Li L (2014) Text mining for drug-drug interaction. Methods Mol Biol (Clifton, N.J.) 1159:47–75
54. Bae H, Ji H, Konstantinov K, Sluyter R, Ariga K, Kim YH, Kim JH (2025) Artificial intelligence-driven Nanoarchitectonics for smart targeted drug delivery. Adv Mater (Deerfield Beach, Fla) 37(42):e10239
55. Heydari S, Masoumi N, Esmaeeli E, Ayyoubzadeh SM, Ghorbani-Bidkorpeh F, Ahmadi M (2024) Artificial intelligence in nanotechnology for treatment of diseases. J Drug Target 32(10):1247–1266
56. Mazumdar H, Khondakar KR, Das S, Halder A, Kaushik A (2025) Artificial intelligence for personalized nanomedicine; from material selection to patient outcomes. Expert Opin Drug Deliv 22(1):85–108
57. Rao L, Yuan Y, Shen X, Yu G, Chen X (2024) Designing nanotheranostics with machine learning. Nat Nanotechnol 19(12):1769–1781
58. Ori MO, Ekpan FM, Samuel HS, Egwuatu OP (2024) Integration of artificial intelligence in nanomedicine. Eurasian J Sci Technol 4(2):88–104
59. Sarvat Z, Nadim R(2025) The convergence of nanotechnology and artificial intelligence: unlocking future innovations. Recent Innov Chem Eng 18(2):85–99
60. Alshamrani M (2022) IoT and artificial intelligence implementations for remote healthcare monitoring systems: a survey. J King Saud Univ—Computer Inf Sci 34:4687–4701
61. Solangi NH, Mazari SA, Mubarak NM, Karri RR, Rajamohan N, Vo DN (2023) Recent trends in MXene-based material for biomedical applications. Environ Res 222:115337
62. Barati F, Avatefi M, Moghadam NB, Asghari S, Ekrami E, Mahmoudifard M (2023) A review of graphene quantum dots and their potential biomedical applications. J Biomater Appl 37(7):1137–1158
63. Bikash M, Himanshu S, Tamanna K, Ajay P (2024) Artificial intelligence: a catalyst for breakthroughs in nanotechnology and pharmaceutical research. Int J Pharm Sci Nanotechnol 17:4

AI–Nano Strategies in Smart Healthcare Systems: Opportunities and Challenges

Mohammed Amine Lafraxo, Mohammed Dalli, and Hinde Hami

Abstract The convergence of artificial intelligence and nanotechnologies represents a major driving force in the development of smart healthcare systems, particularly in the field of real-time health status monitoring. This chapter presents a bibliometric study aimed at mapping the evolution, intellectual structure, and emerging themes of research focusing on AI–nano strategies applied to healthcare. A corpus of 113 articles published between 2014 and 2025 was retrieved from the Scopus database and analyzed using the *Bibliometrix* package in RStudio. The findings show a steady increase in scientific output organized around key conceptual pillars like nanotechnology, artificial intelligence, and human-centered applications. Keyword analysis, thematic trends, and co-occurrence network analysis reveal a strong orientation toward the development of nano-integrated devices, biosensing techniques, personalized medicine, and connected healthcare systems. These findings provide a comprehensive and structured overview of the field, enabling the identification of both consolidated research axes and emerging directions likely to shape the future of intelligent monitoring and healthcare systems.

Keywords Artificial intelligence · Nanotechnology · Smart healthcare systems · Real-time monitoring · Personalized medicine · Bibliometric analysis

1 Introduction

Over the past ten years, advances in artificial intelligence (AI), nanoscale technologies, and the widespread use of digital health infrastructures have all contributed to the acceleration of the digital transformation of healthcare systems. Smart healthcare systems now rely on the integration of big data, connected devices, decision-support

M. A. Lafraxo · M. Dalli (✉)
Higher Institute of Nursing Professions and Health Techniques, Oujda, Morocco
e-mail: m.dalli@ump.ac.ma

M. A. Lafraxo (✉) · H. Hami
Laboratory of Biology and Health, Faculty of Science, Ibn Tofail University, Kenitra, Morocco
e-mail: lafraxo.ma@gmail.com; mohammedamine.lafraxo@uit.ac.ma

J. Chen (ed.), *AI-Driven Nanotechnology in Precision Medicine and Healthcare*, Smart Nanomaterials Technology, https://doi.org/10.1007/978-981-92-1690-1_2

platforms, and emerging technologies aimed at improving disease prevention, diagnosis, treatment, and patient follow-up. The combination of AI and nanotechnologies is a significant lever for organizational and biomedical innovation. Nanotechnologies offer previously unheard-of capabilities for continuous physiological and biological parameter monitoring, ultra-sensitive detection, and the miniaturization of medical devices. The integration of nanosensors into wearable, implantable, or environmental devices enables real-time measurement of key biomarkers, including electrophysiological signals, metabolic parameters, inflammatory markers, and early molecular variations associated with specific pathologies. When coupled with wireless communication technologies and digital health platforms, these nano-integrated solutions facilitate fine-grained, continuous, and contextualized data collection, surpassing the limitations of traditional point-based clinical approaches. Within this framework, artificial intelligence plays a central role by ensuring the analysis, interpretation, and valorization of large volumes of heterogeneous data generated by these connected devices. Machine learning and deep learning algorithms enable the identification of complex patterns, early detection of physiological anomalies, and the support of predictive models of health status evolution. This synergy between AI and nanotechnologies thus contributes to the emergence of intelligent and proactive monitoring systems capable of continuously supporting patients, enhancing care personalization, and strengthening data-driven clinical decision-making. Collectively, these advances position AI–nano strategies as a structural pillar of contemporary smart healthcare systems, particularly in the areas of disease prevention, long-term monitoring of chronic conditions, and personalized medicine.

This chapter's goal is to conduct a bibliometric analysis of the scientific literature on AI–nano approaches used in smart healthcare systems, with an emphasis on real-time health status monitoring via nano-integrated and connected devices. In particular, the analysis looks at how annual scientific output has changed, finds the most highly cited articles, uses a keyword word cloud to explore dominant themes, examines keyword co-occurrence networks, and creates a thematic mapping to conceptually organize the field.

2 Methodology

2.1 Study Design and Rationale

This study adopts a descriptive and analytical bibliometric design aimed at mapping and analyzing the evolution of scientific production addressing the convergence between artificial intelligence and nanotechnologies within the field of smart healthcare systems. Bibliometrics is a well-established quantitative approach for exploring the structure, dynamics, and trends of a given research field based on indexed scientific publications. It enables, in particular, the identification of temporal developments, dominant themes, and conceptual relationships among published studies. This methodological choice is especially relevant in an emerging and interdisciplinary domain such as AI–nano strategies applied to digital health.

2.2 Data Source and Search Strategy

Bibliographic data were extracted from the Scopus database, selected for its broad multidisciplinary coverage and the quality of its metadata, which are widely used in bibliometric studies. The search strategy was designed to capture publications located at the intersection of artificial intelligence, nanotechnologies, and applications in healthcare and smart care systems. The following query was applied to the TITLE-ABS-KEY field (title, abstract, and author keywords):

TITLE-ABS-KEY ((“artificial intelligence”) AND (“nanotechnology”) AND (“health” OR “healthcare” OR “healthcare system” OR “smart healthcare” OR “digital health” OR “connected health”)).

To reflect the recent emergence of artificial intelligence technologies and nano-integrated applications in healthcare, the analysis period was restricted to publications published between 2014 and 2025.

2.3 Selection Criteria and Corpus Constitution

The corpus construction process was carried out in several successive steps (see Fig. 1). Initially, applying the search query in Scopus identified 670 documents. After restricting the publication period to 2014–2025, the number of references was reduced to 632 documents.

In a second step, additional selection criteria were applied to ensure methodological rigor and corpus homogeneity. Only publications written in English and classified as research articles were retained. This choice was intended to ensure the comparability of the analyzed documents and to focus on peer-reviewed studies. After applying these filters, the final corpus analyzed consisted of 113 articles.

2.4 Analytical Framework and Tools

The bibliometric analysis was conducted using RStudio software (version 2023.12.1 + 402) and the Bibliometrix package, through its graphical interface Biblioshiny (version 5.0). These tools provide a comprehensive set of functions dedicated to the quantitative analysis of scientific literature and the visualization of bibliometric networks.

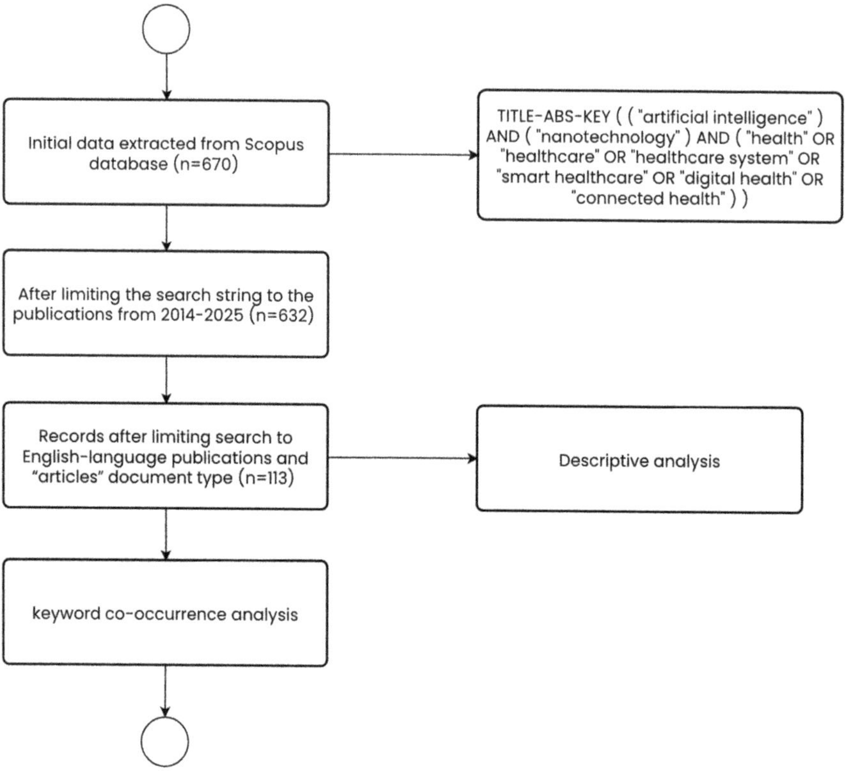

Fig. 1 Study selection flowchart following PRISMA guidelines

3 Results

3.1 Descriptive Analysis

3.1.1 Temporal Evolution of Scientific Production

An analysis of the yearly scientific output pertaining to AI–nano tactics in smart healthcare systems shows a highly exponential growth trajectory, which is typical of a developing field of study going through a significant turning point. The observed period, which spans from 2015 to 2025, can be divided into three distinct phases that demonstrate how quickly this interdisciplinary field is developing (see Fig. 2).

The first phase of latency and emergence extends from 2015 to 2018, during which the volume of publications remains marginal, with one to two articles per year. This period corresponds to the conceptual foundations of the field and is characterized by pioneering explorations and isolated proof-of-concept studies. It shows how two different scientific communities, nanotechnology and artificial intelligence,

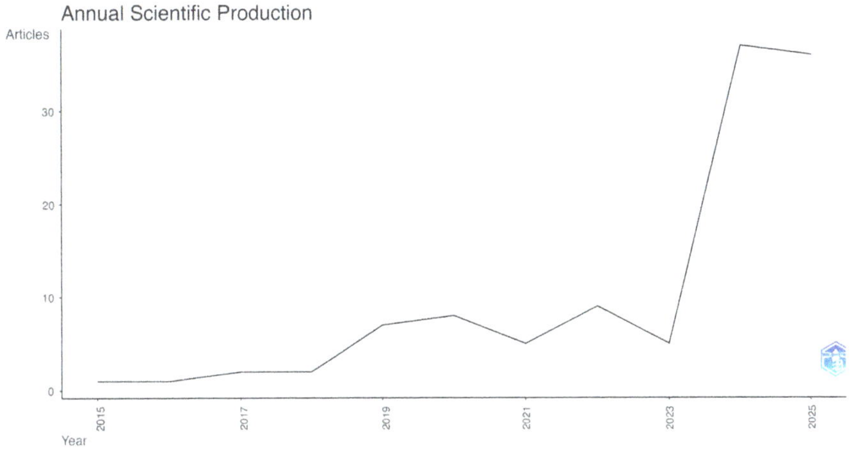

Fig. 2 Annual scientific production (2014–2025)

have begun to tentatively come together around the problems of digital health. This intersection is truly novel, as evidenced by the lack of publications before 2015, which was made possible by concurrent developments in deep learning and nanoscale engineering at the start of the decade (see Fig. 2).

The second phase of experimental growth and consolidation spans the period from 2019 to 2023 (see Fig. 2). During this phase, a significant yet fluctuating increase in the number of publications is observed, ranging from five to nine articles per year, with a notable peak in 2022. These dynamics indicate the increasing confirmation of the potential of AI–nano integration and the expansion of its uses. The expansion seen starting in 2019 corresponds with the development of algorithms and their use in solving progressively challenging biomedical issues. The annual fluctuations, including slight declines in 2021 and 2023, suggest a phase of methodological adjustment and thematic refinement. Moreover, it is plausible that the COVID-19 pandemic, occurring between 2020 and 2022, acted as a catalyst by accelerating research on rapid diagnostic tools, monitoring technologies, and personalized medicine, domains that naturally align with these converging technologies (see Fig. 2).

In 2024, the third and final phase—which was the most important and defined the current state of research—begins. With 37 publications in 2024 alone and an estimated 36 publications in 2025, it is marked by a dramatic increase in scientific productivity that is more than four times higher than the previous peak (see Fig. 2). This quantitative change clearly indicates that a tipping point has been reached. It shows that the field has advanced past the experimental validation stage and entered a period of extensive and varied research. This sudden shift is explained by a number of convergent factors, such as the expansion of application domains toward regenerative medicine, chronic disease monitoring, and healthcare cyber-physical systems, the development of a strong body of proof-of-concept evidence that has drawn significant funding, and the increased accessibility of software tools and fabrication platforms (see Fig. 2).

3.1.2 Most Globally Cited Documents

The foundational and high-impact studies that have shaped the field of AI–nano strategies applied to intelligent healthcare systems are highlighted by the analysis of the most widely cited papers. The three main thematic axes of the most highly cited articles are intelligent wearable health monitoring devices, self-powered nano-integrated sensors, and point-of-care diagnostics.

The article by Berg et al. (2015) [1], published in ACS Nano, emerges as the most influential document in the corpus, with 321 citations. It presents a smartphone-based portable microplate reader for point-of-care ELISA testing (see Fig. 3). This work represents a seminal reference by demonstrating the feasibility of low-cost, connected, and portable diagnostic devices, foreshadowing the later integration of artificial intelligence for automated result analysis and clinical decision-making. Its significant influence highlights how mobile and nano-integrated technologies have shaped the development of intelligent diagnostic systems. The studies by Zhao et al. (2021) [2] and Wang et al. (2020) [3], published in Advanced Materials, illustrate the rise of multifunctional intelligent sensors for ultra-sensitive health monitoring (see Fig. 3). Moreover, the article published by Zhao et al. (195 citations) introduces a bio-inspired electronic skin combining photonic and electronic functions for visual and self-powered sensing, while Wang et al. (165 citations) present a self-powered angle sensor with nanoradian resolution, designed for applications in robotics and personalized medicine (see Fig. 3). These studies demonstrate the increasing significance of next-generation intelligent devices that can provide high-precision data that AI algorithms can use for functional monitoring and rehabilitation. Articles published in Nano Energy and ACS Nano, particularly those by Kim et al. (2021) [4], Das et al. (2019) [5], Li et al. (2017) [6], Wen et al. (2019) [7], and Du et al. (2022) [8], reveal a dominant research axis centered on triboelectric nanogenerators and self-powered sensors (see Fig. 3). These studies, each accumulating between 90 and 174 citations, demonstrate the potential of flexible, stretchable, and self-healing nanomaterials for the development of intelligent wearable devices capable of continuous monitoring without reliance on external power sources (see Fig. 3). Their strong bibliometric visibility reflects the strategic interest in energy autonomy, identified as a key driver for the sustainability of intelligent healthcare systems (see Fig. 3). Furthermore, the article by Tsikala-Vafea et al. (2020) [9], with 111 citations, occupies a distinctive position due to its pandemic context. It provides a synthesis of emerging technologies mobilized for the study, diagnosis, and treatment of COVID-19, illustrating the capacity of AI–nano strategies to respond rapidly to major public health crises. This work emphasizes the importance of integrating intelligent technologies into public health and emergency management approaches, while also highlighting challenges related to large-scale deployment (see Fig. 3). The article by Kumar et al. (2022) [10], published in Production and Operations Management (109 citations), stands out for

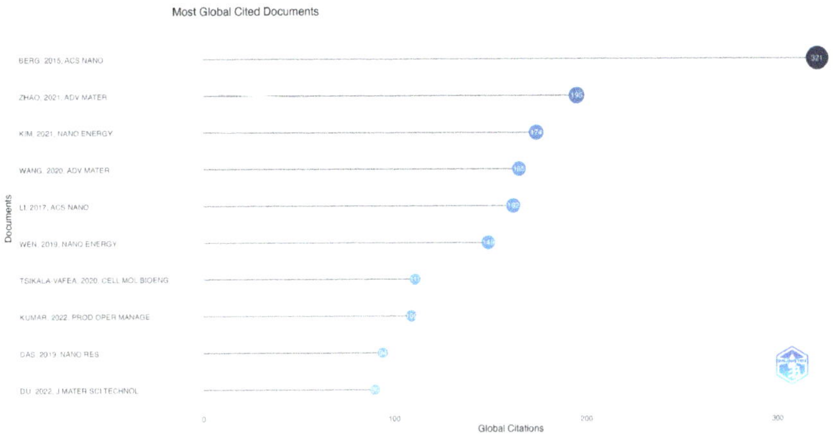

Fig. 3 Most globally cited documents

its focus on cybersecurity and complex systems management (see Fig. 3). Although less directly related to nanotechnologies, its high citation count reflects the growing importance of data security, system resilience, and supply chain management issues in the context of intelligent healthcare systems integrating AI and connected devices.

3.1.3 Thematic Evolution

A keyword cloud illustrating the frequency of occurrence of the most frequently used terms in the articles of the examined corpus, which focuses on the convergence of nanotechnologies, artificial intelligence, and healthcare applications, is shown in the following figure (see Fig. 4). Each term's size corresponds to how frequently it appears, indicating its relative significance within the reviewed scientific literature. The terms "nanotechnology" and "artificial intelligence" are the most frequently used, indicating their pivotal and organizing roles in the examined field of study. Their high visibility suggests that most studies are clearly situated at the intersection of these two domains, confirming the applicability of the search approach and the selected theme scope. The strong focus of research on human-centered applications, especially in relation to health, medicine, and healthcare systems, is reflected in the highly prevalent term "human." This prominence suggests that AI–nano strategies are predominantly explored from a biomedical and clinical perspective rather than a purely technological one. In addition, the visibility of terms such as "machine learning," "personalized medicine," "diagnosis," "drug delivery system," and "biosensing techniques" highlights several major application axes. These keywords reflect the growing use of machine learning algorithms for analyzing data generated by nano-integrated devices, as well as the development of solutions oriented toward personalized medicine, assisted diagnosis, and advanced drug delivery systems. The presence of terms such as "telemedicine," "health care,"

Fig. 4 Keywords frequency analysis

"public health," "internet of things," and "medical technology" underscores the integration of AI–nano strategies within intelligent and connected healthcare ecosystems, enabling continuous monitoring, remote follow-up, and device interoperability.

The transition to proactive, data-driven, digital models of care is strengthened by this trend. Lastly, the emergence of more specialized terms like "nanoparticles," "nanomedicine," "biosensors," and "regenerative medicine" illustrates the field's interdisciplinary richness and the variety of technological approaches used. These phrases refer to the increasing spread of applications of AI-supported nanotechnology in various areas of the healthcare spectrum.

The figure presents the temporal evolution of the dominant research themes identified through keyword analysis over the 2014–2025 period. Each point represents the emergence and relative frequency of a term over time, while bubble size reflects the importance of the theme within the analyzed scientific literature (see Fig. 5). The results highlight a progressive structuring of the field around central thematic cores. The terms "artificial intelligence," "nanotechnology," and "human" emerge as persistent cross-cutting themes, showing continuous presence and strong intensity in recent years. Their dominance, particularly pronounced from 2021–2022 onward, underscores the growing maturity of research integrating AI and nanotechnologies into human-centered and healthcare applications. In addition, the emergence and consolidation of the term "machine learning" in the most recent years reflect a methodological shift toward more specific algorithmic approaches within the broader field of artificial intelligence. This evolution suggests increasing specialization of the techniques employed for the analysis and exploitation of data generated by nano-integrated devices. Application-oriented themes such as "drug delivery system" and "nanoparticle" show a marked rise in visibility during the most recent period (2023–2025). This trend reflects growing interest in intelligent nanomedicine solutions that combine nanoparticle design with AI models to optimize targeted drug delivery and enable real-time therapeutic monitoring. Other themes, including "health care," "medical technology," and "devices," display a more sustained presence over time, indicating their structuring role as broad applications and technological frameworks.

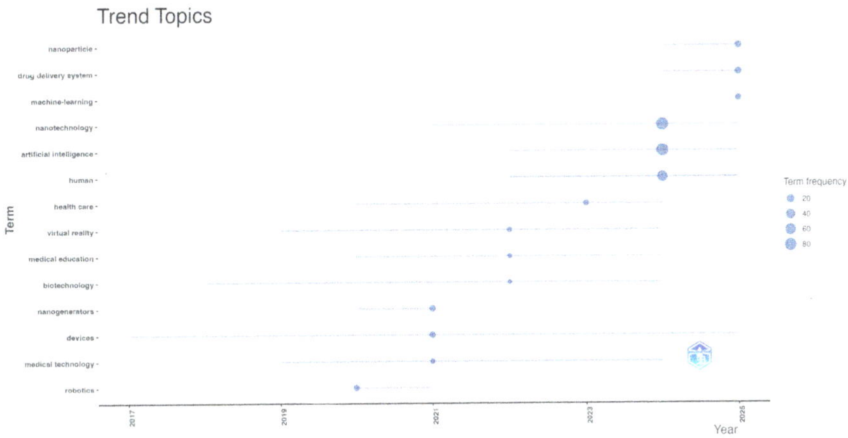

Fig. 5 Thematic evolution over time

Their continuity highlights the embedding of AI–nano strategies within existing healthcare systems and medical infrastructures.

3.2 Keyword Co-occurrence Network Analysis

The keyword co-occurrence network derived from the analyzed corpus highlights the conceptual relationships and the intellectual structure of the research field devoted to AI–nano strategies applied to intelligent healthcare systems. Nodes correspond to keywords, with their size reflecting frequency of occurrence, while links indicate co-occurrences between terms within the same publications. Colors differentiate distinct thematic clusters, revealing coherent subdomains (see Fig. 6).

The network is structured around a strongly dominant central core organized around the term “Artificial intelligence,” which exhibits the highest centrality and connects all thematic areas. This term is closely associated with “human” and “machine learning,” highlighting a strong orientation toward human-centered AI applications supported by machine learning approaches for data analysis and decision support in healthcare. The cluster located on the left side of the Fig. 6 groups keywords such as “biosensors,” “biosensing techniques,” “electrochemical analysis,” “enzyme-linked immunosorbent assay,” “cyclic voltammetry,” “point-of-care testing,” and “diagnosis.” This thematic set highlights a research axis focused on the development of nano-integrated devices and advanced analytical techniques for real-time health monitoring and diagnosis. The density of internal connections reflects strong scientific coherence around nanotechnology applications in biosensing and medical diagnostics. The main cluster, shown in red, encompasses terms such as “Personalized medicine,” “Drug delivery system,” “Nanomedicine,” “Telemedicine,” “Health care,” “Medical technology,” “Biotechnology,” and “Quality of life”. This grouping

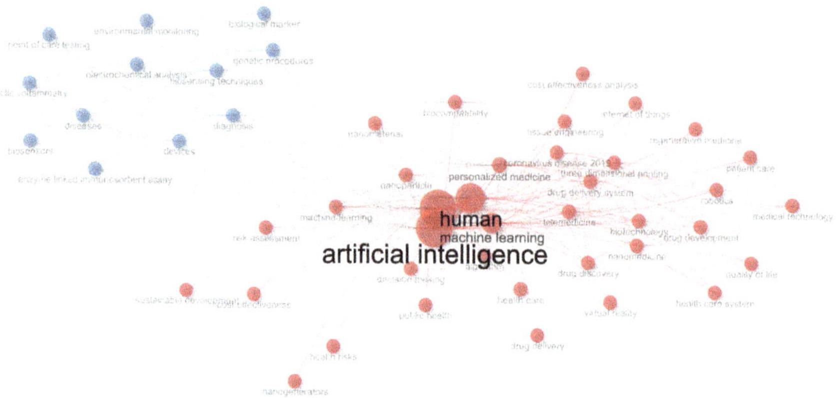

Fig. 6 Keyword co-occurrence network analysis

illustrates the integration of AI–nano strategies into clinical applications and intelligent healthcare systems, with an emphasis on personalized medicine, targeted drug delivery, and patient care. The strong interconnections among these terms indicate a mature research field in which nanotechnologies are combined with AI to enhance the efficiency, personalization, and continuity of care.

A thematic map derived from keyword co-occurrence analysis is shown in the following figure. It is organized along two complementary bibliometric dimensions: density (vertical axis), which shows the degree of internal development and conceptual maturity of the identified themes, and centrality (horizontal axis), which reflects the significance and degree of interconnection of a theme within the overall scientific field (see Fig. 7).

In the upper-right quadrant, motor themes appear that are both highly central and strongly developed. Notably, this quadrant includes the terms "nanotechnology," "artificial intelligence," and "human". Their positioning reflects their structuring role within the field of intelligent healthcare systems integrating AI–nano strategies.

These themes constitute the dynamic core of current research, emphasizing the importance of a human-centered approach in which artificial intelligence and nanotechnologies are leveraged to improve diagnosis, clinical decision-making, and quality of care. Another subset, including "diagnosis," "nanoparticle," and "devices," is also located among the motor themes, indicating that the development of nano-assisted devices for medical diagnosis, often coupled with AI algorithms, represents a research axis that is both mature and strategically important.

Themes like "personalized medicine," "drug delivery system," and "tissue engineering" occupy an intermediate position near the map's center, with high centrality and moderate density. This placement alludes to transversal themes that are presently uniting and serving as links between basic developments in nanotechnology and clinical applications based on artificial intelligence. They illustrate the emergence of personalized medicine grounded in intelligent targeted drug delivery systems and

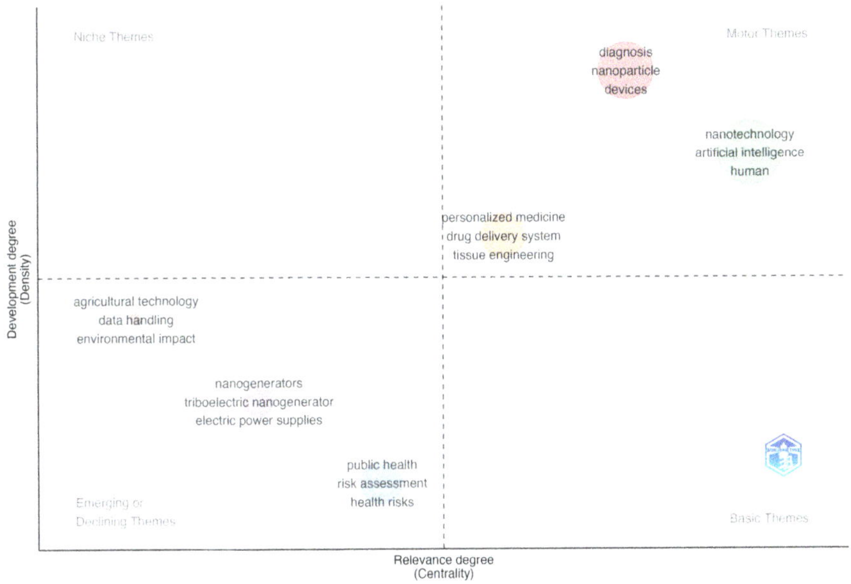

Fig. 7 Thematic map

tissue engineering supported by advanced computational tools. The lower-left quadrant groups emerging or potentially declining themes, such as "nanogenerators," "triboelectric nanogenerator," and "electric power supplies." Although these themes exhibit low centrality, they point to innovative research avenues, particularly in the area of energy autonomy for intelligent medical devices, which may gain importance in the future as wearable and implantable technologies continue to evolve. Within the same quadrant, terms such as "agricultural technology," "data handling," and "environmental impact" suggest still-marginal connections to the core field, while opening interdisciplinary perspectives related to sustainability, data management, and the environmental impact of nano-intelligent technologies.

4 Discussion

This bibliometric analysis demonstrates the increasing organization and scientific development of the field of study focused on methods integrating nanotechnologies and artificial intelligence in intelligent healthcare systems. The corpus's temporal evolution shows a noticeable increase in scientific output starting in 2021–2022, which is consistent with the growth of connected devices, nanosensors, and machine learning algorithms used in healthcare [2–4]. This dynamic reflects a shift in the field toward integrated, human-centered approaches, in which nanomaterials are no longer viewed solely as passive components but as active elements of intelligent systems capable of generating, transmitting, and analyzing biological data in real time [1, 6, 7].

The persistent predominance of the terms "artificial intelligence," "nanotechnology," and "human" confirms that the literature is structured around a techno-clinical paradigm aimed at improving diagnosis, personalized monitoring, and medical decision-making. The most highly cited studies in the corpus clearly illustrate this orientation, particularly through the development of self-powered, flexible, and portable biosensing devices designed for continuous monitoring of physiological parameters [2, 4, 5, 8]. The rise of triboelectric nanogenerators, intelligent e-skins, and sensors based on 2D or hybrid materials highlights the convergence of advanced nanotechnologies with embedded artificial intelligence, enabling the overcoming of limitations associated with conventional medical devices [3, 11].

Co-occurrence and thematic structuring analyses reveal that the motor themes in the field primarily concern nano-assisted diagnostics, intelligent devices, and integrated clinical applications. The high centrality of terms related to biosensing techniques, electrochemical analysis, point-of-care testing, and devices highlights the central role of miniaturized nanosensors in real-time health monitoring. This trend is strongly supported by major contributions focusing on portable diagnostic platforms, often coupled with smartphones or connected systems, capable of providing rapid, low-cost, and widely deployable solutions [1, 12, 13]. The integration of machine learning algorithms enables the exploitation of continuous data streams generated by these devices, improving early disease detection, clinical outcome prediction, and medical decision support [14, 15].

Real-time health monitoring through connected, nano-integrated devices thus emerges as one of the structuring axes of the field. The portable and implantable sensors described in the corpus enable continuous tracking of complex physiological signals, such as blood pressure, neurological tremors, metabolites, or inflammatory biomarkers, under real-life conditions [4, 11, 13, 16–18]. When coupled with artificial intelligence architectures, these devices facilitate a transition toward predictive and proactive medicine, in which anomalies can be detected before the onset of severe clinical symptoms. This approach also paves the way for improved management of chronic and neurodegenerative diseases, as well as personalized remote patient monitoring, thereby reducing the burden on healthcare systems [3, 9, 19, 20].

At the same time, the intermediate yet structuring position of themes such as "personalized medicine," "drug delivery systems," and "tissue engineering" reflects the emergence of an intelligent personalized medicine paradigm. Recent studies in the corpus show that AI plays an increasingly important role in the design, optimization, and evaluation of nanoparticle-based targeted drug delivery systems, enhancing therapeutic efficacy while minimizing adverse effects [21, 22]. This convergence of computational modeling, functional nanomaterials, and clinical applications strengthens the ability of healthcare systems to provide treatments tailored to individual biological profiles [23].

Despite these advances, the results highlight persistent imbalances in the structuring of the field. The peripheral positioning of themes such as "public health," "risk assessment," and "health risks" indicates that population-level, preventive, and evaluative dimensions remain underrepresented compared with the dominant technological developments. Although some studies explicitly address issues of risk governance, nanomaterial toxicology, and long-term safety [24, 25], these approaches remain a minority within the overall corpus. This observation suggests that the systematic integration of risk governance frameworks, safe-by-design principles, and regulatory assessment has not yet reached the same level of maturity as the technological innovations themselves.

This study presents several limitations that should be considered when interpreting the results. First, the restriction of the corpus to English-language publications may have led to the exclusion of relevant studies published in other languages, particularly regional research from non-English-speaking countries, where the development and adoption of AI–nano strategies in healthcare may follow specific trajectories. Second, the exclusive use of the Scopus database, although providing extensive multidisciplinary coverage and high-quality metadata, may not include all relevant publications indexed in other databases such as Web of Science, PubMed, or specialized repositories. This methodological choice may thus influence the overall representativeness of the analyzed corpus. Third, limiting the analysis to peer-reviewed journal articles, while ensuring scientific rigor and methodological homogeneity, may underrepresent certain applied research or emerging technological innovations disseminated as conference proceedings, technical reports, or institutional documents, which sometimes precede formal publication in scientific journals. Finally, it should be noted that bibliometric analysis relies exclusively on the quantitative and relational characteristics of the published literature. It does not directly assess the clinical effectiveness, operational feasibility, or real-world impact of nano-integrated devices and AI algorithms in healthcare settings. Therefore, the results should be interpreted as a mapping of research dynamics rather than an evaluation of technological or clinical performance.

Despite these limitations, this study provides a robust and structured overview of the scientific landscape regarding AI–nano strategies applied to intelligent healthcare systems. It offers valuable insights for guiding future research and supporting the development of innovative solutions for real-time monitoring and personalized care.

5 Conclusion

The structure, development, and scientific dynamics of the field of study devoted to methods integrating nanotechnologies and artificial intelligence in intelligent healthcare systems are highlighted in this bibliometric analysis. The findings show a steady increase in scientific output over the previous ten years, which is indicative of growing interest in novel approaches meant to enhance diagnosis, real-time monitoring, and individualized treatment. The field is currently dominated by central and mature themes related to AI, nanotechnologies, nano-integrated diagnostic devices, and personalized medicine, according to keyword analysis, co-occurrence networks, and thematic mapping. The development of more proactive, predictive, and patient-centered healthcare systems is made possible by this technological convergence, especially when it comes to data-driven clinical decision-making and chronic illness monitoring.

References

1. Berg B et al (2015) Cellphone-based hand-held microplate reader for point-of-care testing of enzyme-linked immunosorbent assays. ACS Nano 9(8):7857–7866. https://doi.org/10.1021/acsnano.5b03203
2. Zhao Y et al (2021) Bioinspired multifunctional photonic-electronic smart skin for ultrasensitive health monitoring, for visual and self-powered sensing. Adv Mater 33(45):2102332. https://doi.org/10.1002/adma.202102332
3. Wang Z et al (2020) A self-powered angle sensor at Nanoradian-resolution for robotic arms and personalized Medicare. Adv Mater 32(32):2001466. https://doi.org/10.1002/adma.202001466
4. Kim JN, Lee J, Lee H, Oh IK (2021) Stretchable and self-healable catechol-chitosan-diatom hydrogel for triboelectric generator and self-powered tremor sensor targeting at Parkinson disease. Nano Energy 82:105705. https://doi.org/10.1016/j.nanoen.2020.105705
5. Das PS, Chhetry A, Maharjan P, Rasel MS, Park JY (2019) A laser ablated graphene-based flexible self-powered pressure sensor for human gestures and finger pulse monitoring. Nano Res 12(8):1789–1795. https://doi.org/10.1007/s12274-019-2433-5
6. Li T et al (2017) From dual-mode triboelectric Nanogenerator to smart tactile sensor: a multiplexing design. ACS Nano 11(4):3950–3956. https://doi.org/10.1021/acsnano.7b00396
7. Wen D-L et al (2019) Printed silk-fibroin-based triboelectric nanogenerators for multi-functional wearable sensing. Nano Energy 66:104123. https://doi.org/10.1016/j.nanoen.2019.104123
8. Du Y, Wang X, Dai X, Lu W, Tang Y, Kong J (2022) Ultraflexible, highly efficient electromagnetic interference shielding, and self-healable triboelectric nanogenerator based on Ti3C2T*x* MXene for self-powered wearable electronics. J Mater Sci Technol 100:1–11. https://doi.org/10.1016/j.jmst.2021.04.078
9. Tsikala Vafea M et al (2020) Emerging Technologies for use in the study, diagnosis, and treatment of patients with COVID-19. Cell Mol Bioeng 13(4):249–257. https://doi.org/10.1007/s12195-020-00629-w
10. Kumar S, Mallipeddi RR (2022) Impact of cybersecurity on operations and supply chain management: emerging trends and future research directions. Prod Oper Manag 31(12):4488–4500. https://doi.org/10.1111/poms.13859
11. Venugopal K, Panchatcharam P, Chandrasekhar A, Shanmugasundaram V (2021) Comprehensive review on triboelectric Nanogenerator based wrist pulse measurement: sensor fabrication and diagnosis of arterial pressure. ACS Sens 6(5):1681–1694. https://doi.org/10.1021/acssensors.0c02324
12. Zhan J, Wang S, Li X, Zhang J (2025) Molecular engineering of functional DNA molecules toward point-of-care diagnostic devices. Chem Commun 61(22):4316–4338. https://doi.org/10.1039/D5CC00338E
13. Shen C, Xi X, Wu D, Guo X, Su Y, Liu R (2025) Active-matrix extended-gate field-effect transistor array for simultaneous detection of multiple metabolites. Biosens Bioelectron 267:116787. https://doi.org/10.1016/j.bios.2024.116787
14. Agrahari V, Choonara YE, Mosharraf M, Patel SK, Zhang F (2024) The role of artificial intelligence and machine learning in accelerating the discovery and development of nanomedicine. Pharm Res 41(12):2289–2297. https://doi.org/10.1007/s11095-024-03798-9
15. Omidian H (2025) AI-powered breakthroughs in material science and biomedical polymers. J Bioact Compat Polym 40(2):161–174. https://doi.org/10.1177/08839115241308202

16. Lafraxo MA, Remaida A, Achachi H, Zarouali AB, El Madhi Y, Hami H (2025) Artificial intelligence and machine learning in connected health: a bibliometric analysis of evolving trends within society 5.0. In: 2025 AI-driven smart healthcare for society 5.0, pp 7–12. https://doi.org/10.1109/IEEECONF64992.2025.10963175
17. Lafraxo MA, Soulaymani A, Hami H (2024) Decadal trends in smart healthcare: a bibliometric review of Smart City innovations. In: 2024 Mediterranean smart cities conference (MSCC), pp 1–6. https://doi.org/10.1109/MSCC62288.2024.10697027
18. Lafraxo MA, Douelfiqar I, Remaida A, El Madhi Y, Soulaymani A, Hami H (2025) Burnout syndrome and coping strategies among nursing staff: Modeling and prototyping a smart prevention system using the internet of medical things. In: 2025 international conference on ambient intelligence in health care (ICAIHC), pp 1–8. https://doi.org/10.1109/ICAIHC64101.2025.10957041
19. Lafraxo MA et al (2023) Building a recommender system to predict the shape of bacteria in urine Cytobacteriological examination using machine learning. Int J Online Biomed Eng (IJOE) 19(13). https://doi.org/10.3991/ijoe.v19i13.36185
20. Lafraxo MA et al (2025) Advancing resilience in healthcare professionals: a multifaceted IoMT-based strategy to combat burnout. Eur Psychiatry 68(S1):S780–S780. https://doi.org/10.1192/j.eurpsy.2025.1582
21. Kamal IR, El-atty SMA, El-Zoghdy SF, Soliman RF (2025) Intelligent deep learning model for targeted cancer drug delivery. Sci Rep 15(1):19068. https://doi.org/10.1038/s41598-025-96149-6
22. Kapoor DU et al (2024) AI-driven design and optimization of nanoparticle-based drug delivery systems. Sci Eng Health Stud 24010003–24010003. https://doi.org/10.69598/sehs.18.24010003
23. Cai Y et al (2024) Nanotechnology for the diagnosis and treatment of liver cancer. Int J Nanomedicine (IJN) 19:13805–13821. https://doi.org/10.2147/IJN.S490661
24. Linkov I et al (2018) Comparative, collaborative, and integrative risk governance for emerging technologies. Environ Syst Decis 38(2):170–176. https://doi.org/10.1007/s10669-018-9686-5
25. Subramanian V et al (2016) Sustainable nanotechnology decision support system: bridging risk management, sustainable innovation and risk governance. J Nanopart Res 18(4):89. https://doi.org/10.1007/s11051-016-3375-4

AI-Driven Personalized Nanomedicines: An Updated Overview

Hitesh Malhotra, Preeti Garg, Chirag Madan, Jyoti, and Rupesh K. Gautam

Abstract The field of nanomedicine is swiftly developing by solving healthcare challenges and providing more practical solutions. However, these convoluted systems face many hurdles concerning the specifications for functionality, scalable production, characterization, quality assurance, and clinical application. In addition to addressing the hurdles of designing nanomedicines for specific functions and creating systems capable of scalable production and quality control, utilizing machine learning (ML) and/or artificial intelligence (AI) methods will enable faster and more precise assessment of data, develop patterns, and predict outcomes, leading to the development of viable nanomedicines. The creation of nanomedicine for diagnostic and therapeutic purposes can be enhanced by AI technology via its ability to process vast quantities of data and recognize patterns within that data. By using ML/AI, multi-omics data, medical imaging (MRI, CT, etc.), and clinical data can be fused to produce adaptive treatment protocols in real time and shorten the length of time necessary to discover solutions by eliminating the need for trial-and-error experimentation. AI also provides an excellent example of an individualized dosage approach for each patient and demonstrates, through recent developments in convolutional and graph neural networks, the extent to which AI can aid both structural modeling and medical imaging. The chapter also discusses the ethical, legal, and data governance issues that come with using AI-driven nanomedicine, such as prejudice, data privacy, and fair access. In order to realize genuinely individualized, AI-powered nanotherapeutics for global healthcare applications, the chapter ends by highlighting the revolutionary possibilities of interdisciplinary collaboration, open-source innovation, and upcoming technologies like quantum computing and digital twins. Thus, with an emphasis on their uses in discovery, evaluation, production, and

H. Malhotra
Faculty of Pharmaceutical Sciences, PCTE Group of Institute, Ludhiana, Punjab, India

P. Garg · C. Madan · Jyoti
Hindu College of Pharmacy, Sonipat, Haryana, India

R. K. Gautam (✉)
Centre for Pharmacology, Amity Institute of Pharmacy, Amity University Uttar Pradesh, Noida, India

J. Chen (ed.), *AI-Driven Nanotechnology in Precision Medicine and Healthcare*, Smart Nanomaterials Technology, https://doi.org/10.1007/978-981-92-1690-1_3

clinical trials, the chapter explores the possibilities of AI and ML in nanomedicine product development.

Keywords Artificial intelligence · Drug · Machine learning · Nanomedicine · Nanoparticle design · Personalized medicine · Targeted drug delivery

1 Introduction

Nanomedicine is a rapidly developing field that is expected to grow significantly in order to promote the development of healthcare services. The US market for nanomedicine, which includes nanoparticles (NPs), nanotubes, nanodevices, and nanoshells, was estimated by Grand View Research to be worth USD 91.1 billion in 2023. Small molecule medications, biologics, and vaccines are just a few of the many therapies that can be delivered via nanomedicines. In particular, nanomedicines can interact with cells and cross-intricate physiological barriers, such as the blood–brain barrier, due to their small size and advantageous surface characteristics [1].

Nanomedicine is defined as the involvement of nanotechnology for therapeutic goals and has developed as a rapidly growing domain that promotes drug delivery. Nanoscale paradigms include polymeric nanoparticles, liposomes, and inorganic nanoparticles that lead to precise targeting of therapeutic agents, decrease adverse events, and can overcome biological barriers. Nanomedicine enables superior therapeutic outcomes and agent efficacy by controlling the timing and location of release from nanocarriers. In this context, personalized medicine is a practice predicated on tailoring therapy to an individual's unique genetic and physiological characteristics and is particularly relevant in difficult-to-treat diseases, such as cancer, where interindividual responses to a particular agent are common. A key goal of integrating nanomedicine into personalized medicine is to design drug delivery systems tailored to the patient-specific biomarker, tumor type, or gene-expression profile [2].

Advanced technological capabilities, such as AI and ML, provide a promising opportunity to use such technologies throughout many stages of the development process of nanomedicine. AI is a collection of technologies that can replicate some aspects of human thought; it can learn, reason, and evaluate solution options to solve complex problems. ML is an aspect of AI where computer scientists create an algorithm based on data that was collected previously, and on the experience(s) that a machine has used to create the algorithm to model, predict, and operate on complex systems with the data that has been collected, through the machine's process of learning from prior data collected [3]. AI now supports decision-making in biomedical research by allowing the use of machine learning algorithms to analyze and develop patterns within large datasets that would take humans years or decades to analyze. Recently, the application of AI technology has enabled researchers to develop innovative nanoparticle formulations that predict how nanoparticles behave in the human body and will assist with determining which patients will benefit from the nanoparticle treatments. Through these innovative technologies, researchers will be able to develop effective, safe, and timely therapies. Two core tenets are forming the foundations of nanomedicine today through AI technologies used to improve nanomedicine. First, the biological characteristics of a nanoparticle play a dominant role in regulating the effect, i.e., how a NPs behave in vivo, how it passes through and enters cells, and how it releases its therapeutic payloads. For example, NPs will have different mobility based on their size, shape, and surface charge. Simple equations cannot adequately represent these relationships, but machine learning is quite good at uncovering hidden patterns in variable data. Second, there is increasing interest in using patient data—such as immunological markers, gene expression, or tumor characteristics—to inform the design of the nanoparticle. This makes it possible to identify the formulation that is most likely to be effective for a certain subgroup of patients, a process known as patient stratification. Through virtual testing, AI models may sometimes show how altering a formulation would affect performance in real time, speeding up the design process [4].

Kolluri and colleagues, for instance, concentrated on the background of AI and ML as well as various tools that might be applied in the pharmaceutical industry. The AI/ML workflow typically begins with building or gaining access to a sizable and varied historical data bank, cleaning and organizing the data, making sure that the labels and annotations are accurate, identifying the assumptions and correlational limitations, and choosing a model. Following quality control and preparation, half of the data is used to validate and verify the model to prevent overfitting, and the other half is used as a training set to build the model. Lastly, best practices should be incorporated into the AI/ML process to ensure that the training data is of the highest quality possible, raising the possibility that the model's goals will be successfully met [5].

With the growing need for advanced and creative resolutions in the pharmaceutical industry, AI is being integrated into nanomedicine. Combining AI's capabilities, i.e., computational capabilities and pattern recognition, will enhance the potential and precision of nanotechnology. Clinical studies, diagnostics, and therapeutics are the areas where numerous breakthroughs have occurred. Researchers and medical professionals can use AI opportunities for targeted delivery, personalized medicine, and better therapeutic outcomes by involving AI in nanoscale processes. By enhancing the effectiveness of research and development procedures, AI helps healthcare organizations to reduce medication development costs [6]. Designing experiments and predicting the pharmacokinetics and toxicity of drug candidates are made easier by machine learning techniques. This capability makes it easier to prioritize and optimize lead compounds, reducing the need for resource-intensive animal testing and lowering costs. The pharmaceutical industry has a bright future for advancing drug development processes and enhancing patient care if it continues to engage in and investigate AI. Advanced AI-nanotechnology-based systems function as active pharmaceutical agent carriers or therapies. Nanocrystals, liposomes, lipid nanoparticles, PEGylated polymeric nanodrugs, polymers, protein-based, and metal-based nanoparticles are examples of commercially available choices. It is imagined that an automated medical technology will be approved, go on sale, and be used by qualified medical practitioners. AI is viewed as a helpful tool that reduces errors, improves precision, and lessens the workload for doctors [7].

AI is crucial to nanomedicine because it can analyze large amounts of biological and chemical data, speeding up the process of developing new drugs. Anticipating how a patient will react to particular treatments and enhancing drug delivery methods for more precise targeting and lower toxicity make personalized medicine possible. AI also aids in picture processing for accurate diagnosis and real-time monitoring using nanosensors. AI's integration with nanomedicine generally drives advancements in precision medicine, patient care, and therapeutic efficacy. "AI's spellbinding role in nanomedicine" refers to the fascinating collaboration between AI and nanomedicine, where AI-driven techniques and algorithms are employed to revolutionize healthcare at the nanoscale level. Precision medicine with hitherto unheard-of accuracy and efficacy could greatly benefit from this collaboration [8].

The field of scientific medicine transitioned during the twentieth century from a logical approach to one based on molecular science. This transition has led to many major advances in human health and medicine (e.g., antibiotics), which, when used correctly, can reduce bacterial infections due to their ability to fight pathogens (bacteria) at the molecular level. Many of the advances made in our understanding of human biology at a molecular level have been made possible through advances made in areas such as Genomics, Proteomics, and Bioinformatics. The determination of our Human Genome has enabled us to create small instruments to study and repair individual cells to maintain ongoing health. The application of nanoparticles has increased tremendously over the last few years, leading to their extensive application within many areas of clinical usage. Due to the innovative design of these nanoparticles, they have also enabled us to overcome some of the limitations of traditional therapies and provide us with more effective pathways through the complex biological barriers that exist within many different patients and diseases. Precision therapies have become more widely applied to help reduce patient heterogeneity and to provide patients with increased therapeutic effectiveness through the creation of personalized therapeutic approaches [9]. However, most of the studies done on nanoparticles have focused on creating uniform delivery systems without the required customization.

The discovery, production, pre-clinical, and clinical development of nanomedicines are being accelerated by AI and ML, as we have briefly covered in this chapter. The possible drawbacks of using AI/ML techniques in the creation of nanomedicine products have also been discussed. In order to finally shorten the time and expense of developing nanomedicine products, the AI/ML tools assist with screening and analyzing massive datasets, modeling, prediction, process optimization, and decision-making. This chapter examines how the development of nanomedicines for individualized and focused therapies is being revolutionized by artificial intelligence. The development of nanoparticles has historically relied heavily on trial and error and guessing, which hinders progress and raises costs. AI has made it possible to get closer to data-driven decision-making, where models may direct formulation decisions, forecast therapy results, and provide customized treatments for each patient. Researchers are starting to combine AI algorithms with intelligent nanomaterials to develop precise drug delivery systems for conditions like cancer, neurological diseases, and cardiovascular disease. We demonstrate how AI is not only improving nanomedicine but also making personalized therapy a feasible objective in the near future by examining recent research in this rapidly developing field.

2 Integration of AI in Nanotechnology

2.1 Discovery and Design

AI-enabled drug discovery has been linked to a variety of medicinal uses, including biologics and small molecules. However, AI has only lately (over the last decade) been used in the development of nanomedicine formulations. Historically, the identification of innovative nanomedicine formulations has relied heavily on trial and error. In the last decade, there have been several attempts to incorporate AI technology into the development of nanomedicine formulations, particularly when developing formulations with specific properties such as the ability to target organs and tissues, safety, efficacy, drug loading, release kinetics, pharmacokinetics, biodistribution, and particle size distribution. Figure 1 visualizes the schematic representation of the use of artificial intelligence (AI) in the drug discovery process [10].

For predicting chemical interactions and finding new nanomedicines with desirable features using different deep learning algorithms, research in this area has so far mostly concentrated on investigating large molecule medications and excipients (such as lipids and polymers). Achieving high drug-load in nanoformulation products, particularly when composed of hydrophobic drugs, has historically posed significant obstacles; however, increased understanding of how machine learning (ML) guided high-throughput experimentation can support prediction of the self-assembly of hydrophobic drugs through the use of polyfunctional polymers and/or nanoscale molecular dyes, particularly in identifying rapidly and uniformly the application of nanomedicine that has the potential for high aggregated drug load capacity across many independent applications. In addition, ML and AI are being used to optimize the formulation development of long-acting nanomedicine products. The utilization of ML-based algorithms to analyze the imaging data to project long-term drug release, overall stability, and in vivo performance has facilitated the accelerated development of nanomedicine formulation products [11].

Advancements in mRNA and LNPs are one area where AI/ML-based approaches have made substantial progress. The classic discovery of ionizable lipids that enable efficient mRNA delivery has relied on time-consuming in vitro library screening. Li et al. have more recently developed a machine learning-based virtual screening platform to identify novel ionizable lipids with high transfection efficiency. To identify lipid compositions that enable LNPs to reach cells or organs other than the liver and lymph nodes, researchers have also developed an AI/ML technique that synergistically integrates deep learning with combinatorial chemistry.

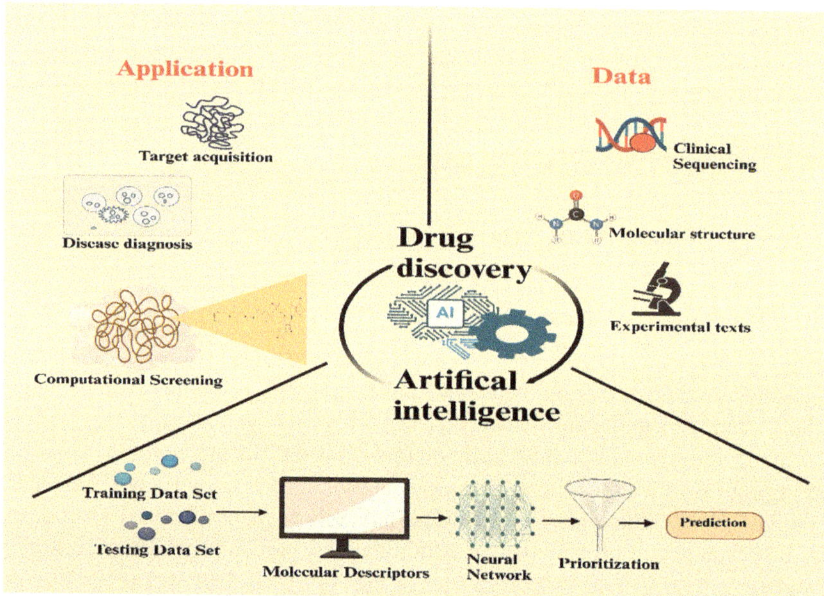

Fig. 1 A schematic illustration of how AI is employed in drug discovery. It shows the workflow starting from data collection—such as clinical sequencing, experimental testing, and molecular structure analysis—leading to the application of AI algorithms and neural networks. This method can be used for disease diagnosis, target identification, and computational screening. The diagram emphasizes AI's transformative impact on medical research and healthcare

It's noteworthy that several centralized AI-based web platforms, like Formulation AI and AI-Guided Ionizable Lipid Engineering (AGILE), have also been developed recently by merging existing excipient and pharmaceutical data [12]. The first formulation-based platform, AGILE, was developed using the GNN algorithm to specifically represent the extremely adaptable yet simple molecular structures of ionizable lipids. In order to build and construct LNPs that carry mRNA to muscle and immune cells, it has been utilized to predict novel ionizable lipids. Several models based on over ten distinct algorithms were developed for formulation AI to choose the best-fit model. Formulation AI makes it possible to develop a formulation by simply entering basic drug and excipient data. The majority of work in this field, particularly the application of AI in formulation design, relies on internal data generated from preformulation experiments or data derived from molecular computations (e.g., molecular descriptors, intermolecular distances, and energies) due to the limited availability of data.

Notably, the majority of published data related to LNPs' in vitro performance is provided on heat maps or bar graphs and frequently lacks a standard control, making data processing and comparisons difficult for AI scientists. This demonstrated how important it is for scientists to take the meta-analysis into account when publishing study findings. The quality of the data is equally crucial since it frequently significantly affects the model's quality. Data audits and reviews can help create a healthy, impartial model by ensuring that the training data is representative, diversified, and bias-free [13].

AI can be used in nanomedicine design to predict drug-excipient interactions in addition to designing new lipids and other excipients. Protein or peptide-based targeting moieties or therapeutic payloads can be found more quickly thanks to AlfaFold and AlfaFold2's effective prediction of protein or peptide structures and their interactions. Using such payloads and targeting moieties as basic building blocks, new nanomedicine products can then be engineered. AI and ML demonstrate efficient cellular internalization and target binding, and thus aid in screening the peptide architectures. Incorporating such structures within nanomedicine using AI-assisted peptide engineering and formulations can achieve improved targeting efficiency [14].

2.2 *Manufacturing, Characterization, and Quality Check*

AI-based approaches like deep learning provide unparalleled access to vast amounts of data that can be processed arguably better than traditional prediction methods. Such prediction capability has fueled dramatic advancement in nanomedicine production and quality control. Main applications explored in reference literature include formulation development, identification of key process parameters (CPPs) or critical quality attributes (CQAs), process optimization, or process validation for quality assurance by process analytical technology (PAT). In optimizing formulation for rational liposome design with optimal size distribution and Polydispersion Index (PDI), Rebollo et al. employed the combination of an artificial neural network (ANN) with designs of experiments (DoE). In this study, only those parameters that significantly influenced formulation outcome, such as flow rate, were input into the ANN for the prediction of liposome size distribution and Polydispersion Index. It is important to note that ANN was trained for prediction from experimental data sources not organized by pre-designed experiments, indicating that AI can devise predictions from available data that need not be pre-designed for experiments. Using ANN, only one prediction model could be developed for PDI and size distribution, which was impossible for DoE alone [15].

The in vivo activity of NPs is significantly affected by CQAs, including particle size, surface charge, and composition. Reproducible manufacturing of NPs is crucial but complex because small changes in process parameters can significantly change these CQAs. The reproducibility of manufacturing processes can be enhanced notably by integrating cutting-edge analytical technologies with AI-based data analysis and ML algorithms that offer real-time analysis. Apart from that, the use of AI will enhance our understanding of the CPPs concerning the CQAs in the nanomedicine sector, which further reduces the risks associated with any quality issues in a product. Many approaches related to AI/ML, for example, ANNs, DL, support vector machines (SVM), genetic algorithms, and multiple linear regression, have also been utilized in the nanomedicine manufacturing process [16].

Continuous manufacturing has increasingly incorporated AI and ML tools. In pharmaceutical continuous manufacturing, the PAT initiative employs in-line, at-line, and online sensors to measure CQAs and CPPs in real time. This approach facilitates error correction, process condition adjustments to prevent batch failures, and supports a more agile manufacturing paradigm. The integration of PAT with continuous production generates substantial multivariate data. ML-based techniques, including ANNs, are capable of adapting and modifying model behavior to enhance ongoing performance and can process large datasets in real time. These capabilities support adaptive control, autonomous decision-making, and the advancement of digital twins and smart factories [17].

The Multi-Input Single-Output (MISO) model, with particle size variables modeled from one neural network and PDI from another, performed better than the MIMO model with two neural networks. During development, training, and testing phases, errors were monitored. The mean relative error between predicted and actual values was reduced in the optimized model that used molecular descriptors. The results showed that algorithmic models can accurately predict the CQAs of liposomes and other nanoformulations. These AI/ML-based investigations are strongly advised for both creating development platforms and cutting down on the time and resources needed to optimize nanoformulations, which will ultimately speed up the clinical development process [18].

The pattern recognition and prediction capabilities of AI can have major applications in nanomedicine manufacturing to enhance prediction, enable automation of PAT approaches, identify product defects and out-of-specifications, and support real-time monitoring of product physicochemical properties. As demonstrated with tablets, the performance and dependability of PAT can be greatly enhanced by the incorporation of sophisticated data analytic capabilities driven by AI. ANNs and a digital twin approach for milling operations have been used by Davidopoulou et al. to forecast and improve nanosuspension production process results. In order to supplement the experimental dataset and mimic the dynamic comminution profile inside the actual mill, an ANN was specifically integrated into data sampling, model deployment, and curve fitting. This allows for the estimation of process outcomes, such as particle size profile, at any moment during milling. All things considered, even if they are still in their infancy, AI and ML-based techniques in conjunction with PAT will be essential instruments for tracking, managing, and optimizing a wide range of variables in real time to guarantee the repeatable production of nanomedicines, as illustrated in Fig. 2 [19].

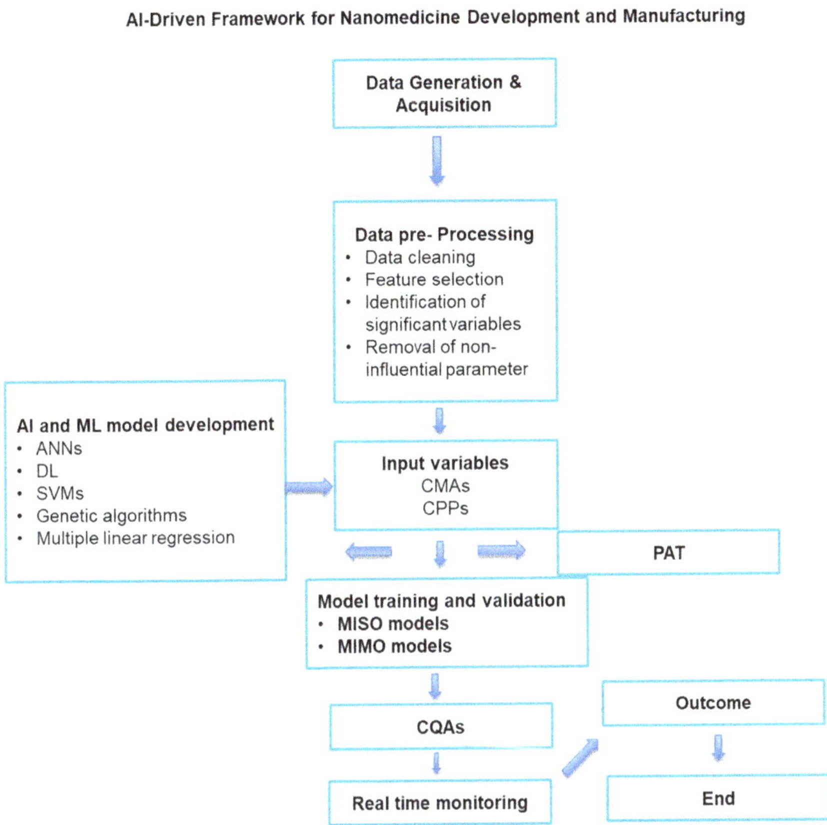

Fig. 2 AI-driven framework for nanomedicine development and manufacturing

2.3 *Illustrations and Justification*

The ideas and innovations described are not merely theoretical; they have already been applied in practical settings. A machine learning model to forecast the effectiveness of LNP formulations was created by a group of researchers at Johns Hopkins University. Research and development of new nanoparticle designs will be carried out using the technology. The most efficient particles for targeting tumors were found thanks to the AI system's ability to precisely forecast cellular absorption and shape in breast cancer models. AI promotes research on brain tumors by involving nanoparticle interaction with tumor tissues and overwhelming the BBB and thus led to the development of therapeutic advancements in glioblastoma treatment, enhancing efficacy and precision. [20].

3 Nanomedicine in Personalized Therapy

A key advantage of nanomedicine is the shift from uniform treatments to individualized care based on each patient's unique biology. This personalization is crucial for complex conditions such as cancer, heart disease, and brain disorders, where patients may respond differently to the same therapy. AI plays a key role in enabling this transition by analyzing clinical data, patient biomarkers, and genetic profiles [21].

3.1 Targeted Delivery

Nanomedicine offers significant opportunities for targeting specific molecular pathways in various diseases. AI plays a pivotal role in predicting optimal formulation parameters, including size, surface characteristics, and drug loading factors for each application. In cancer therapy, AI has facilitated the development of nanoparticles that respond to the unique tumor microenvironment. For example, in oral cancer, researchers have engineered nanocarriers that release therapeutic agents only in the presence of specific enzymes found within the tumor microenvironment. In brain cancer, AI has enabled the prediction of optimal nanoparticle designs for targeted drug delivery to glioblastoma cells, minimizing impact on healthy brain tissue (visualized in Fig. 3) [22].

Artificial intelligence is currently utilized to develop lipid-based nanocarriers that can target sites of cholesterol accumulation or deliver anti-inflammatory agents. These nanocarriers have demonstrated significant potential in dyslipidemia treatment. AI-driven design is also advancing the management of neurological and autoimmune disorders. Researchers are investigating how nanoparticles can be engineered to target inflamed regions of the brain and deliver therapeutics for conditions such as multiple sclerosis and Alzheimer's disease.

3.2 Biomarker Identification and Immune Modulation

The key to personalized nanomedicine in the future is to understand what makes every individual biologically unique. Scientists use AI to sift through massive amounts of data to find biomarkers, simulate immune system reactions, model patient responses to specific nanoparticle formulations, and predict patient responses to specific nanoparticle formulations. For example, biomarker discovery is critical in the creation of tailored vaccinations and glioblastoma-specific nanocarriers. AI algorithms study cancer genomes to discover specific mutations or neoantigens that can be used as targets. In one study, AI assisted in the selection of patient-specific antigen profiles, which facilitates the delivery of individualized cancer vaccines via

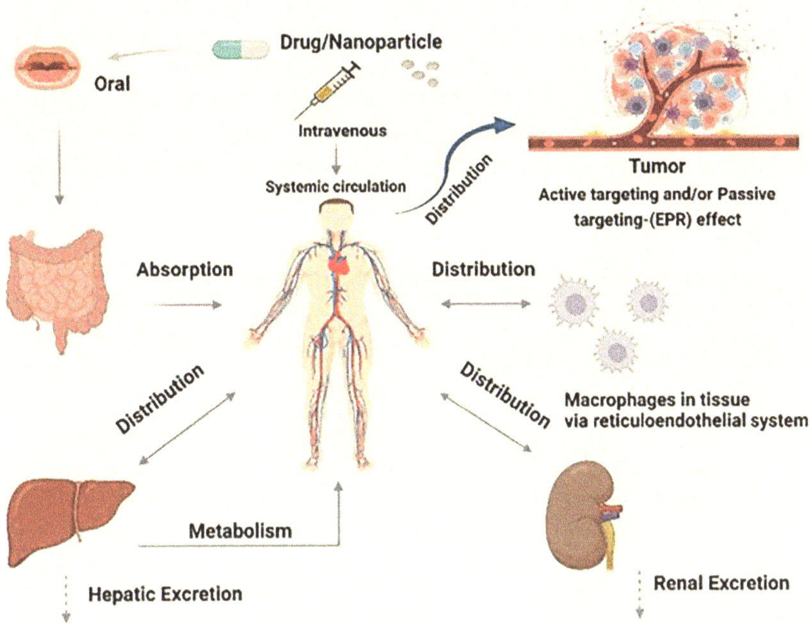

Fig. 3 Precision-based target delivery

lipid nanoparticles to induce immune system reactions particular to each individual's tumor type [23]. In the study of brain cancer, specifically glioblastoma, researchers have applied AI to identify receptors that are overexpressed in particular patient samples. Based on these patient-specific characteristics, scientists have designed particular nanoparticles.

AI is also greatly impacting the modulation of the immune system. While just carrying drugs in the body is sufficient in nanoparticles, they can be made more intelligent so as to train or modulate the immune system. In the treatment of cancers through the immune system, AI has assisted in the development of intelligent nanomedicines. These drugs work by reducing the inflammation associated with the treatment method while triggering an accurate response from T-cells. These techniques involve training the computer using biological data about the immune system. They also involve the best way to inhibit or activate a given biological pathway. Most importantly, it helps the scientist predict how the drug will behave in the patient. While using machine learning concepts, the machine uses the composition of the nanoparticle and the health background of the patient. This helps the machine predict the effectiveness and toxicity levels in the human body. Johns Hopkins scientists have used machine learning to predict the performance levels of differently structured lipid nanoparticles in response to varied biological conditions. This has increased the success levels of mRNA delivery. Other models focus on finding the appropriate candidate for the new nanomedicine therapies. As the technology advances, AI will also help in advancing nanomedicine from a laboratory setting into a personalized treatment option. This is the right treatment at the right time for the right patient [24]. AI closes the gap between Big Data and Personalized Medicine. It enables predictions for individuals, the identification of biomarkers, and the guidance of the development of pharmaceuticals. In assisting scientists in developing safer, more effective, and more intelligent medications for individuals, we move closer to the reality of Personalized Medicine.

4 Practical Application of AI and Nanomedicine

4.1 Cancer Vaccine Design

Designing a cancer vaccine that is truly effective for every individual patient is one of the greatest hurdles in immunotherapy. Since every patient has a tumor that is genetically specific to them, a traditional vaccine would not be as effective. AI is now going to be the most handy tool for researchers in their quest to find the best way to immunize, per the unique tumor profile of the patient. Recently, Dubey et al. [4] have developed an AI-based approach that supports the creation of patient-specific mRNA and DNA cancer vaccines (illustrated in Fig. 4). Their method predicts tumor antigens most likely to generate a powerful immune response by combining biological datasets such as patient genomics and proteomics with machine learning. These are called neo-antigens, which are unique to the tumor and make excellent therapeutic targets. After recognizing probable neo-antigens, the algorithm employs deep learning to optimize the vaccination sequences. This strengthens factors such as mRNA stability and its capability to get into cells. Furthermore, by estimating the level of recognition of specific antigens by T-cells, the system simulates the patient's immune response in advance, thus offering researchers a pool of the most promising vaccine candidates for that particular patient. The study showcases not only the assistance with delivery methods but also the control of what is supplied as the main support, which is the case in the AI-driven personalized nanomedicine as the first example. It's a great example of how data science can be applied to the enhancement of cancer treatment and the creation of more intelligent, patient-specific drugs [4].

4.2 Therapy for Breast Cancer

The large design space (various polymers, particle sizes, surface changes, and drug loadings can drastically alter how nanoparticles interact with target cells) is one of the main obstacles to developing effective nanoparticle-based therapeutics. It is expensive, time-consuming, and frequently ineffective to manually test every formulation. To speed up the creation of nanocarriers for breast cancer therapy, a group of researchers invented a pioneering AI-guided platform that combines high-throughput

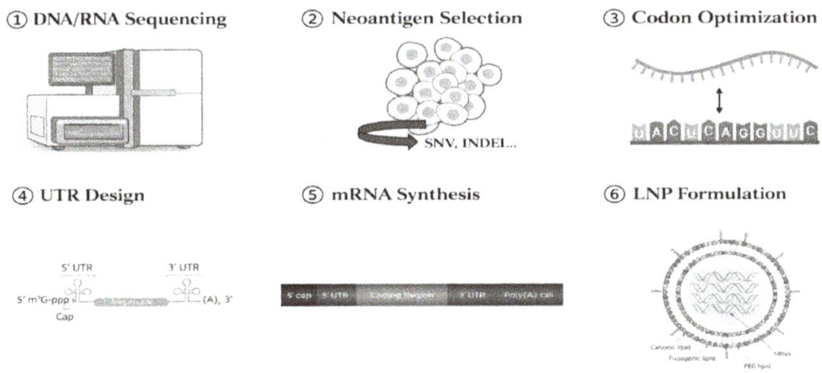

Fig. 4 Cancer vaccine designing via DNA/RNA sequencing through AI

imaging, active machine learning, and microfluidic nanoparticle production (visualized in Fig. 5). They specifically worked on the development of nanoparticles made up of PLGA-PEG for selective uptake in the human breast cancer cell line MDAMB-468. The traditional trial-and-error procedures were replaced by an iterative machine learning procedure. The process started with a learning set of various nanoparticle materials and how they were taken up by cells. The model identified the next set of promising nanoparticle compositions to synthesize and test. After each test cycle, new knowledge was cycled back into the model. The process allowed researchers to sample how the ingredients came together. After just two rounds of this process of finding the best nanoparticle composition, the biggest advantage was a 15-fold increase in cellular uptake in days compared to months [25].

This demonstrates how AI can initiate the nanomedicine development that enhances its ability to scale and adapt to new cell types/therapy targets. Thus, inclusion of data-driven practice into nanomedicine, predominantly in oncology, leads to enhanced and efficient delivery and individualized therapy. The case also depicts the involvement of experimental techniques, with AI transitioning into experiment design and decision-making.

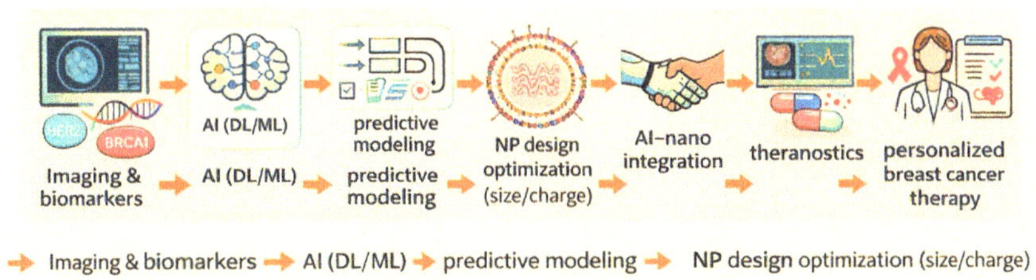

Fig. 5 AI-driven nanomedicine for breast cancer, depicting progression from imaging and biomarker data by AI-based studies

4.3 LNP for Gene Delivery

As per the study conducted by Hai Quan Mao, AI has the potential to transform nanomedicine by increasing machine learning methods for the design of lipid-based nanoparticles (LNPs) for gene delivery. This might be able to solve the problem of LNPs by forming effective and cell-type-specific transfection by considering the broad usage of LNPs in mRNA vaccine delivery. Over 1000 different LNP formulations tested on six different cell types, which included brain, kidney, and cancer cells, were combined and curated by the researchers from the high-throughput screening tests. They devised a strategy to train ML algorithms like ensemble-based classifiers on this data to predict the best compositions of nanoparticles for targeted delivery. The models made the process of formulation refinement faster by showing lipid component combinations and design rules that were not obvious but improved gene transfer efficiency in specific cells. This technique not only led to a scalable, predictive framework for the design of next-generation LNPs that are suitable for different therapeutic use cases, but it also reduced the experimental burden usually related to nanoparticle screening [26]. Because it demonstrates how AI can convert empirical trial-and-error techniques into logical, data-driven design strategies, the study is highly relevant to personalized medicine and AI-integrated nanomedicine pipelines.

5 Advanced Drug Delivery Systems

Smart delivery systems for drugs have been introduced in nanomedicine as a paradigm shift in the way that drugs are carried, from passive carriers to intelligent systems that can respond to the physiological cues. The systems are designed to release the drugs that heal, through both time and place, and the release is to a large extent determined by the local changes in the physiological parameters such as pH, temperature, and/or the body's levels of specific enzymes. It is predicted by the authors Akbarzadeh et al. [27] that AI can design an array of multifunctional nanocarriers that can instantly adapt under the influence of physiological or therapeutic stimuli, thus improving the accuracy, timing, and customization of the patient's therapy [27]. The use of AI to create the materials' properties, simulate the drug release properties, and integrate wearable or implantable biosensors that allow the treatment based on real-time feedback of the local conditions has all been indicated by recent findings.

5.1 Nanocarriers

Smart drug delivery systems stand out because they can release medication only when certain signals show up—things like changes in pH, temperature, or enzyme levels inside the body. These nanocarriers function as responders to the fluctuations that occur in the body using a pre-programmed trigger system. When a drug delivery system utilizes these types of carriers, the drug will only be released at the appropriate organ and time. Therefore, the drug will perform its intended function at the targeted site of action, which will substantially reduce the potential for adverse effects and toxicity, as well as improve overall treatment efficacy.

pH and Temperature-Responsive Systems: Tissue in proximity to cancerous tissue or inflammation is typically slightly more acidic compared to normal tissue. This acidity is detected by AI-derived nanocarriers. When these carriers detect changes in pH, they can change in shape/size or release their therapeutic contents on demand. In addition to these types of responses, some types of these carriers may also respond to thermally induced stimuli, which are beneficial for the use of hyperthermia treatments. By utilizing ML algorithms, researchers can optimize the formulation of these smart carriers and investigate what minor alterations will result in more effective delivery of drugs.

Enzyme-Triggered Carriers-The following approach makes use of AI to design nanocarriers that are specifically activated by enzymes that are produced in high quantities in affected tissues. Hyaluronidase and matrix-metalloproteinases (MMP's) are two examples of enzymes that are often produced in much greater quantities when there is a bacterial infection and/or when cancer is present. When utilizing AI to detect combinations of these metabolically produced metabolic signals, researchers can design nanoparticles that deliver therapeutic drug compounds only when the presence of these specific enzymes occurs. The specificity in which these newly developed systems operate is evidenced by their success in managing oral and lung cancer models where treatment is delivered exactly where the spike in enzyme production occurs, thus limiting the toxicity to surrounding, healthy tissue while delivering the therapeutic agent(s) directly to those infected with this particular cancer type. In addition, the potential to develop future delivery systems utilizing AI provides an additional opportunity to better tailor the treatment regimen to meet the individual needs of specific patients in managing their particular, complex, and prolonged medical conditions such as cancer.

Stimuli Sensitive Materials-Intelligent nanocarriers are made of materials that can respond to a broad range of external stimuli and biological conditions, and the ability of these materials to undergo means developed through the application of intelligent carrier systems allows for the ability to engineer systems that will react directly to their environment. The structure, permeability, or rate of release of a pharmaceutical agent from a nanocarrier may be modified by incorporating smart materials into nanocarrier systems. Physiologically speaking, the stimuli for nanocarrier systems can be both internal and external stimuli, with examples of internal stimuli being changes in pH, concentration of enzymes, and redox potential in the diseased tissue. Examples of external stimuli include exposure to light, magnetic fields, and ultrasound, to name a few. For example, some polymeric materials can expand and/or decompose when exposed to acidic environments such as those found in tumors, thereby allowing for targeted drug delivery.

AI has great potential for developing and improving responsive materials to obtain predictive results based on specific product compositions and polymer structures. For instance, the role of AI in the field of nanomedicine for lung cancer management involves designing and manufacturing enzyme-sensitive polymer-lipid hybrid nanoparticles that enhance the efficacy of technologies by providing enhanced permeation into tumors, thus enabling greater specificity of drug activation. The combination of responsive materials and AI-enabled design procedures can facilitate the development of flexible medications that will elicit a different response for each patient, thereby providing the foundation for a truly personalized nanomedicine. Furthermore, responsive polymeric materials may also be tailored to target specific tissues, treat particular diseases, or address unique biomarker profiles of individuals. By utilizing AI-augmented design methodologies, it is possible to design and develop responsive materials that will deliver the appropriate therapeutic agents at precisely the right times and locations of the body for maximum effectiveness [25, 27].

5.2 *Multi-functional Nanocarriers*

Multi-functional nanocarriers are multi-tasking platforms used for the simultaneous performance of therapeutic and diagnostic functions and represent the next generation of Smart Delivery Systems. Multi-functional nanocarriers can function as theragnostic delivery systems (i.e., they target with drugs, sense with imaging technologies, and deliver therapeutics by incorporating targeted Diagnostic Devices). Artificial intelligence is dramatically changing how theragnostics are designed, as it can process large amounts of data from chemistry, materials science, PK, and human biology to identify the optimal combinations of materials, targeting ligands, and release triggers for a specific therapeutic goal. AI will enable us to prototype designs in silico, thereby allowing us to evaluate hundreds of proposed designs before testing in the lab instead of relying solely on empirical testing. By using AI, researchers can now use in silico prototypes to simulate and evaluate many different nanocarrier designs before creating physical prototypes in the laboratory. Through this method, researchers can create prototypes quickly without waiting for experimental results; therefore, this technology is particularly advantageous for personalized medicine. For example, individualized releases can be initiated by the expression of certain biomarkers, dosages can be adjusted based on metabolic differences, and feedback systems (e.g., wearable or implantable biosensors) for adaptive therapy can be built into multifunctional nanocarriers. The combination of AI and intelligent nanotechnology has brought us much closer to achieving real-time, data-based therapies that will adapt as the patient's health changes over time [28].

5.3 Wearable and Implantable Sensors

Incorporating biosensors into nanocarrier systems and creating closed-loop therapy systems represents one of the most exciting developments in smart delivery methods for pharmaceuticals. These devices continuously monitor and collect real-time physiological data to facilitate real-time adjustments to both dosage and release profiles, as well as timing and frequency of medication administration throughout the therapy process. Patients may be cared for by clinicians over time without needing to visit regularly for physical examinations (i.e., the application of clinical judgment) while providing a greater degree of precision in treating the patient's condition and minimizing the risk of adverse drug reactions than would be possible under the existing standard-of-care practice. For example, when a nurse makes a decision based on glucose levels, inflammatory markers, or specific cancer biomarkers, the biosensor connected to the wearable will wirelessly or through physical connection to the automated medication delivery system tells the system what to do with the medication, i.e., to dispense the drug on demand or to stop dispensing it altogether, based on the response of the individual to treatment once the monitored value for the biomarker reaches the threshold specified. More intricate versions of closed-loop therapy systems have also been developed that use artificial intelligence software with predictive capabilities to help prevent long-term morbidity and improve treatment outcomes for chronic diseases by deploying microdevices capable of sensing and dispensing drugs under computer control [29]. This allows for proactive therapy modifications rather than reactive dosage.

AI will play a critical role in utilizing this overwhelming amount of data generated by wearable sensors since it will enable real-time adjustments to the parameters of the delivery system [30]. Examples include the early detection of a flaring or recurring disease through monitoring, adjustment of unfair interference noise during delivery/sensor use, and modified treatment regimens based on machine learning predictive analytics. AI-enabled sensors integrated into wearable devices have been suggested as a new method of providing targeted drugs for the treatment of specific cancers. Immunological markers present in the blood of cancer patients can be used to provide localized delivery of anticancer drugs at optimal points during the immune cycle. When combined with AI-enabled sensor data, drug distribution becomes adaptable to individual patients and creates a framework that will enable physicians and AI systems to develop therapeutic regimens that will continuously adapt to each patient's changing needs through AI's predictive capabilities and nanomedicine's precise and highly targeted delivery.

6 Ethical and Safety Considerations in AI-Enabled Nanomedicine

As the rapidly developing field of AI-enabled nanomedicine grows, the ethical/safety concerns associated with its increasing use should be addressed. Addressing issues surrounding the ethical, regulatory, and safety considerations needs to take place along with the increase in the number of applications of AI and nanotechnology for use in medicine. Furthermore, these new technologies can generate new challenges associated with safety and ethics. AI-powered use of robotics in nanotechnology can modulate healthcare by introducing personalized treatments as well as providing them much more efficiently. Accordingly, it is vital to ensure that while including these technologies in the healthcare system, it should be executed ethically and equitably. For instance, promoting maximum benefit while limiting associated risks requires an equilibrium between informed consent, safety, regulation, transparency, and a framework for ethical decision-making [31]. Numerous ethical/safety issues are related to the use of AI-based nanomedicine products.

6.1 Legal and Ethical Issues

AI nanomedicine faces many legal and ethical difficulties, such as (i) Informed Consent and Autonomy, (ii) Privacy and Security, (iii) Accountability and Transparency, (iv) Bias, (v) Intellectual Property Rights, (vi) Regulatory Backgrounds, and (vii) Monitoring and Accountability. Security and privacy will require consideration while developing a process to provide consent for the collection and use of data and for the development of data protection laws. Informed Consent and Autonomy are two of the most important aspects of developing clear protocols for obtaining patient consent to use their information and protecting their Autonomy. The need to balance IP protection with the need to create an environment conducive to continuing research for developing and promoting new treatments via the AI nanomedicine industry will determine how IPRs in AI nanomedicine will be defined. Regulatory Systems (RS) must be created and revised to ensure ongoing safety and efficacy while also providing accountability for ongoing assessment and addressing the emergence of New Ethical Issues and Risks associated with AI nanomedicine. Additionally, there is currently a debate regarding the classification of artificial intelligence (AI) technology into either an established category of Law or into a separate category specifically for AI technology with its own unique characteristics and legal implications.

6.2 Safety Considerations

AI and nanotechnology pose significant safety hazards when applied to therapy. Lack of accountability raises concerns about the use of AI systems (AIs) in medical practice without validation or confirmation for safety risks. The responsibility to "not harm" should rest on all stakeholders, including doctors and AI developers/engineers, and ethical considerations for system safety should involve them both because they share the common goal of preventing patient harm [32].

6.3 Risk Assessment

AI can spot both the pros and cons of the products developed by other AI. Besides, through this process, the ethical issues that might come up can already be understood and dealt with before the product is launched. AI's role in medical decision-making is very wide and complicated. In order to avoid the possibility of privacy violations, it is important to pinpoint and evaluate the risk of unauthorized access to confidential health data using encryption techniques, access control systems, and data storage practices. There are indeed many ways to help patients to have a better understanding of the applicability and consequences of AI, but informed consent is still the major problem. The issue is not only the chance of adverse side effects or unforeseen reactions from the nanoparticles used in the medical treatments, but it also raises significant ethical dilemmas. In addition, one of the major factors influencing the safety of nanomaterials is the assessment of their toxicity and safety profiles. Hence, measures should be taken that will ensure the AI assistance does not go beyond the level of expertise of the medical professionals involved in the decision-making process, since in this case, the AI over-reliance may only cause the human participation to be reduced in the area of medicine. Moreover, the situation where access to AI-empowered healthcare technologies is not equal can make global health inequalities even worse [33].

6.4 Monitoring and Guidelines

Ethical dilemmas related to the use of AI in nanomedicine include real-time monitoring of algorithms regarding any relationship to patient outcomes, data security assessment, transparency about the data used to develop AI, and verification of all stakeholders regarding the ethical obligations arising from AI deployment in healthcare. Several avenues may be pursued to support the ethical use of AI in healthcare, including recognizing and mitigating biases related to AI, safeguarding patient privacy, and ensuring the explainability of AI algorithms developed for clinical decision-making. Finally, a regulatory compliance framework describes the steps necessary for obtaining all required regulatory approvals of AI-enabled nanomedicine while sustaining ongoing compliance with relevant laws and standards [34].

6.5 Standards and Regulations

Data Privacy Standards, Algorithmic Fairness Standards, Explainability and Transparency Standards, Informed Consent Guidelines, and Safety and Toxicity Evaluation Criteria are the five standards for the creation and application of AI nanomedicine products. By guaranteeing concern for bias in AI algorithms, ensuring informed consent to the use of AI and nanotechnology in healthcare (Informed Consent Guidelines), and offering transparency and explainability regarding how AI algorithms make medical decisions on behalf of patients (Explainability and Transparency Standards), these standards aim to provide equitable access to healthcare for various populations. The Compliance with Existing Healthcare Regulations verifies that the use of AI-assisted medical products complies with current healthcare regulations and describes the regulatory procedures for their approval and oversight. Continuous Monitoring and Reporting Requirements call for routine audits of AI algorithms' performance in order to assess and resolve any ethical concerns [35].

6.6 Interoperability

Patient-centric interoperability, data exchange standards, consent management across systems, cross-institutional collaboration, transparent data usage policies, the impact on vulnerable populations, continuity of care, cybersecurity, interoperability standards for AI models, and education and training for healthcare professionals are all important ethical issues in AI-enabled nanomedicine. People can effectively access and share their health information thanks to patient-centric interoperability, which is coherent with the ethical norms of autonomy and informed consent. Standards for data interchange promote collaboration, close information gaps, and improve transparency. Patient consent preferences must be consistently respected through potent consent management across systems to uphold ethical integrity. Collaboration across institutions advances ethical best practices, increases knowledge, and enhances healthcare outcomes. To safeguard ethical data practices and promote patient confidence in AI-enabled nanomedicine, transparent data usage policies are crucial. Continuity of care is an ethical requirement because effective interoperability facilitates smooth patient data circulation. Lastly, creating standards for AI models guarantees consistent, dependable performance and interoperability.

6.7 Transparency and Comprehensibility

The rapidly developing field of AI-enabled nanomedicine calls for the ethical and responsible integration of several approaches. Algorithmic transparency, clear explanations of AI decisions, patient-focused communication, user-friendly interfaces, access to algorithmic insights, comprehensive documentation of AI models, education for healthcare professionals on AI, management of uncertainties and limitations, public awareness campaigns, and ethical considerations during design and development are important elements. By enabling patients and healthcare professionals to comprehend the decision-making process, transparent algorithms promote confidence. To make informed decisions and get patient consent, AI decisions must be clearly explained. Clinicians can make morally sound decisions by having access to thorough insights produced by AI systems. Healthcare professionals who receive AI education are better prepared to handle moral dilemmas and appropriately incorporate AI into clinical practice. By educating the public about AI's role in healthcare, public awareness campaigns promote autonomy and foster trust. Incorporating ethical considerations into the design and development of AI-enabled nanomedicine guarantees transparency and comprehensibility while reducing potential problems. For human experts and AI to work together effectively, it is essential to comprehend the reasoning behind AI recommendations [36].

7 Limitations of AI/ML in Nanomedicine Product Development

The time and expense needed for the discovery, production, PK assessment, clinical design, and testing of nanomedicine products has significantly decreased due to the application of AI and ML-based techniques. AI/ML-based models have certain drawbacks despite their many benefits. In their discussion of AI's limits in drug discovery, Benders and Cortes-Ciriano emphasized the significance of data quality for forecasts and success rates. It is crucial to make sure that the underlying training data for AI/ML models is substantial, varied, high-quality, and bias-free. Bias and low predictive power may result from little and insufficiently diversified training data. Because the data might not match the pattern that the algorithm was trained on, it is possible to overlook a clinical candidate that is a perfect fit. The algorithm must be founded on precise assumptions and correlation limitations to avoid biases. Because bias can result in erroneous predictions and uneven performance, machine learning algorithms should routinely assess and correct bias in clinical testing. Patients, including minorities, women, and low- and middle-income nations, are not being adequately represented in the clinical trial design. So, with regard to the ethical and security concerns, sponsors must inform the patients about the advantages and disadvantages of sharing data via AI/ML models, allowing patient permission and retention during trials. Besides, financial restrictions and limited access to the infrastructure prevent the organization from utilizing AI/ML tools in their product development [37, 38].

8 Future Perspectives

The advancement of AI-enabled nanomedicine will necessitate the integration of novel computational and biological technologies alongside artificial intelligence. Such technological convergence is expected to substantially enhance the capabilities of intelligent and personalized nanomedicine platforms. Quantum algorithms have the potential to accelerate molecular simulations in nanomedicine, facilitating the prediction of nanoparticle–biological interactions with unprecedented precision. This advancement would enable researchers to investigate extensive formulation spaces, including billions of possible nanocarrier compositions in silico, thereby expediting the identification of optimal drug delivery strategies. When combined with AI-driven nanomedicine platforms, the IoMT can facilitate continuous, patient-specific feedback loops that support adaptive therapy. The integration of AI with these advanced technologies is expected to transform nanomedicine from static formulations to dynamic, responsive, and predictive therapeutic systems, while also enhancing computational capacity. The future of nanomedicine hinges on implementing interconnected, AI-powered ecosystems that deliver highly customized treatments that adapt to each patient's changing state. By gradually learning from each treatment cycle, the approach improves therapeutic outcomes while reducing side effects. These AI-driven, customized nanomedicine ecosystems will require robust data infrastructure, privacy-preserving analytics, and strong interoperability standards to integrate laboratory, clinical, and patient-generated data [39].

Clinical results, proven AI algorithms, and standardized nanoparticle characterization data can all be centralized on open-source platforms. Over time, these frameworks may support international nanomedicine innovation ecosystems that link academic research with biotech startups, pharmaceutical firms, and public health organizations [40]. AI-powered nanomedicine represents a transformative approach with the potential to address health inequities in low-income countries. Integrating local epidemiological data, artificial intelligence analytics, and resource-optimized delivery strategies may enable historically marginalized populations to benefit from advanced nanomedicine. AI-driven precision health models can tailor nanomedicine interventions to local needs. For example, these models may recommend cost-effective, thermally stable nanocarriers for regions lacking cold-chain storage or adapt treatment regimens based on local disease prevalence and available diagnostic capabilities [41]. Precision health models can provide regionally appropriate and effective treatment recommendations by incorporating local genetic, clinical, and public health data. Additionally, AI-powered algorithms can monitor shifts in disease epidemic trends, optimize resource allocation, and anticipate treatment demand. These tools also enable primary health centers to implement preventive strategies by targeting high-risk areas with nanomedicine-based therapies, thereby improving efficiency and reducing resource waste. Precision health models can guide cost-effective nanomedicine deployment strategies, ensuring that investments in state-of-the-art treatments have the greatest possible impact on population health [38].

9 Conclusions

In order to better anticipate and optimize the discovery process, development, scalable production, and in vivo pharmacology of nanomedicines before clinical trials, artificial intelligence (AI) is set to play a crucial role. Using the vast quantity of PK and biodistribution data accessible in the literature, a number of ML-based techniques can specifically improve the PK/PD features of NPs to aid their clinical translation. Furthermore, AI-integrated PAT can enhance process understanding, which benefits CQAs from a Quality by Design (QbD) standpoint that is essential to regulatory success. This is because the reproducible manufacture of nanomedicines is crucial. The FDA is actively creating its AI regulatory framework to guarantee the ethical and successful integration of AI in clinical studies and commercial development.

In general, the development of nanomedicine will be more successful, economical, and transferable for wider clinical applications when AI and ML techniques are incorporated. However, it is crucial to make sure the model is validated for precise prediction of nanomedicine properties when utilizing innovative AI-enabled modalities. It is also crucial to have a procedure for data quality control. These mitigation techniques can be utilized in the pharmaceutical industry's research, production, and clinical development processes to address biases brought on by ML-enabled prediction models. All in all, the product development scientists, ML engineers, data scientists, robotics engineers, regulatory agencies, AI researchers, and clinicians must work in perfect collaboration so that the gradual and advantageous application of this technology in the product development processes is made possible, and thus the integration of AI/ML methods in nanomedicine research is opened up.

The use of AI in medicine has been associated with the running of clinical trials, personalized treatment, and other innovations on the laboratory front, which made the practice of nanomedicine much faster. In this chapter, the application of AI across the nanomedicine pipeline is discussed. The areas of application include clinical trial optimization, adaptive treatments, nanoparticle generation, and smart drug delivery systems. Some of these applications have already demonstrated that AI is not merely a computational tool but rather a revolutionary facilitator that can speed up the discovery, enhance the support from specific patients, and create better interventions, giving increasing precision. The association of AI with technologies such as Internet of Medical Things and quantum computing will enable precision in medicine to reach the highest possible level. Thus, the AI-powered future of medicine will be made by the involvement of clinicians, scientists, and regulators, as well as across different disciplines such as materials science, computational biology, and data science.

References

1. U.S. Nanomedicine Market Size, Share & Trends Analysis Report By Application (Drug Delivery), By Indication (Clinical Oncology, Infectious Diseases), By Molecule Type, By Region, And Segment Forecasts, 2024–2030. Grand View Research
2. Sedky NK, Braoudaki M, Mahdy NK, Amin K, Fawzy IM, Efthimiadou EK, Youness RA, Fahmy SA (2023) Box–behnken design of thermo-responsive nano-liposomes loaded with a platinum(iv) anticancer complex: evaluation of cytotoxicity and apoptotic pathways in triple negative breast cancer cells. Nanoscale Adv 5(19):5399–5413
3. Thapa RK, Kim JO (2023) Nanomedicine-based commercial formulations: current developments and future prospects. J Pharm Investig 53(1):19–33
4. Dubey A, Raju F, Lobo CL, Gs R, Hebbar S, Shetty A, Kumar P, El-Zahaby SA (2024) Formulation and characterization of RBCS-coated carboplatin-loaded nano-liposomal formulation for managing breast cancer. Drug Dev Res 85(8):70019
5. Kolluri S, Lin J, Liu R, Zhang Y, Zhang W (2022) Machine learning and artificial intelligence in pharmaceutical research and development: a review. AAPS J 24(1):19
6. Hamet P, Tremblay J (2017) Artificial intelligence in medicine. Metabolism 69:36–40
7. Meetoo DD (2009) Nanomedicine: the revolution of the big future with tiny medicine. Br J Nurs 18(9):1201–1206
8. Dawoud MHS, Mannaa IS, Abdel-Daim A, Sweed NM (2023) Integrating artificial intelligence with quality by Design in the Formulation of lecithin/chitosan nanoparticles of a poorly water-soluble drug. AAPS PharmSciTech 24(6):169
9. Adir O, Poley M, Chen G, Froim S, Krinsky N, Shklover J, Shainsky-Roitman J, Lammers T, Schroeder A (2020) Integrating artificial intelligence and nanotechnology for precision cancer medicine. Adv Mater 32(13):1901989
10. Hersh AM, Alomari S, Tyler BM (2022) Crossing the blood-brain barrier: advances in nanoparticle technology for drug delivery in neuro-oncology. Int J Mol Sci 23(8):4153
11. Dordevic S, Gonzalez MM, Conejos-Sanchez I, Carreira B, Pozzi S, Acurcio RC, Satchi-Fainaro R, Florindo HF, Vicent MJ (2022) Current hurdles to the translation of nanomedicines from bench to the clinic. Drug Deliv Transl Res 12(3):500–525
12. Suriyaamporn P, Pamornpathomkul B, Patrojanasophon P, Ngawhirunpat T, Rojanarata T, Opanasopit P (2024) The artificial intelligence-powered new era in pharmaceutical research and development: a review. AAPS PharmSciTech 25(6):188
13. Bender A, Cortes-Ciriano I (2021) Artificial intelligence in drug discovery: what is realistic, what are illusions? Part 1: ways to make an impact, and why we are not there yet. Drug Discov Today 26(2):511–524
14. Chou WC, Chen Q, Yuan L, Cheng YH, He C, Monteiro-Riviere NA, Riviere JE, Lin Z (2023) An artificial intelligence-assisted physiologically based pharmacokinetic model to predict nanoparticle delivery to tumors in mice. J Control Release 361:53–63
15. Rebollo R (2022) Microfluidic manufacturing of liposomes: development and optimization by design of experiment and machine learning. ACS Appl Mater Interfaces 14(35):39736–39745
16. Singh AV, Varma M, Laux P, Choudhary S, Datusalia AK, Gupta N, Luch A, Gandhi A, Kulkarni P, Nath B (2023) Artificial intelligence and machine learning disciplines with the potential to improve the nanotoxicology and nanomedicine fields: a comprehensive review. Achieves Toxicol 97(4):963–979
17. Mendes BB, Zhang Z, Conniot J, Sousa DP, Ravasco JMJM, Onweller LA, Lorenc A, Rodrigues T, Reker D, Conde J (2024) A large-scale machine learning analysis of inorganic nanoparticles in preclinical cancer research. Nat Nanotechnol 19(6):867–878
18. Zhu M, Zhuang J, Li Z, Liu Q, Zhao R, Gao Z, Midgley AC, Qi T, Tian J, Zhang J, Kong D, Tian J, Yan X, Huang X (2023) Machine-learning-assisted single-vessel analysis of nanoparticle permeability in tumour vasculatures. Nat Nanotechnol 18(6):657–666
19. Hoseini B, Jaafari MR, Golabpour A, Momtazi-Borojeni AA, Karimi M, Eslami S (2023) Application of ensemble machine learning approach to assess the factors affecting size and polydispersity index of liposomal nanoparticles. Sci Rep 13(1):18012

20. Firouzabadi FB, Oryan SH, Sheikhha MH, Kalantar SM, Javed A (2019) Preparation and evaluation of a novel liposomal Nano-formulation in metastatic cancer treatment studies. Cell J 21(2):135–142
21. Cheng Y, Zhao P, Wu S, Yang T, Chen Y, Zhang X, He C (2018) Cisplatin and curcumin co-loaded nano-liposomes for the treatment of hepatocellular carcinoma. Int J Pharm 545(1–2):261–273
22. Valinezhadi N, Dehghan G, Yaghoubzad-Maleki M, Mohammadi M, Alizadeh AA, Hamishehkar H (2025) Liposome-assisted combination chemotherapy improves the antiproliferation and anti-angiogenesis response of cisplatin in breast cancer; experimental and computational study. J Chemother:1–18
23. Hasnain MS, Nayak AK (2019) Natural polysaccharides in drug delivery and biomedical applications, 1st Edn. Academic Press
24. Mendiratta S, Hussein M, Nasser HA, Ali A, a. A. (2019) Multidisciplinary role of mesoporous silica nanoparticles in brain regeneration and cancers: from crossing the blood–brain barrier to treatment. Part Part Syst Charact 36(9)
25. Zheng R, Wang Y, Song S, Liu W, Liu Y, Wang D, Liu L (2024) Development and efficacy evaluation of nanoliposomes targeting CAFs–LCSCs communication for hepatocellular carcinoma treatment. Drug Deliv 31(1):2273355
26. Chen L, Wang Y, Xu H, Shi M, Sun M (2023) Nano-liposomal zein hydrolysate for improved apoptotic activity and therapeutic index in lung cancer treatment. Drug Deliv 30(1):2204874
27. Akbarzadeh A, Rezaei-Sadabady R, Davaran S, Joo SW, Zarghami N, Hanifehpour Y, Samiei M, Kouhi M, Nejati-Koshki K (2013) Functionalized liposomes as drug nanocarriers for active targeted cancer therapy: a systematic review. J Control Release 170(3):346–362
28. Emtiazi G, Khodaverdi E, Afzalzadeh R (2022) Comparative study of nano-liposome and nanoniosome for delivery of Achillea millefolium essential oils: development, optimization, characterization, and their cytotoxicity effects on cancer cell lines and antibacterial activity. J Sci Food Agric 102(9):3863–3873
29. Li W, Zhang H, Assaraf YG, Zhao K, Xu X, Xie J, Yang D, Chen Z-S (2021) Gemcitabine and Pin1 siRNA co-delivery with fucoidan-coated nano-liposomes for therapy of pancreatic cancer. Int J Biol Macromol 183:1767–1778
30. Ebrahimi M, Fathi M, Ahmadian S (2020) Co-administration of curcumin and bromocriptine nano-liposomes for induction of apoptosis in lung cancer cells. J Cell Physiol 235(12):9150–9163
31. Hajhouji F, Ganau M, Helene C, Romano A, Gubian A, Proust F, Chibbaro S (2019) Rare encounters: primary pineal malignant melanoma with lepto-meningeal spread. Case report and literature review on management challenges and outcomes. J Clin Neurosci 65:161
32. Li F, Ruijs N, Lu Y (2022) Ethics & AI: A systematic review on ethical concerns and related strategies for designing with AI in healthcare. AI 4:28
33. Stone P, Brooks R, Brynjolfsson E, Calo R, Etzioni O, Hager G, Hirschberg J, Kalyanakrishnan S, Kamar E, Kraus S, Leyton-Brown K (2022) Artificial intelligence and life in 2030: the one hundred year study on artificial intelligence. arXiv 2211:6318
34. Schicktanz S, Welsch J, Schweda M, Hein A, Rieger JW, Kirste T (2023) AI-assisted ethics? Considerations of AI simulation for the ethical assessment and design of assistive technologies. arXiv 2305:566
35. Wasti S, Lee IH, Kim S, Lee JH, Kim H (2023) Ethical and legal challenges in nanomedical innovations: a scoping review. Front Genet 14:1163392
36. Zhang M, Zhang F (2023) AI-assisted prediction of cancer nanomedicine efficacy using a physiologically-based pharmacokinetic model. Cancer Nanotechnol 14:1
37. Naik GG, Jagtap VA (2024) Two heads are better than one: unravelling the potential impact of artificial intelligence in nanotechnology. Nano TransMed 3:100041
38. Center for Drug Evaluation and Research, U.S.F.D.A (2023) Artificial intelligence in drug manufacturing
39. Sun M, Xu H, Li Y, Wang Y, Shi M (2023) Nano-drug co-delivery system of natural active ingredients and chemotherapy drugs for cancer treatment: a review. Drug Deliv 30(1):2207489

40. Vora LK, Gholap AD, Jetha K, Thakur RRS, Solanki HK, Chavda VP (2023) Artificial intelligence in pharmaceutical technology and drug delivery design. Pharmaceutics 15(7):1916
41. Das KP (2022) Nanoparticles and convergence of artificial intelligence for targeted drug delivery for cancer therapy: current progress and challenges. Front Med Technol 4:1067144
42. Administration, U.S.F.D. (2023) Using artificial intelligence and machine learning in the development of drug and biological products—discussion paper and request for feedback
43. Chen C, Ng DYW, Weil T (2020) Polymer bioconjugates: modern design concepts toward precision hybrid materials. Prog Polym Sci 105:101241

Intelligent Nanotechnology for Cancer Management: Prediction, Screening, and Diagnosis

Ekambaram Gayathiri, Jayashree Shanmugam, Nagarajan Sankaranarayanan, Palanisamy Prakash, Sonia Kesavan, Aruna Sharmili Sundararaj, Mahendrakumar Mani, and Sekar Vijayakumar

Abstract Nanotechnology and artificial intelligence are a game-changing model for oncology, facilitating unprecedented accuracy in cancer prediction, screening, and diagnosis. In-depth review: This novel in-depth literature survey consolidates the latest robust accomplishments of intelligent nanotechnology systems for diagnosing adversaries at the platform via incorporating smart nanomaterials in association with machine learning-based algorithms to facilitate precise diagnosis and thereby support personalized therapy of cancers. We focus on essential nanomaterials, such as gold nanoparticles (AuNPs), Quantum Dots (QDs), liposomes, and polymeric and carbon-based agents, which provide picomolar sensitivity in biomarker detection and further allow multimodal imaging for MRI, CT, and PET. Deep learning methods such as Convolutional Neural Networks (CNNs), vision transformers, and

E. Gayathiri
Department of Plant Biology and Plant Biotechnology, Guru Nanak College (Autonomous), Chennai, India

J. Shanmugam · A. S. Sundararaj
Department of Biotechnology, Stella Maris College (Autonomous), Chennai, Tamil Nadu, India

N. Sankaranarayanan
Department of Chemistry, National Institute of Technology Manipur, Imphal, India

P. Prakash
Department of Biotechnology, Vinayaka Missions Kirupananda Variyar Engineering College, Vinayaka Missions Research Foundation (Deemed to be University), Salem, Tamil Nadu, India

S. Kesavan
Department of Biochemistry, Ethiraj College for Women (Autonomous), Chennai, Tamil Nadu, India

M. Mani (✉)
Department of Biotechnology, Guru Nanak College (Autonomous), Chennai, Tamil Nadu, India

S. Vijayakumar
College of Material Science and Engineering, Huaqiao University, Engineering Research Center of Environment-Friendly Functional Materials, Ministry of Education, Xiamen, People's Republic of China

J. Chen (ed.), *AI-Driven Nanotechnology in Precision Medicine and Healthcare*, Smart Nanomaterials Technology, https://doi.org/10.1007/978-981-92-1690-1_4

ensemble methods have led to diagnostic systems that could confidently attain 94–99% accuracy in cancer detection and classification for several types of cancers. Ultramodern machine learning methodologies for discovering biomarkers, integrating multi-omics data, and predicting responses to treatments translate complex heterogeneous cancer information into clinically actionable knowledge. Nonetheless, there are great obstacles to clinical translation, such as scalability of manufacturing, generalizability of algorithms between different populations, regulatory burden, and the necessity to consider biocompatibility. FDA oversight under current regulatory frameworks sets pathways for approval; however, the lack of standardization in the characterization of nanomaterials and AI model validation is still a concern. Future trends are biodegradable nanomaterials, explainable artificial intelligence, learning federations for privacy-preserving algorithm development, and combinations with immunotherapy strategies. Here, we review recent literature describing the potential of advanced nanotechnology systems for earlier cancer detection and precision medicine delivery, whereby smart nanotechnologies use bio-mimicking body signal processing capabilities to afford accurate diagnosis and tailored therapeutics (precision oncology), and we will discuss key research priorities and obstacles that require collaboration among researchers (from different disciplines) as well as with regulators, clinicians, and policy makers to achieve the full spectrum of these transformative technologies in mainstream clinical applications for cancer management.

Keywords Nanotechnology · Artificial intelligence · Machine learning · Cancer diagnosis · Nanomaterials · Deep learning · Precision medicine · Biomarkers · Clinical translation · Radiomics

1 Introduction

Cancer is one of the primary causes of mortality worldwide. Nanotechnology has become a revolutionary tool in oncology, offering the potential to achieve previously unattainable accuracy for the detection, assessment, and treatment of cancer. At the heart of nanotechnology in cancer treatment are the distinctive physicochemical properties of nanomaterials, such as controllable size, large surface area-to-volume ratio, and tunable surface chemistry that allow tailor-made targeting toward cancer biomarkers and interactions with the tumor microenvironment [1]. From drug carriers to smart and multifunctional theranostic platforms: A way of being between two worlds nanomaterials that go from passive carriers to active, multicomponent architectures able to perform dual functions (diagnosis and therapy), a concept known as "theranostics." These nanosystems can also be developed to diagnose cancer at the molecular level using ultrasensitive biosensors for circulating tumor cell morphologies, nucleic acid, and disease-related proteins in biological fluids. Nanotechnology with artificial intelligence: From the viewpoint of technology, more than 1 decade ago, the integration of nanotechnology was now followed by another strong side—artificial intelligence. This results in a new era for oncology, where smart nanomaterials containing AI algorithms are capable of data processing speed, which allows analysis and prediction for decision modeling for individual treatment [2, 3].

1.1 Physicochemical Characteristics and Targeting Mechanisms

The efficacy of nanotechnology for cancer therapy is mainly dependent on specific physicochemical properties of synthetically engineered nanomaterials. Due to their tunable size of 1–100 nm, nanomaterials can penetrate biological barriers and selectively accumulate into the tumor tissues via passive targeting by the Enhanced Permeability and Retention (EPR) effect [4]. High drug encapsulation capacity is allowed by their large surface area-to-volume ratio as well as surface modification with targeting ligands, antibodies, and homing peptides for active targeting. By tuning the surface chemistry, nanoparticle charge/hydrophobicity/biodegradability can be fine-tuned to improve cancer cell uptake and reduce undesirable effects from non-cancerous cells. These features together allow nanomaterials to preferentially accumulate in the tumor and lower systemic toxicity when compared with conventional chemotherapy, resulting in a higher therapeutic index and better patient tolerability [5].

Nanoparticle-based cancer cell-specific targeting can be achieved through passive and active mechanisms. Passive targeting takes advantage of the EPR effect, which refers to the fact that NPs tend to accumulate in tumors by default, owing to leaky blood vessels and nonfunctional lymphatic drainage in the tumor microenvironment. Manipulation for these particles was improved by active targeting through functionalization with tumor-specific ligands, for instance, folate receptor, transferrin receptor, or antibodies against tumor-associated antigens. This combination approach to tumor targeting can substantially enhance drug concentrations within the target tissue while retaining low systemic exposures, which in turn minimize toxicity elsewhere, leading to a broadening of the therapeutic window [6].

1.2 Transition from Passive to Intelligent Theranostic Systems

Advancement of nanomaterials from traditional passive drug carriers to smart and multifunctional theragnostic systems has proved a fundamental paradigm transition for the treatment of cancer. Earlier generation nanoformulations primarily act as passive carriers for delivery by depending on EPR effects for tumor accumulation. The recent smart nanomaterials have the sensitivity to specific tumor microenvironmental markers, including acidic pH, increased oxidative stress, overexpression of an enzyme, or increased temperature that can trigger them and lead to drug release or diagnostic effects [3]. These stimuli-responsive nanosystems provide controlled and localized drug releases at the site of action, which prevents systemic exposure to a loaded drug yet maintains the therapeutic concentration locally. Multi-stimuli-responsive features of nanomaterials enable the recognition of multiple characteristics of TME at the same time, which contributes to high specificity and low side effects [7].

Theranostic-imaging platforms that can diagnose the tumor and deliver therapeutic drugs allow for real-time visualization of drug delivery and monitoring of response to treatment, as well as adjusting the target volume according to the response of the individual patient. Such multifunctional systems integrate imaging probes like fluorophores, magnetic resonance contrast agents, or radioactive isotopes with chemotherapeutic drugs, immunotherapy drugs, or gene therapy vectors on the same nanoparticle platform. This integrated process allows clinicians to check drug biodistribution, evaluate the targeting efficiency of a tumor, evaluate the treatment response directly, and manage therapy using quantifiable imaging biomarkers [8]. The generation of biodegradable theragnostic nanoplatforms also decreases issues associated with long-term toxicity as well as environmental hazards while preserving therapeutic effectiveness [9].

1.3 Computer-Assisted Diagnosis and Prognosis of Cancer with AI

The combination of AI with nanotechnology-borne cancer management systems yields unprecedented gains in early detection, molecular profiling, and prognostic prediction. These machine learning algorithms analyze multiple layers of data collections from nano-biosensors, medical images, genomic information, or clinical data to recognize these "big data" contacts even in their complexity for all features/classes that are known and unknown predictors of cancer presence and subtype, aggressiveness, or therapy response. Deep learning models using training from high-resolution image data combined with their information about biomarkers obtained by nanoparticles approach the diagnostic accuracies in various cancer types over 95%, which are much higher than traditional diagnostics [10]. AI-based solutions make possible the real-time control of treatment strategies by continuously dynamic biomarkers through smart nanosensors, which underlines truly personalized precision medicine strategies that personalize therapies according to the characteristics of individual patient tumors and responses to treatments [11].

Moreover, the machine learning algorithms can be used to optimize nanoparticle design parameters such as size, shape, surface charge, and functionalization with a focus on improving tumor tropism and therapeutic potency, or, in other words, creating an iterative computation predictions-to-experiments loop where computational predictions guide nanomaterials design leading toward new training data generation for algorithm refitting. Multi-omics data ranging from genomic to proteomic, metabolomic, and imaging can now be integrated using artificial intelligence into comprehensive predictive models that accommodate the full complexity of cancer heterogeneity [12]. This integrated approach substantially increases prognostic precision compared with single-modal analysis and allows us to prospectively find high-risk patients for treatment resistance or aggressive disease course. By connecting smart nanomaterial systems with sophisticated AI designs, oncologists can progress toward genuinely predictive and preventive cancer care that may be capable of finding malignancies at molecular levels or stages well before traditional imaging tools become sensitive enough to see tumors [13].

2 Smart Nanomaterials for Cancer Diagnosis

2.1 Nanotechnology-Based Biosensing Platforms

Smart nanomaterials belong to one type of engineered NPs with the ability to respond to cancer targets due to their selective molecular recognition. Gold nanoparticles (AuNPs) are one of the most popular types of nanomaterials used in various anatomic and cancer studies because of their unique optical characteristics, surface plasmon resonance behavior, and biocompatibility [14]. These materials display the effect of localized surface plasmon resonance, which is useful for either absorption and/or scattering of light at certain wavelengths for detection or photothermal application (Fig. 1).

Quantum dots are another class of excellent nanomaterials for tumor diagnosis due to their tunable fluorescence, high photostability, and low photobleaching as compared with traditional organic fluorophores [15]. Among the emerging theragnostic agents, carbon quantum dots have been regarded as excellent candidates having both low toxicity and superior fluorescence properties for simultaneous bioimaging, photoacoustic imaging, and photodynamic therapy. Due to solar irradiation-induced reactive oxygen species production, these structures are thus suitable for non-invasive treatment of cancer and the possibility of real-time monitoring of the tumor [16].

Functionalized CNs have displayed enormous potential, especially for cancer detection using highly sensitive biosensors, which can detect a variety of cancer biomarkers [17]. The biocompatibility and diagnostic efficacy have been improved considerably through recent developments in both covalent and non-covalent functionalization strategies. These functional nanotubes make possible electrochemical, optical, and field-effect transistor-based sensing systems for fluorescence imaging, Magnetic Resonance Imaging (MRI), Computed Tomography (CT), and ultrasonic imaging, enhancing contrast, resolution power, and diagnostic specificity [18].

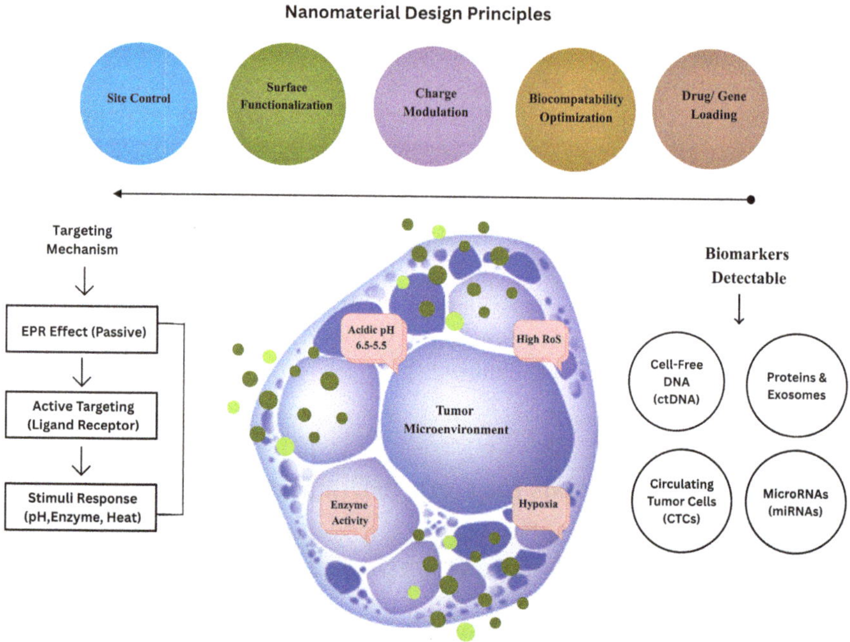

Fig. 1 Overview of nanomaterial design

2.2 Contrast Agents and Molecular Imaging

Nanomaterials have greatly affected the field of medical imaging through the design of innovative contrast agents that are adapted for different imaging techniques. Superparamagnetic Iron Oxide Nanoparticles (SPIONs) are efficient contrast agents for MRI and can help early detection of liver cancer, as well as other soft tissue cancers [19]. These nanoparticles selectively accumulate in the tumor due to the enhanced permeability and retention effect, which offers better spatial resolution and sensitivity than conventional gadolinium agents.

Metal-conjugated graphene nanohybrids are emerging multipurpose theragnostic platforms integrating diagnosis by imaging and therapy. Metallic nanoparticles, such as Fe_3O_4, Au, Ag, TiO_2, and Pd-particle-modified graphene materials, show enhanced electronic modulation/electrochemical behavior for use as a biosensor in addition to better biocompatibility [20]. In vitro studies show size- and dose-dependent inhibition of cancer cell viability at nontoxic concentrations with excellent imaging contrast for early detection of cancer.

2.3 Detection of Circulating Tumor Cells and Liquid Biopsy

Circulating tumor cells and circulating tumor DNA have been identified as important blood-based biomarkers in cancer liquid biopsy for non-invasive early diagnosis and monitoring [21]. The novel smart nanomaterial-triggered systems have revolutionized CTCs using the design of tailored nanoplatforms with multivalent recognition elements. DNA networks-based diagnostic system, which can differentiate cancer patients from healthy individuals with high sensitivity and selectivity, on account of ultralong DNA having aptamers bound specifically to Epithelial Cell Adhesion Molecule (EpCAM) in the membrane of circulating tumor cells.

3 Types of Cancer and Their Specific Applications of Intelligent Nanotechnology

Cancer is a collection of diverse diseases with respect to multiple biological pathways, origins, and prognostic courses. Therefore, intelligent nanotechnology for cancer must be developed based on molecular signature, sample availability, and treatment mode in cancer. This review discusses how nanotechnology in combination with Artificial Intelligence (AI) can be used for prediction, screening, and diagnosis in different cancer types, summarizing more general technical approaches as well as tumor-specific alterations [22].

3.1 Breast Cancer

Breast cancer is the most clinically studied cancer for nano-enabled diagnostics, driven by an existing screening infrastructure and distinct molecular subtypes. With the introduction of nanotechnology-enabled molecular assays, ultrasensitive detection of circulating tumor-derived factors has enhanced breast cancer risk prediction [23]. Proteomic profiling of exosomes and methylation panels of liquid biopsies (circulating tumor DNA—ctDNA) can detect oncogenic alterations at a very early stage, even before their detection on radiologic studies. These molecularly amplified features, in conjunction with AI-based risk models, strengthen the conventional predictors, like mammographic density, hormonal status, and family history. Longitudinal nano-biomarker trajectories are integrated into machine learning models to achieve improved individual risk stratification and prognosis prediction [24]. Conventional mammography is insensitive in women with dense breast tissue. Nanoparticle MR contrast agents and nano-enhanced mammographic agents increase lesion conspicuity and signal-to-noise ratios. Simultaneously, handheld nano-biosensors have the potential for low-cost detection of multiplexed plasma protein and miRNA panels for point-of-care screening, particularly in resource-limited or rural settings [25]. Surface-Enhanced Raman Scattering (SERS)-based multiplex immunoassays provide specific molecular fingerprints of ER, PR, and HER2 status. AI-driven image and spectral analysis in nanoparticle-targeted imaging and biosensing technologies enables molecular subtypes to be accurately determined, assists in tumor biopsy localization, and guides personalized treatment plans [25].

3.2 Lung (Non-small Cell and Small Cell) Cancers

Lung cancer remains the leading cause of death from cancer because it can be diagnosed very late. Nanotechnology-based nanosensors for mutation-specific ctDNA and spectrophotometric exosomal RNA have shown promise in the detection of early lung cancer in high-risk cohorts, such as heavy smokers. Time to progression: using longitudinal nano-biomarker data and AI models for better tracking and intervention of premalignant lesions into malignant lung cancer at a much earlier step [26]. New approaches of pooled ctDNA testing (whole genome and targeted approach) have been developed, which have improved sensitivity for mutation detection as well as specificity. Low-Dose Computer Tomography (LDCT) has been the standard of care for screening, but it suffers from a high burden of false positives and radiation. NP-based contrast agents enhance the imaging sensitivity, and an AI-based CAD system reduces the false positive rate and discovers sub-centimeter nodules. In the absence of CT infrastructure, nano-biosensors based on liquid biopsy might be employed as a complementary or alternative platform for screenings. Nanopore and nanogap sequencing have the potential to enable rapid, low-input detection of actionable driver mutations (e.g., EGFR, ALK, KRAS) from blood. AI integration into the radiomic features of nano-liquid biopsy products allows non-invasive molecular phenotyping and minimizes the need for tissue biopsies [27].

3.3 Colorectal Cancer

Colorectal Cancer (CRC) is especially suitable for nano-enabled screening due to the accessibility of biofluids and current population-based screening programs. In recent years, the detection of ctDNA, exosomal microRNAs, and protein biomarkers based on nanoparticles in developed stool and blood assays has shown much improved sensitivity to early-stage or precancerous adenomas [28]. These transportable nano-biosensors offer a scalable alternative to the colonoscopy, especially for low-screened populations. Broad multi-omics exosome-based assays combined with analysis of cfDNA methylation have been demonstrated to reach a high diagnostic performance (AUC 0.99) for CRC detection [29]. Liquid-based biopsy exosomes for diagnosis in yCRC have been attractive even at the age of 20–35 years. Local and systemic nanoparticle contrast agents appear to offer potential improvements in the visualization of dysplastic lesions at colonoscopy. Histopathologic identification of the probed target in SERS using AI-assisted real-time reading improves adequate diagnosis and facilitates on-the-spot clinical choice [30].

3.4 Prostate Cancer

The dilemma in the management of prostate cancer lies predominantly with the detection of non-lethal disease, rather than significant cancers. These can include exosomal-protein-panel-type nanoassays and urine-based detection of tumor-derived nucleic acids, in addition to Prostate-Specific Antigen (PSA)-based diagnostics, which are expected to be more specific and less misdiagnosed [31]. Nano-biomarker profiles integrated with clinical features enrich AI models for aggressive disease detection. Portable graphene oxide nanostructure QCM biosensors have been developed for point-of-care PSA and PCA3 detection. PSMA-targeted nanoparticles increase MRI sensitivity in the molecular imaging of PSMA expression. AI-based radiomics further step toward a finer differentiation between lesions and their threshold between active surveillance and the need for intervention. Based on IGZO transistors and electrolyte gates, an extended-gate sensor has been readily developed to sense the PSA enzymatically at those ultralow ranges within a fairly short-time scale in real time [32].

3.5 Pancreatic Cancer

Pancreatic cancer presents late, with a poor prognosis, and so begs for comprehensive surveillance. It is therefore necessary to explore sensitive nano-biosensors that can be used for early diagnosis of pancreatic cancer by detecting minimal amounts of ctDNA and tumor-derived exosomes. AI models integrating nanobiomarkers and genetic susceptibility, diabetic status, and chronic pancreatitis history might help to diagnose early for high-risk populations [33]. Exosome-liquid biopsies in the nanoliter volume level, according to a dual co-recognition mode of two tumor-specific biomarkers, can be further reduced to an LLOD as low as 78 pg/mL. PancC patient cfDNA methylation analysis at a fragment level has proved to be more sensitive than that of a classical CG-methylation region-based test. Nanoparticles are used as contrast agents for the imaging of cancer and inflammation resulting from pancreatitis. Further, we can explore the SERS-enhanced biopsy tests for AI analysis, which might be possible to improve the sensitivity of biopsy [34].

3.6 Hepatocellular Carcinoma (HCC)

There is usually an underlying chronic liver disease in cases of HCC, which warrants the use of effective screening techniques. Methods of NANO-liquid biopsy in ctDNA methylation detection and exosomal protein could be potentially non-invasive methods for early detection of cancerous signals. AI/ML models integrating nanodata with liver damage clinical parameters contribute to improving risk stratification for HCC at an early stage. The quantification of circulating tumor DNA (ctDNA) has demonstrated its predictive value for Disease-Free Survival (DFS) in patients with resectable Hepatocellular Carcinoma (HCC), as evidenced in a cohort of 21 patients. The sensitivity of detection for small tumors (especially in cirrhotic liver) may be increased by ultrasound follow-up using nano-contrast agents and AI image analysis. Deep learning radiomics models and MRI-based analysis are highly accurate in prognosis prediction for unresectable HCC [35].

3.7 Ovarian and Gynecologic Cancers

Ovarian carcinoma lacks an accepted population-based screening test and is detected with advanced disease. Multiplexing platforms, like those that quantify circulating protein signatures and exosomal biomarkers, have improved sensitivity and specificity compared to single-marker testing [36]. These nano-biomarker panels based on AI classifiers have potential for early detection. Exosome proteomics analysis revealed that ZNF587B was a novel biomarker related to tumor stage. For these reasons, the intraoperative use of nano-powered enhanced imaging to aid in the visualization of tumors and determinations of staging and cytoreductional results is an effective procedure. SERS LIF droplet microfluidic devices are developed for a high-throughput analysis of cancer exosomes with a high sensitivity [37].

3.8 Cervical Cancer

Screening is a crucial step in preventing cervical cancer. Decentralized Point-of-Care (POC) Human Papillomavirus (HPV) test by nanotechnology: The nanotechnology-supported HPV-POC device enables fast on-site testing. Commercially available lateral flow immunoassays, such as OVAL's OncoE6, offer a straightforward method for obtaining results using either a nasal lozenge or cervical swab sample, providing an immediate bedside readout within 30 seconds. Nano-biosensor triage of abnormal cytology restrains the inflation of specificity and prevents unnecessary colposcopy. AI-based techniques have been very promising for cervical cytology interpretation. Computer-aided design with AI on nano-enhanced cytology images coupled biomolecular markers for optimized colposcopy and biopsy [38].

3.9 Reciprocal Regulation in HNSCC (Head and Neck Squamous Cell Carcinoma)

HNSCC is a well-accessible biofluid, such as saliva. Saliva and plasma nano-biosensors for the tumorigenic exosomes and ctDNA from tumors have potential use in non-invasive screening of at-risk people (smokers and HPV-positive parties). Composite AI models, including both nano-biomarker and behavior-based risk factors, improve early detection. Lateral Flow Assays (LFAs) based on SERS at POC have also been reported for the detection of two GATK-known ctDNAs down to attomolar levels [39]. Intraoperative margin assessment and guidance of biopsies is also considered with the help of nanoparticle-delivered optical probes in combination with SERS-guided imaging to reduce the occurrence of resection.

3.10 Melanoma and Skin Cancers

Cutaneous malignancies are located at the surface and are easy to reach; therefore, these types of cancers represent excellent targets for cutaneous nano-AI imaging techniques. Active nano-imaging patches and target nanoparticles further enhance dermal contrast. Community screening programs can achieve better sensitivity and specificity with the help of image classification using AI. Molecular nano biosensors, nanoparticle-assisted dermoscopy, and machine learning have been demonstrated to assist in differentiating malignant and benign lesions [40].

3.11 CNS (Brain) Tumors

The blood-brain barrier is unique to brain tumors. Notably, ultrahigh-sensitive nano-biosensors to the tumor-origin exosomes and ctDNA in blood or Cerebrospinal Fluid (CSF) can enable non-invasive molecular profiling. Nanoparticulate MRI contrast agents and AI-assisted post-processing may provide benefits in tumor delineation and planning. Machine learning models built on top of radiomic features combined with clinical covariates have also been reported to achieve improved performance in neuroblastoma risk stratification and survival prediction [41].

3.12 Hematologic Malignancies (Leukemia, Lymphoma, Myeloma)

Liquid biopsy techniques are inherently applicable to hematological malignancies. Based on nanoparticles and Raman technology, immunocapture-SERS can achieve rapid detection of circulating malignant cells and nucleic acids. AI models trained on nano-biomarker longitudinal data are able to facilitate MRD (Minimal Residual Disease) monitoring and relapse prediction. The clinical utility of single-molecule ctDNA detection according to whole-genome-based assays is demonstrated in treatment response and survival prediction. Subtypes for therapy are identified rapidly using nanoflow cytometry, single-cell nanoassays, and nanopore sequencing [42].

3.13 Pediatric Cancers

Everything in pediatric oncology requires it to be safer, even if it is more toxic. Detection of very small sample volumes from blood or urine with high sensitivity has been achieved using nano-biosensors. Models adapted to age-specific baselines provide precise surveillance and reduce unnecessary testing. Safety in the long term is most important, and biodegradable and well-defined NP platforms are extremely desirable for pediatric applications [43].

3.14 Rare and Tumor-Agnostic Cancers

Rare cancers struggle with fewer screening options and data. Pan-cancer nano-affixed liquid biopsy panels identify quantifiable gene fusions as well as pan-oncogenomic signatures for organ-of-origin prediction in rare histologic types. Federated AI models trained on data from multiple institutions perform as well as or even better than AI models trained on data from any single institution—all without the need for underlying raw data to be shared between the institutions. Additionally, reciprocal analysis of multi-omics exosome (proteomic, genomic, and transcriptomic; R-MOE) has performed well for pan-cancer screening [44].

4 Application of Machine Learning Techniques in Tumor Comprehension

4.1 Data-Driven Techniques for Biomarker Identification

Since genes interact with each other, Wei et al. proposed a prediction model of differential expression combining gene interaction in human data with gene expression array data from yeast through a maximal margin classifier (Support Vector Machine) [22]. In recent years, Machine Learning (ML) methods have been widely used to discover cancer biomarkers from high-dimensional multi-omics data. Epigenetic analysis with artificial intelligence has discovered that DNA methylation, histone modification, and non-coding RNA regulate the expression of genes through mechanisms that are independent of gene sequences and thereby bring us a clue to the mechanism(s) underlying cancer progression [45]. More sophisticated machine learning approaches, such as random forests, CNNs, and autoencoders, have detected epigenetic signatures with great accuracy to discriminate cancer from normal tissues. Machine learning models, which incorporate biomarkers as features, have shown great potential in early cancer diagnosis and prognosis prediction. These strategies combine protein markers with clinical surrogates and reportedly improve the overall accuracy for risk stratification and treatment choice [46].

4.2 Multi-omics Integration and AI

Multi-omics data will allow "cancer cell signals" that show the type of cancer to be precisely extracted, supporting both patient diagnosis and treatment decisions. Artificial intelligence-based multi-omics data integration has transformed the understanding of the heterogeneity and complexity of the tumor microenvironment in leiomyosarcoma. Combining genomics, epigenomics, transcriptomics, proteomics, and metabolomics shows cooperative interaction between molecular strata and enhances the performance of metastasis prediction over monolayer modalities [47]. This integrative method makes it possible to discover new biomarkers and therapeutic targets not otherwise found by conventional analysis. Explainable Artificial Intelligence (AI) approaches have improved confidence in AI-generated cancer prediction models by providing interpretable models and linking them to clinical domain knowledge [48]. For clinical adoption, clinicians must understand how AI makes decisions in patient care (Fig. 2; Table 1).

5 AI-Driven Cancer Diagnosis and Management

5.1 Applications of Deep Learning in Cancer Imaging

Deep learning significantly alters cancer diagnosis by enabling automatic feature extraction of medical images and optimizing their identification via Convolutional Neural Networks (CNNs) and Vision Transformers [57]. These models correctly distinguish between malignant and benign lesions and classify specific cancer subtypes. The combination of attention mechanisms in conjunction with sophisticated architectures is allowing hybrid models to attain high accuracy in detecting cancer from medical images [58]. Ensemble learning models that involve a combination of multiple deep learning architecture parameters have been shown to achieve better diagnosis performance. Transfer learning methods, by maintaining the high

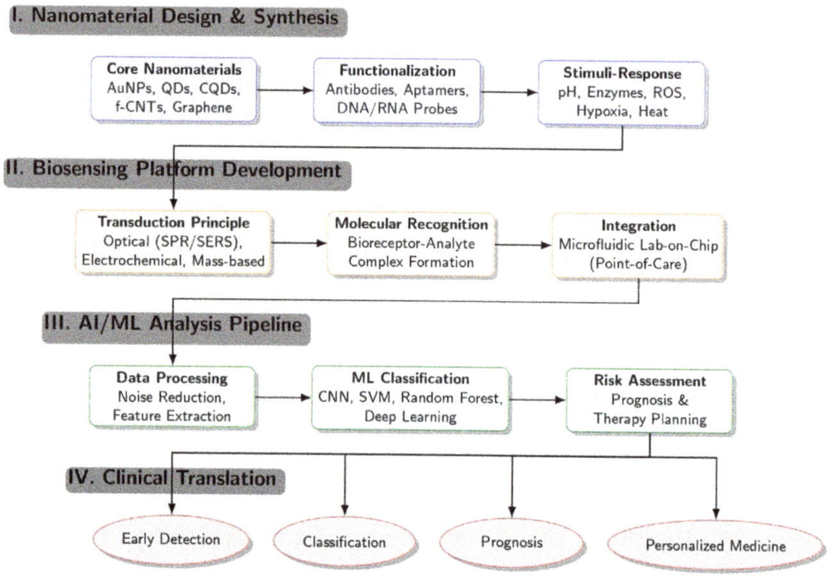

Fig. 2 Nanotechnology-based biosensing platforms for cancer diagnosis

rate of diagnostic accuracy and reducing the computational load, achieved the fine-tuning of pre-trained models over a cancer-specific dataset. Deep convolutional neural networks have attained high accuracy in histopathological image analysis with optimized hyperparameters [59] (Table 1).

5.2 *Radiomics and Predictive Modeling*

Radiomics, which is defined as the high-throughput extraction of quantitative imaging characteristics combined with Machine Learning (ML), has been considered as an exciting tool for precision cancer medicine. Deep learning algorithms, which are trained on radiomics features of MP imaging, achieved high prediction accuracy of treatment response [60]. Dynamic models of prediction based on repeated imaging have had the advantage of better performance in risk prediction.

5.3 *Prediction of Treatment Response and Personalized Medicine*

Machine learning-based predictive models of response to neoadjuvant chemotherapy, immunotherapy, and targeted agents have demonstrated potential increases in clinical

Table 1 Comparison of nanomaterials, AI solutions, and clinical application in the management of cancer

Nanomaterial type	Key properties	Cancer applications	AI/ML approaches	Clinical stage	References
Gold nanoparticles	Localized surface plasmon resonance, excellent optical properties, biocompatibility	Biosensing, photothermal therapy, contrast-enhanced imaging	CNN, SVM, SERS image analysis	Phase I–II trials	[49]
Quantum dots	Tunable fluorescence, minimal photobleaching, superior photostability	Fluorescence imaging, photoacoustic imaging, theranostics	Deep learning, transfer learning	Preclinical-Phase I	[50]
Liposomes	High drug encapsulation, PEGylation reduces opsonization, tunable size	Chemotherapy delivery, targeted therapy, combination therapy	Radiomics, ML prediction models	FDA-approved	[51]
Polymeric nanoparticles	Biodegradable, stimuli-responsive release, versatile functionalization	Targeted drug delivery, immunotherapy, and gene therapy	Deep learning for characterization	Phase I–II trials	[52]
Iron oxide nanoparticles	Superparamagnetic properties, excellent MRI contrast, and magnetic hyperthermia	MRI imaging, magnetic hyperthermia, targeted delivery	CNN-based image segmentation	Phase I–II	[53]
Carbon nanotubes	Exceptional tensile strength, electrochemical properties, tunable reactivity	Biosensors, electrochemical detection, photothermal therapy	ML feature extraction, classification	Preclinical-Phase I	[54]
Graphene-based hybrids	Large surface area, excellent thermal conductivity, tunable chemistry	Multimodal imaging, photothermal therapy, and drug delivery	Graph neural networks, AI optimization	Preclinical	[55]
DNA/RNA nanostructures	Programmable specificity, aptamer integration, and enzymatic amplification	Circulating tumor cell detection, liquid biopsy	Deep learning for signal interpretation	Clinical validation	[56]

relevance for stratifying patients. These systems interrogate TME features, immune infiltrating cell signatures, and genetic alterations to predict therapeutic responses prior to therapy commencement. The use of AI in precision medicine spans drug discovery and therapy target identification. AI algorithms have identified new drug targets and improved drug cocktails. The combination of AI with the nanoformulation design has resulted in intelligent DDSs that can smartly respond to tumor microenvironment signals, leading to enhanced therapeutic efficacy and reduced side effects [61].

6 Regulatory and Ethical Considerations

6.1 Regulatory Pathways and FDA Approval

There are complex regulatory issues that must also be considered for the clinical translation of nanotechnology-based cancer diagnostics and therapeutics, including safety, efficacy, and manufacturing reproducibility oversight by the FDA. The USFDA approved nanomedicines for the market. Liposomal formulations have well-established regulatory paths for nanoformulations [62]. Nevertheless, combination products that combine several nanoscale scavenging mechanisms of action introduce new regulatory challenges, which need to be analyzed on a case-by-case basis. Regulations that apply to model-based diagnostic systems, however, are fundamentally different compared with the traditional medical device regulations and call for new frameworks addressing algorithm transparency and clinical validation as well as post-market surveillance. Regulatory authorities are increasingly demanding prospective clinical validation across diverse populations and imaging modalities to guarantee external generalizability and safety in patient demographics [63].

6.2 Biocompatibility and Toxicity Considerations

Biocompatibility is an overall challenge for clinical translation of most nanoformulations. Cationic NPs exhibit a higher level of plasma membrane disturbance, mitochondrial and lysosomal damage, and autophagosome generation than anionic NPs [64]. Cellular internalization pathway and intracellular distribution are crucial for nanoparticle cytotoxicity. Besides, the long-term biodegradation routes and accumulation in organs (such as liver, lung, and spleen) are still obscure for a large number of nanomaterials. Standard toxicological testing procedures must be adapted to the particular features of NPs, such as size, shape, surface chemistry, and agglomeration state, and their effects on biological interactions [65].

6.3 Clinical Translation Hurdles and Ethical Guidelines

The clinical translation of intelligent nanotechnology is hindered, on the one hand, by manufacturing scalability, standardization, and cost-effectiveness for mass use in clinics. Shortcomings include poor reproducibility in synthesizing nanoparticles from lot to lot in terms of physicochemical properties and high production costs, which restrict the use of nanomedicine to resource-poor environments [66, 67]. Ethical dimensions of deploying artificial intelligence for cancer diagnostics touch on issues related to algorithmic bias, data privacy, and equitable access to precision medicine technologies. Machine learning models developed from data derived from homogeneous populations are likely to have poor performance when applied to underrepresented racial and ethnic groups, therefore requiring diverse and well-annotated training datasets across different demographic factors. Frameworks for data governance need to reconcile open science principles for algorithm development and patient privacy protection, including security measures [68].

7 Advantages, Challenges, and Outlook

7.1 Key Intelligent Nanotechnology Benefits

The intersection of nanotechnology and AI has never brought so many benefits for cancer control. Improved sensitivity is a key benefit, and nano-biosensors have been used to detect picomolar cancer biomarker levels as well as to enable the detection of circulating tumor cells down to the single-cell count [23]. This sensibility has the potential to perform cancer diagnosis at an earlier stage than, for example, conventional means and allow certain therapies, such as adjuvant therapy of HDIL-2. Conventional types of treatment include surgery followed immediately by radiation therapy. Multiplexing allows different combinations of multiple biomarkers to be quantified within each single test sample, allowing for complete tumoral characterization with reduced invasiveness. Artificial intelligence facilitates the rapid analysis of data derived from multiple biomarkers, recognizing complex trends and sets of biomarkers that can predict treatment response and clinical response. Theranostic nanoplatforms, integrating diagnostic and therapeutic properties in one system, provide the capacity to record drug delivery and treatment responses that promote the optimization of a therapeutic regimen [11].

7.2 Current Challenges and Limitations

Although there have been important advances, several challenging issues are still limiting intelligent nanotechnology for cancer treatment in clinical practice. Data diversity in machine learning models is a major bottleneck, with algorithms trained on certain imaging protocols, patient cohorts, and cancer types losing performance when generalized to external populations [8]. The feature extraction, image preprocessing, and algorithm development are within the standardization process, currently still insufficient, with different research groups performing in an inconsistent manner, which will restrict the reproducibility and clinical validation. Scalable manufacturing and cost reduction are major obstacles for clinical translation. Bars to the uptake of learning systems in clinical practice include a lack of transparency, clinician distrust regarding the opaqueness of algorithmic decision-making, inadequate technical expertise among practicing physicians, and reticence about workflow disturbances [13]. Development of machine learning algorithms is also complicated by a lack of data and class imbalance, which is further exacerbated for rare cancers due to small patient cohorts hampering the size of the training dataset. External validation of AI models in different geographical regions, health systems, and patient populations is also lacking, with the need for large-scale prospective studies before applications are introduced to a clinical setting [18].

7.3 Directions for Future Work and Open Issues

The next generation of intelligent nanotechnology is expected to involve the design of real "smart" nanomaterials, which can autonomously respond to tumor microenvironment stimuli. The coupling of sophisticated genetic tools with nanotechnology promises site-targeted therapeutic strategies at the tumor level, introducing new therapeutic alternatives [10].

Explainable artificial intelligence approaches are likely to be included in the development of AI to ensure that clinicians understand diagnostic and predictive algorithms before they are used in clinical practice. Multimodal Data Fusion using images, genomics, proteomics, metabolomics, and clinical data together using deep learning in multimodally encoded patient stratification models robustly representing the complexity of cancer [6]. Therefore, the future direction of biomedical NPs will be biodegradable nanomaterials produced by green methods with minimized environmental effects and enhanced biocompatibility [13]. A promising approach includes the combination of nanotechnology with immunotherapy, where nanoparticles can be engineered to modify tumor immune microenvironment traits and to increase the effectiveness of checkpoint inhibitors. Feedback closure by diagnostic nanosensors and therapeutic nanocarriers will achieve automated, personalized spinodal treatment selections employing real-time biomarker surveillance [11].

8 Conclusion

When fusing intelligent nanotechnology with artificial intelligence, this would stand for a transformative cancer management from the classical diagnostic and therapeutic modalities to precision oncology systems that can monitor the molecular level of cancer and even predict individual responses to patient-specific treatment exposures. Nanotechnology and machine learning algorithms merged into smart nanomaterial-based diagnostic systems with higher sensitivity and specificity than standard disaggregated methods. Early diagnosis by nano-biosensors facilitates the intervention of disease at earlier stages, which leads to better treatments and patient survival. Explainable artificial intelligence methods can improve clinical adoption by offering interpretable diagnostic support based on clinically meaningful characteristics. However, before we can bring these MRD tests to the clinic, we must overcome significant hurdles in the validation process. Manufacturing scale-up and standardization need to be developed so that affordable production of biocompatible nanoformulations is possible. The generalization of the machine learning model across other populations and healthcare systems remains to be validated in prospective external validation studies in multiple geographic locations and patient populations. The regulatory landscape must be transformed to account for novel features of combination products consisting of nanomaterials combined with artificial intelligence. Ethical issues related to algorithmic bias, privacy, and equitable access to precision medicine technologies require a coordinated effort from researchers, clinicians, industry, and policymakers. The research in the future needs to focus on making a biodegradable nanomaterial, which is better for the environment and has less toxicity. The multi-omics data integration, along with modern AI architectures, will certainly contribute to a well-defined patient stratification covering cancer heterogeneity. Closed-loop feedback between diagnostic nanosensors and therapeutic nanovehicles will provide a new line of Personalized/Adaptive (P/A) regimes that are optimized through real-time monitoring of sensitive molecular biomarkers. Rigorous clinical studies on intelligent nanotechnology systems in a wide variety of cancer patients and types will facilitate clinical translation, leading to improved patient outcomes and decreased mortality worldwide.

References

1. Gupta V, Kumar D, Gupta S, Tanwar R, Jaiswal NK, Islam MM, Singh S, Choudhary N, Gowri S, Webster TJ, Faiyazuddin M (2026) Nanotechnology-driven cancer therapies for precision oncology: advances and clinical outlook. Taylor & Francis. https://www.tandfonline.com/doi/full/10.2147/IJN.S568254?src=
2. Kurul F, Turkmen H, Cetin AE, Topkaya SN (2025) Nanomedicine: how nanomaterials are transforming drug delivery, bio-imaging, and diagnosis. Next Nanotechnol 7:100129. https://doi.org/10.1016/j.nxnano.2024.100129

3. Majumder J, Minko T (2020) Multifunctional and stimuli-responsive nanocarriers for targeted therapeutic delivery. Expert Opin Drug Deliv 18(2):205–227. https://doi.org/10.1080/17425247.2021.1828339
4. Lan H, Jamil M, Ke G, Dong N (2023) The role of nanoparticles and nanomaterials in cancer diagnosis and treatment: a comprehensive review. Am J Cancer Res. https://pmc.ncbi.nlm.nih.gov/articles/PMC10767363/
5. Shirazi AN, Vadlapatla R, Koomer A, Nguyen A, Khoury V, Parang K (2025) Peptide-based inorganic nanoparticles as efficient intracellular delivery systems. Pharmaceutics 17(9):1123. https://doi.org/10.3390/pharmaceutics17091123
6. Kashyap BK, Singh VV, Solanki MK, Kumar A, Ruokolainen J, Kesari KK (2023) Smart nanomaterials in cancer theranostics: challenges and opportunities. ACS Omega 8(16):14290–14320. https://doi.org/10.1021/acsomega.2c07840
7. Hapuarachchige S, Artemov D (2020) Theranostic pretargeting drug delivery and imaging platforms in cancer precision medicine. Front Oncol 10:1131. https://doi.org/10.3389/fonc.2020.01131
8. Parthasarathy N, Karthikeyan L, Fabiola M, Vivek R (2025) Recent progress in multifunctional theranostic inorganic nanomaterials for cancer diagnostic imaging and therapy. Next Nanotechnol 8:100212. https://doi.org/10.1016/j.nxnano.2025.100212
9. Yasir M, Mishra R, Tripathi AS, Maurya RK, Shahi A, Zaki MEA, Hussain SAA, Masand VH (2024) Theranostics: a multifaceted approach utilizing nano-biomaterials. Discov Nano 19(1):35. https://doi.org/10.1186/s11671-024-03979-w
10. Wang M, Chang W, Zhang Y (2025) Artificial intelligence for the diagnosis and management of cancers: potentials and challenges. MedComm 6(11):e70460. https://doi.org/10.1002/mco2.70460
11. Johnson KB, Wei W, Weeraratne D, Frisse ME, Misulis K, Rhee K, Zhao J, Snowdon JL (2020) Precision medicine, AI, and the future of personalized health care. Clin Transl Sci 14(1):86–93. https://doi.org/10.1111/cts.12884
12. Wang Q, Liu Y, Li C, Xu B, Xu S, Liu B (2025) Machine learning-enhanced nanoparticle design for precision cancer drug delivery. Adv Sci 12(30):e03138. https://doi.org/10.1002/advs.202503138
13. Shirzad M, Shaban M, Mohammadzadeh V et al (2025) Artificial intelligence-directions. BioNanoSci 15:354. https://doi.org/10.1007/s12668-025-01980-w
14. Tarantino S, Caricato AP, Rinaldi R, Capomolla C, De Matteis V (2023) Cancer treatment using different shapes of gold-based nanomaterials in combination with conventional physical techniques. Pharmaceutics 15(2):500. https://doi.org/10.3390/pharmaceutics15020500
15. Zdobnova T, Lebedenko E, Deyev S (2011) Quantum dots for molecular diagnostics of tumors. Acta Nat. https://pmc.ncbi.nlm.nih.gov/articles/PMC3347596/
16. Tavan M, Yousefian Z, Bakhtiar Z, Rahmandoust M, Mirjalili MH (2025) Carbon quantum dots: multifunctional fluorescent nanomaterials for sustainable advances in biomedicine and agriculture. Ind Crop Prod 231:121207. https://doi.org/10.1016/j.indcrop.2025.121207
17. Sengupta J, Hussain CM (2023) CNT and graphene-based transistor biosensors for cancer detection: a review. Biomolecules 13(7):1024. https://doi.org/10.3390/biom13071024
18. He Y, Hu C, Li Z, Wu C, Zeng Y, Peng C (2022) Multifunctional carbon nanomaterials for diagnostic applications in infectious diseases and tumors. Mater Today Bio 14:100231. https://doi.org/10.1016/j.mtbio.2022.100231
19. Wang W, Shang S, Wang Y, Xu B (2024) Utilization of nanomaterials in MRI contrast agents and their role in therapy guided by imaging. Front Bioeng Biotechnol 12:1484577. https://doi.org/10.3389/fbioe.2024.1484577
20. Aboushoushah SFO (2025) Iron oxide nanoparticles enhancing magnetic resonance imaging: a review of the latest advancements. J Sci Adv Mater Devices 10(2):100875. https://doi.org/10.1016/j.jsamd.2025.100875
21. Cheng F, Su L, Qian C (2016) Circulating tumor DNA: a promising biomarker in the liquid biopsy of cancer. Oncotarget 7(30):48832–48841. https://doi.org/10.18632/oncotarget.9453

22. Wei T, Liao B, Ackermann BL, Jolly RA, Eckstein JA, Kulkarni NH, Helvering LM, Goldstein KM, Shou J, Estrem ST, Ryan TP, Colet J, Thomas CE, Stevens JL, Onyia JE (2005) Data-driven analysis approach for biomarker discovery using molecular-profiling technologies. Biomarkers 10(2–3):153–172. https://doi.org/10.1080/13547500500107430
23. Samathoti P, Kumarachari RK, Bukke SPN, Rajasekhar ESK, Jaiswal AA, Eftekhari Z (2025) The role of nanomedicine and artificial intelligence in cancer health care: individual applications and emerging integrations—a narrative review. Discov Oncol 16(1):697. https://doi.org/10.1007/s12672-025-02469-4
24. Raina N, Rani R, Kanojia M, Mathias A, Paudel KR, Garg A, Tuli HS, Rahman ATMM, Subramaniyan V, Gupta M (2025) Nanotechnology-driven treatment strategies for breast cancer: recent advances and innovations. Oncol Res 33(10):2787–2831. https://doi.org/10.32604/or.2025.066624
25. Bamodu OA, Chung C, Pisanic TR (2023) Harnessing liquid biopsies: exosomes and ctDNA as minimally invasive biomarkers for precision cancer medicine. J Liq Biopsy 2:100126. https://doi.org/10.1016/j.jlb.2023.100126
26. Smolsky J, Kaur S, Hayashi C, Batra S, Krasnoslobodtsev A (2017) Surface-enhanced RAMAN scattering-based immunoassay technologies for detection of disease biomarkers. Biosensors 7(1):7. https://doi.org/10.3390/bios7010007
27. Habbab FM, Bédard ELR, Joy AA, Alam Z, Abraham AG, Roa WHY (2025) Early detection of lung cancer: a review of innovative milestones and techniques. J Clin Med 14(21):7812. https://doi.org/10.3390/jcm14217812
28. Cherukuri SP, Kaur A, Goyal B, Kukunoor HR, Sahito AF, Sachdeva P, Yerrapragada G, Elangovan P, Shariff MN, Natarajan T, Janarthanan J, Richard S, Venkatesaprasath SP, Karuppiah SS, Iyer VN, Helgeson SA, Arunachalam SP (2025) Artificial intelligence-enhanced liquid biopsy and radiomics in early-stage lung cancer detection: a precision oncology paradigm. Cancer 17(19):3165. https://doi.org/10.3390/cancers17193165
29. Beniwal S, Lamo P, Kaushik A, Lorenzo-Villegas D, Liu Y, MohanaSundaram A (2023) Current status and emerging trends in colorectal cancer screening and diagnostics. Biosensors 13(10):926. https://doi.org/10.3390/bios13100926
30. Mannucci A, Balaguer F, Yamada Y, Nagasaka T, Toiyama Y, Okugawa Y, Martí-Gallostra M, Jiménez-Toscano M, Vidal-Tocino R, Jiménez F, Perea J, Quintero E, Boland C, Cavestro GM, Goel A, Melone S, Orgaz JAR, Altamarina EC, García EG et al (2025) An exosome-based liquid biopsy for the detection of early-onset colorectal cancer: the ENCODER multicenter study. Gastroenterology 170(2):330–343. https://doi.org/10.1053/j.gastro.2025.08.013
31. Liao C, Wu Z, Lin C, Chen X, Zou Y, Zhao W, Li X, Huang G, Xu B, Briganti GE, Qi Y, Wang X, Zeng T, Wuethrich A, Zou H (2023) Nurturing the marriages of urinary liquid biopsies and nano-diagnostics for precision urinalysis of prostate cancer. Smart Med 2(1):e20220020. https://doi.org/10.1002/smmd.20220020
32. Britto JSJ, Guan X, Tran TKA, Lei Z, Bahadur R, Patel V, Zhang X, Wong SL, Vinu A (2024) Emerging multifunctional carbon-nanomaterial-based biosensors for cancer diagnosis. Small Sci 4(3):2300221. https://doi.org/10.1002/smsc.202300221
33. He W, Cui J, Wang X, Siu RHP, Tanner JA (2025) Early-stage pancreatic cancer diagnosis: serum biomarkers and the potential for aptamer-based biosensors. Molecules 30(9):2012. https://doi.org/10.3390/molecules30092012
34. Yu Z, Yang Y, Fang W, Hu P, Liu Y, Shi J (2023) Dual tumor exosome biomarker co-recognitions based nanoliquid biopsy for the accurate early diagnosis of pancreatic cancer. ACS Nano 17(12):11384–11395. https://doi.org/10.1021/acsnano.3c00674
35. Yang W, Nguyen R, Safri F, Shiddiky MJA, Warkiani ME, George J, Qiao L (2025) Liquid biopsy in hepatocellular carcinoma: CTDNA as a potential biomarker for diagnosis and prognosis. Curr Oncol Rep 27(6):791–802. https://doi.org/10.1007/s11912-025-01681-3
36. Liberto JM, Chen S, Shih I, Wang T, Wang T, Pisanic TR (2022) Current and emerging methods for ovarian cancer screening and diagnostics: a comprehensive review. Cancer 14(12):2885. https://doi.org/10.3390/cancers14122885

37. Luo B, Que Z, Lu X, Qi D, Qiao Z, Yang Y, Qian F, Jiang Y, Li Y, Ke R, Shen X, Xiao H, Li H, Wu E, Tian J (2023) Identification of exosome protein panels as predictive biomarkers for non-small cell lung cancer. Biol Proced Online 25(1):29. https://doi.org/10.1186/s12575-023-00223-0
38. Gupta R, Gupta S (2023) Point-of-care tests for human papillomavirus detection in uterine cervical samples: a review of advances in resource-constrained settings. Indian J Med Res 158(5 & 6):509–521. https://doi.org/10.4103/ijmr.ijmr_1143_23
39. Brandt A, Thiele B, Schultheiß C, Daetwyler E, Binder M (2023) Circulating tumor DNA in head and neck squamous cell carcinoma. Cancer 15(7):2051. https://doi.org/10.3390/cancers15072051
40. Jadhav LA, Mandlik SK (2024) Nanocarriers in skin cancer treatment: emerging drug delivery approaches and innovations. Nano TransMed 4:100068. https://doi.org/10.1016/j.ntm.2024.100068
41. Luo Q, Yang J, Yang M, Wang Y, Liu Y, Liu J, Kalvakolanu DV, Cong X, Zhang J, Zhang L, Guo B, Duo Y (2025) Utilization of nanotechnology to surmount the blood-brain barrier in disorders of the central nervous system. Mater Today Bio 31:101457. https://doi.org/10.1016/j.mtbio.2025.101457
42. Talotta D, Almasri M, Cosentino C, Gaidano G, Moia R (2023) Liquid biopsy in hematological malignancies: current and future applications. Front Oncol 13:1164517. https://doi.org/10.3389/fonc.2023.1164517
43. Gharehzadehshirazi A, Zarejousheghani M, Falahi S, Joseph Y, Rahimi P (2023) Biomarkers and corresponding biosensors for childhood cancer diagnostics. Sensors 23(3):1482. https://doi.org/10.3390/s23031482
44. Arslan S, Schmidt J, Bass C, Mehrotra D, Geraldes A, Singhal S, Hense J, Li X, Raharja-Liu P, Maiques O, Kather JN, Pandya P (2024) A systematic pan-cancer study on deep learning-based prediction of multi-omic biomarkers from routine pathology images. Commun Med 4(1):48. https://doi.org/10.1038/s43856-024-00471-5
45. Sahoo K, Lingasamy P, Khatun M, Sudhakaran SL, Salumets A, Sundararajan V, Modhukur V (2025) Artificial intelligence in cancer epigenomics: a review on advances in pan-cancer detection and precision medicine. Epigenet Chrom 18(1):35. https://doi.org/10.1186/s13072-025-00595-5
46. Al-Tashi Q, Saad MB, Muneer A, Qureshi R, Mirjalili S, Sheshadri A, Le X, Vokes NI, Zhang J, Wu J (2023) Machine learning models for the identification of prognostic and predictive cancer biomarkers: a systematic review. Int J Mol Sci 24(9):7781. https://doi.org/10.3390/ijms24097781
47. Peeters F, Cappuyns S, Piqué-Gili M, Phillips G, Verslype C, Lambrechts D, Dekervel J (2024) Applications of single-cell multi-omics in liver cancer. JHEP Rep 6(7):101094. https://doi.org/10.1016/j.jhepr.2024.101094
48. Alum EU (2025) AI-driven biomarker discovery: enhancing precision in cancer diagnosis and prognosis. Discov Oncol 16(1):313. https://doi.org/10.1007/s12672-025-02064-7
49. Zhou H, Yang H, Wang G, Gao A, Yuan Z (2019) Recent advances of plasmonic gold nanoparticles in optical sensing and therapy. Curr Pharm Des 25(46):4861–4876. https://doi.org/10.2174/1381612826666191219130033
50. Trivedi R, Malode D, Umekar M, Shidhaye S, Khobragade R, Raut N (2025) Graphene quantum dots: synthesis, applications, and future directions in bioimaging and cancer therapy. Next Nanotechnol 9:100326. https://doi.org/10.1016/j.nxnano.2025.100326
51. Gabizon A, Ohana P, Amitay Y, Gorin J, Tzemach D, Mak L, Shmeeda H (2021) Liposome co-encapsulation of anti-cancer agents for pharmacological optimization of nanomedicine-based combination chemotherapy. Cancer Drug Resist 4(2):463–484. https://doi.org/10.20517/cdr.2020.87
52. Gagliardi A, Giuliano E, Venkateswararao E, Fresta M, Bulotta S, Awasthi V, Cosco D (2021) Biodegradable polymeric nanoparticles for drug delivery to solid tumors. Front Pharmacol 12:601626. https://doi.org/10.3389/fphar.2021.601626

53. Pucci C, Degl'Innocenti A, Gümüş MB, Ciofani G (2022) Superparamagnetic iron oxide nanoparticles for magnetic hyperthermia: recent advancements, molecular effects, and future directions in the omics era. Biomater Sci 10(9):2103–2121. https://doi.org/10.1039/d1bm01963e
54. Thirumalaisamy R, Nimithap S, Selvankumar T, Manikandan R, Vadivel P (2026) Emerging roles of carbon nanotubes in cancer therapy, diagnosis and targeted drug delivery: current insights and toxicity considerations. Chem Phys Impact 101003. https://doi.org/10.1016/j.chphi.2026.101003
55. Singh R, Rawat H, Kumar A, Gandhi Y, Kumar V, Mishra SK, Narasimhaji CV (2024) Graphene and its hybrid nanocomposite: a metamorphoses elevation in the field of tissue engineering. Heliyon 10(13):e33542. https://doi.org/10.1016/j.heliyon.2024.e33542
56. Zhu Y, Li W, Lan F, Chen S, Chen X, Zhang X, Zhang Y (2023) DNA nanotechnology in tumor liquid biopsy: enrichment and determination of circulating biomarkers. Interdiscip Med 2(1). https://doi.org/10.1002/inmd.20230043
57. Jiang X, Hu Z, Wang S, Zhang Y (2023) Deep learning for medical image-based cancer diagnosis. Cancer 15(14):3608. https://doi.org/10.3390/cancers15143608
58. Yao Y, Lilhore UK, Simaiya S, Aldossary SM, Tekeste LG, Ghith E, Abdallah HA (2025) Advancing skin cancer diagnosis with deep learning and attention mechanisms. Sci Rep 16(1):3610. https://doi.org/10.1038/s41598-025-33597-0
59. Varzaneh ZA, Mousavi SM, Khoshkangini R, Khaliji SMM (2025) An ensemble model based on transfer learning for the early detection of Alzheimer's disease. Sci Rep 15(1):34634. https://doi.org/10.1038/s41598-025-22025-y
60. Parekh V, Jacobs MA (2016) Radiomics: a new application from established techniques. Expert Rev Precis Med Drug Dev 1(2):207–226. https://doi.org/10.1080/23808993.2016.1164013
61. Lai C, Chen C, Chang T (2025) Predicting pathological complete response following neoadjuvant therapy in patients with breast cancer: development of machine learning-based prediction models in a retrospective study. JMIR Cancer 11:e64685. https://doi.org/10.2196/64685
62. Rodríguez-Gómez FD, Monferrer D, Penon O, Rivera-Gil P (2025) Regulatory pathways and guidelines for nanotechnology-enabled health products: a comparative review of EU and US frameworks. Front Med 12:1544393. https://doi.org/10.3389/fmed.2025.1544393
63. Reis ME, Bettencourt A, Ribeiro HM (2022) The regulatory challenges of innovative customized combination products. Front Med 9:821094. https://doi.org/10.3389/fmed.2022.821094
64. Kyriakides TR, Raj A, Tseng TH, Xiao H, Nguyen R, Mohammed FS, Halder S, Xu M, Wu MJ, Bao S, Sheu WC (2021) Biocompatibility of nanomaterials and their immunological properties. Biomed Mater 16(4):042005. https://doi.org/10.1088/1748-605x/abe5fa
65. Havelikar U, Ghorpade KB, Kumar A, Patel A, Singh M, Banjare N, Gupta PN (2024) Comprehensive insights into mechanism of nanotoxicity, assessment methods and regulatory challenges of nanomedicines. Discov Nano 19(1):165. https://doi.org/10.1186/s11671-024-04118-1
66. Liu D, Yang F, Xiong F, Gu N (2016) The smart drug delivery system and its clinical potential. Theranostics 6(9):1306–1323. https://doi.org/10.7150/thno.14858
67. Javid-Naderi MJ, Shaegh SAM (2025) Advanced microfluidic techniques for the preparation of solid lipid nanoparticles: innovations and biomedical applications. Int J Pharm X 10:100399. https://doi.org/10.1016/j.ijpx.2025.100399
68. Hantel A, Clancy DD, Kehl KL, Marron JM, Van Allen EM, Abel GA (2022) A process framework for ethically deploying artificial intelligence in oncology. J Clin Oncol 40(34):3907–3911. https://doi.org/10.1200/jco.22.01113

Smart Nanomedicines for Cancer Therapy

Gratiela Gradisteanu-Pircalabioru, Anita Visan, and Irina Negut

Abstract The junction of Artificial Intelligence (AI) with nanotechnology has transformed cancer research by enabling the design of intelligent, personalized drug delivery systems. AI-driven models can process complex biological and chemical datasets to predict Nanoparticle (NP) behavior, optimize formulation parameters, and simulate interactions within the Tumor Microenvironment (TME). This data-driven approach allows for the rational design of smart nanomedicines capable of targeted, stimuli-responsive, and controlled drug release. The chapter outlines the fundamental principles of smart nanomedicine design, emphasizing the roles of materials such as polymeric NPs, liposomes, dendrimers, inorganic nanostructures, and hybrid or biomimetic carriers. It further explores how AI can enhance every stage of development, from NPs synthesis and toxicity prediction to molecular docking and therapeutic optimization. Special attention is given to AI-assisted modeling of the TME to create adaptive, patient-specific nanocarriers. Despite these advances, challenges persist in data standardization, model interpretability, large-scale production, and regulatory acceptance. By integrating nanoinformatics, digital twin simulations, and multiomics analytics, AI-driven smart nanomedicine offers a promising pathway toward safer, more efficient, and truly personalized cancer therapies.

Keywords Artificial intelligence · Nanocarriers · Nanoparticle · Personalized cancer therapies · Tumor microenvironment

G. Gradisteanu-Pircalabioru
eBio-Hub Research Center, University Politehnica of Bucharest—CAMPUS, Bucharest, Romania

Research Institute of the University of Bucharest (ICUB), University of Bucharest, Bucharest, Romania

Academy of Romanian Scientists, Bucharest, Romania

G. Gradisteanu-Pircalabioru
e-mail: gratiela.gradisteanu@icub.unibuc.ro

A. Visan · I. Negut (✉)
National Institute for Lasers, Plasma and Radiation Physics, Măgurele, Romania
e-mail: negut.irina@inflpr.ro

A. Visan
e-mail: anita.visan@inflpr.ro

J. Chen (ed.), *AI-Driven Nanotechnology in Precision Medicine and Healthcare*, Smart Nanomaterials Technology, https://doi.org/10.1007/978-981-92-1690-1_5

1 Introduction

Over the last decade, the merging of Artificial Intelligence (AI) and nanotechnology has revolutionized the field of oncology. AI algorithms, in particular, Machine Learning (ML) and Deep Learning (DL), can process vast amounts of biological, chemical, and clinical data to identify patterns, optimize Nanoparticle (NP) design, and predict therapeutic outcomes. This has allowed scientists to move beyond traditional trial-and-error drug development and toward rational, data-driven design of smart nanomedicines. These systems can integrate multiple layers of information, from gene expression and metabolic profiles to imaging and pharmacokinetic data, in order to tailor drug delivery specifically to the needs of each patient [1].

At the same time, the integration of multiomics data (genomics, transcriptomics, proteomics, metabolomics, and radiomics) has opened the door to a more accurate prediction of disease progression and therapeutic response, encouraging the development of a new generation of personalized nanotherapies.

This chapter provides a comprehensive overview of the current landscape, challenges, and future directions of AI-driven smart nanomedicine in cancer therapy. It begins with the fundamentals of smart nanomedicines, outlining their defining characteristics, design principles, and the major types of nanomaterials, such as polymeric NPs, liposomes, dendrimers, inorganic nanostructures, and hybrid or biomimetic carriers, along with their mechanisms of action in tumor targeting and controlled drug delivery. Building on this foundation, the chapter explores how AI enhances nanomedicine design, from optimizing NPs synthesis and predicting toxicity to improving drug loading, release kinetics, and molecular targeting. It further examines recent case studies that demonstrate the impact of AI on real-world nanomedicine development.

Attention is given to the application of AI in modeling the Tumor Microenvironment (TME) to design responsive and personalized nanocarrier systems. AI-driven nanomedicines can be "taught" to recognize tumor heterogeneity, anticipate drug resistance mechanisms, and optimize biodistribution, significantly improving treatment precision and minimizing systemic toxicity. For example, predictive models can simulate how NPs interact with the TME, including factors like pH, hypoxia, and immune infiltration, allowing for smarter, stimuli-responsive designs [2].

A critical discussion of the main challenges and future perspectives related to data quality, model interpretability, large-scale production, and regulatory considerations follows the above section. Together, these sections provide an integrated view of how AI and nanotechnology converge to advance the next-generation precision cancer therapies.

However, despite these remarkable advances, the field is still at an early stage of clinical translation. Many AI-optimized nanomedicines that demonstrate exceptional efficacy in silico or in preclinical models fail to replicate the same results in human patients. This is due to the biological complexity of tumors, interpatient variability, and the dynamic nature of nanoparticle–cell interactions in vivo [3]. The lack of standardized data, inconsistent experimental reporting, and the "black-box" opacity of many AI systems also pose major barriers to reproducibility and trust. In addition, translating computational predictions into scalable and reproducible manufacturing remains an ongoing technical challenge, as small variations in NPs synthesis can lead to unpredictable differences in biological behavior [4]. Equally important are the ethical, regulatory, and societal challenges that accompany the integration of AI into biomedicine [5].

2 Fundamentals of Smart Nanomedicines

2.1 Definition and Characteristics of Smart Nanomedicines

Smart nanomedicines are an advanced generation of nanoscale drug delivery systems designed to intelligently interact with biological environments and respond to specific internal or external stimuli to achieve precise and on-demand therapeutic effects. These nanostructures go beyond traditional passive drug delivery vehicles by integrating responsiveness, selectivity, and adaptability, with the purpose of entering a new era of intelligent therapeutics in oncology [6].

Smart nanomedicines can sense and respond to variations in the physiological microenvironment, such as differences in pH, enzymatic concentration, redox gradients, hypoxia, or temperature, which distinguish tumor tissues from normal cells. They can also be activated by external stimuli, including light, magnetic fields, ultrasound, or electrical signals, allowing researchers and clinicians to remotely control drug activation and release at the target site. This dual responsiveness significantly enhances therapeutic precision and minimizes collateral toxicity to healthy tissues [7].

Structurally, smart nanomedicines are fabricated from an array of organic, inorganic, or hybrid materials, each chosen for its tunable physicochemical and biological properties.

- Organic nanocarriers, such as polymeric NPs, liposomes, micelles, and dendrimers, offer biodegradability, biocompatibility, and flexibility in surface functionalization.
- Inorganic nanostructures such as gold, silica, iron oxide, and quantum dots provide superior mechanical stability, optical properties, and the potential for imaging-guided therapy.
- Hybrid and biomimetic nanocarriers, which combine the advantages of both, often coat synthetic NPs with cell membranes or peptides to evade immune clearance and enhance tumor targeting [8, 9].

The key features that define these next-generation nanomedicines include the following:

(i) Site-Specific Targeting: Smart nanomedicines exploit the Enhanced Permeability and Retention (EPR) effect to passively accumulate in tumor tissues, while active targeting is achieved through the conjugation of ligands such as antibodies, aptamers, or small molecules that selectively bind to overexpressed receptors on cancer cells [8].

(ii) Stimuli-Responsive Drug Release**:** The hallmark of smart nanomedicine is its controlled and intelligent drug release behavior, triggered by endogenous (pH, enzyme, redox) or exogenous (temperature, magnetic field, light, ultrasound) stimuli. This responsiveness enables spatiotemporal control, allowing precise timing and localization of therapeutic action [10].

(iii) Multifunctionality (Theranostics): Smart nanomedicines often combine therapeutic and diagnostic capabilities in a single nanoplatform. These "theranostic" systems enable real-time monitoring of drug distribution and therapeutic response through imaging modalities like MRI, CT, or photoacoustic imaging [11].

(iv) Biocompatibility and Minimal Off-Target Toxicity: The use of biodegradable polymers, lipids, and natural biomaterials ensures low immunogenicity and safe degradation in vivo. Surface engineering (e.g., PEGylation, biomimetic coatings) further enhances circulation time and reduces clearance by the reticuloendothelial system [12].

The integration of AI in the design and optimization of these nanomedicines is further accelerating their evolution. AI algorithms can predict NPs interactions with biological systems, optimize size and surface charge, and model drug release kinetics to improve therapeutic outcomes [13].

2.2 *Key Design Principles: Stimuli-Responsiveness, Targeting, and Controlled Release*

The fundamental design philosophy behind smart nanomedicines is based on their ability to adaptively respond to the biological environment or external triggers and to deliver therapeutic agents with high spatial and temporal precision. This adaptability, derived from their physicochemical engineering, is realized through three interrelated design principles: stimuli-responsiveness, targeting, and controlled release. These principles work synergistically to enhance therapeutic efficacy, minimize systemic toxicity, and achieve precise tumor-specific delivery.

2.2.1 Stimuli-Responsiveness

Stimuli-responsiveness represents the defining feature of smart nanomedicines, enabling them to recognize and react to specific biological or physical cues. This design strategy allows nanocarriers to release their therapeutic payload only upon encountering characteristic tumor conditions or upon external activation, thereby maximizing drug activity at the tumor site while minimizing damage to normal tissues.

Endogenous stimuli originate within the Tumor Microenvironment (TME), which exhibits unique pathological conditions distinct from normal tissues. These include the following:

- pH Gradients: The acidic extracellular pH ($\approx$ 6.5) of tumor tissues and even lower pH ($\approx$ 5.5) inside endosomes/lysosomes can be exploited by incorporating acid-labile linkages or pH-sensitive polymers that trigger drug release upon acidification [10].
- Redox Potential: Tumor cells exhibit high intracellular concentrations of Glutathione (GSH), which can cleave disulfide bonds in redox-sensitive nanocarriers, leading to controlled intracellular drug liberation.
- Enzyme Expression: Overexpressed enzymes such as Matrix Metalloproteinases (MMPs) and cathepsins degrade specific peptide or polymer linkages, allowing site-specific release.
- Hypoxia: Oxygen-deficient tumor regions activate hypoxia-sensitive linkers or prodrugs integrated into nanocarriers, further enhancing selectivity.

Exogenous stimuli, conversely, provide external control over nanomedicine activation and release, allowing clinicians to adjust therapy dynamically. These include the following:

- Temperature: Thermo-responsive polymers (e.g., PNIPAM) undergo phase transitions upon mild hyperthermia, releasing the encapsulated drug.
- Light: Near-Infrared (NIR) or UV light can trigger photo-cleavable bonds or induce local heating for photothermal drug release.
- Ultrasound: Acoustic waves enhance NPs' permeability or cause cavitation, enabling deep tissue drug penetration.
- Magnetic Fields: Magnetic NPs respond to alternating magnetic fields, producing localized heating or mechanical disruption to trigger drug discharge [11].
- The integration of multiple stimulus-responsive mechanisms (dual or multi-responsive systems) allows higher precision and flexibility, facilitating customized therapy regimens for complex cancer pathophysiologies [6].

2.2.2 Targeting Strategies

An equally critical aspect of smart nanomedicine design is targeting specificity, which represents the ability of nanocarriers to accumulate selectively in tumor tissues while sparing normal organs. Targeting can be achieved via passive or active mechanisms, often used synergistically.

Passive accumulation relies on the Enhanced Permeability and Retention (EPR) effect, a phenomenon arising from the abnormal vasculature and poor lymphatic drainage in tumor tissues. NPs, typically sized between 50–200 nm, can preferentially extravasate through leaky capillaries and remain retained within the tumor interstitium. This passive targeting approach has been foundational in the success of several clinically approved nanomedicines such as *Doxil®* and *Abraxane®* [9].

Active targeting involves ligand-mediated recognition of overexpressed receptors on cancer cell membranes. Nanocarriers are functionalized with targeting moieties such as antibodies (anti-HER2, anti-EGFR), peptides (RGD, TAT), small molecules (folic acid, transferrin), or aptamers that bind to their corresponding receptors, promoting receptor-mediated endocytosis and intracellular drug delivery [8].

Advanced strategies combine dual-targeting (e.g., ligand + magnetic guidance) and AI-assisted ligand optimization, allowing enhanced selectivity and reduced off-target uptake [13].

Moreover, biomimetic approaches, where NPs are masked with natural cell membranes (e.g., erythrocyte, platelet, or cancer cell membranes), enable them to evade immune surveillance and exhibit "self-recognition" behavior, significantly prolonging systemic circulation.

2.2.3 Controlled Drug Release

The controlled release mechanism ensures that therapeutic agents are delivered at the optimal rate, dose, and location, maintaining an effective concentration within tumor tissues for extended periods while minimizing systemic exposure.

The release mechanisms in smart nanomedicines are governed by the material composition, drug–matrix interactions, and environmental stimuli. Key release strategies include the following:

- Diffusion-Controlled Release: Drug molecules gradually diffuse through the polymer matrix or lipid bilayer, providing a sustained and predictable release profile.
- Degradation-Triggered Release: Biodegradable materials (e.g., PLGA, chitosan) undergo hydrolysis or enzymatic breakdown, gradually releasing the encapsulated drug as the carrier degrades.
- Stimuli-Induced Membrane Disruption: Environmental triggers such as acidic pH, redox imbalance, or temperature cause conformational changes or pore formation in nanocarrier membranes, leading to burst release at the tumor site [7].

Additionally, mathematical modeling and AI-based predictive algorithms are increasingly being used to simulate and optimize drug release kinetics, allowing for precise prediction of in vivo behavior based on nanocarrier properties and patient-specific tumor parameters [13].

Through this integrated triad of stimuli-responsiveness, targeted delivery, and controlled release, smart nanomedicines offer a paradigm shift from passive drug delivery to dynamic, intelligent, and patient-tailored therapy, ultimately aligning with the goals of precision medicine in oncology.

2.3 *Types of Nanomaterials in Cancer Therapy*

Smart nanomedicines are engineered using a wide variety of nanomaterial platforms, each offering distinct physicochemical, biological, and functional properties that can be tuned for specific cancer therapeutic needs. These nanocarriers serve as modular, multifunctional systems capable of drug encapsulation, controlled release, active targeting, and diagnostic integration. The diversity of materials, ranging from biodegradable polymers to inorganic metals and biomimetic hybrids, enables the design of intelligent systems that combine precision, adaptability, and safety.

2.3.1 Polymeric NPs

Polymeric NPs are among the most versatile and widely studied nanocarriers for cancer therapy. Constructed from biodegradable and biocompatible polymers such as poly(Lactic-co-Glycolic Acid) (PLGA), Polyethylene Glycol (PEG), Polycaprolactone (PCL), and chitosan, these NPs exhibit controlled degradation profiles, tunable surface chemistry, and high drug-loading capacity.

Their structural versatility allows both hydrophobic and hydrophilic therapeutic molecules to be encapsulated or conjugated, achieving sustained and targeted drug release. Functional modifications, such as PEGylation for stealth properties, ligand conjugation for receptor targeting, or incorporation of pH/redox-sensitive linkers, transform these carriers into stimuli-responsive systems capable of precise release within the tumor microenvironment [12].

Recent advances in AI-driven molecular modeling and multiscale simulations have further refined polymeric NPs design. ML algorithms can now predict self-assembly behavior, NPs morphology, and drug–polymer interaction energies, optimizing pharmacokinetic and biodistribution profiles before in vivo testing [13].

Such predictive design approaches are accelerating the development of polymeric nanomedicines that are both mechanistically intelligent and clinically translatable.

2.3.2 Liposomes and Micelles

Liposomes are spherical lipid bilayer vesicles that mimic biological membranes and are capable of encapsulating both hydrophilic drugs (in the aqueous core) and hydrophobic drugs (within the lipid bilayer). Their biocompatibility and capacity for surface modification make them one of the earliest and most successful nanocarriers in clinical oncology, with FDA-approved formulations such as *Doxil®* and *Caelyx®*.

Modern liposomal systems can be engineered to be pH-, thermo-, or enzyme-sensitive, releasing their payload specifically within the acidic or heated tumor microenvironment [14].

Additionally, PEGylation and ligand modification further extend circulation time and enhance active targeting capabilities.

Polymeric micelles, on the other hand, are self-assembled nanostructures formed by amphiphilic block copolymers with hydrophobic cores and hydrophilic shells. These micelles are ideal for poorly water-soluble chemotherapeutic agents such as paclitaxel and doxorubicin. Their nanoscale size (typically 20–100 nm) allows for deep tumor penetration, and their surface hydrophilicity prevents opsonization and clearance by the mononuclear phagocyte system. Importantly, micelles can be designed to be stimuli-responsive, releasing drugs in response to pH, redox potential, or temperature variations within tumor tissues [15].

Together, liposomes and micelles represent soft-matter nanocarriers that merge safety, adaptability, and tunable drug delivery, critical parameters for modern cancer nanomedicine.

Dendrimers

Dendrimers are highly branched, monodisperse, tree-like macromolecules with precisely defined architectures and multiple surface functional groups. Their interior cavities and surface peripheries allow simultaneous drug encapsulation, covalent drug conjugation, and multivalent ligand attachment, providing unprecedented control over drug loading, release kinetics, and targeting specificity.

For cancer therapy, dendrimers such as Poly(Amidoamine) (PAMAM) and Poly(PropyleneImine) (PPI) have demonstrated strong potential for gene and small-molecule delivery, as well as co-delivery of drugs and imaging agents for theranostic applications. Their tunable surface chemistry also supports stimuli-responsive behavior, where pH or redox conditions induce drug release from covalent or electrostatic linkages [10].

Moreover, the modular design of dendrimers facilitates multi-functionalization, allowing them to act simultaneously as carriers, targeting agents, and imaging probes. However, controlling their synthesis complexity and ensuring biocompatibility remain critical challenges for clinical translation.

Inorganic Nanostructures

Inorganic nanostructures such as gold NPs (AuNPs), silica NPs (SiNPs), Iron Oxide NPs (IONPs), and Quantum Dots (QDs) have garnered significant attention for their mechanical stability, tunable surface chemistry, and unique optical and magnetic properties. These nanostructures are inherently suited for dual therapeutic and diagnostic ("theranostic") applications in oncology.

For example, AuNPs are widely employed in Photothermal Therapy (PTT), converting near-infrared light into localized heat to ablate tumor tissues, while iron oxide NPs serve as both drug carriers and Magnetic Resonance Imaging (MRI) contrast agents. Mesoporous Silica NPs (MSNs), with their high surface area and porous structure, allow for high drug-loading efficiency and programmable release behavior, while quantum dots offer fluorescent imaging capabilities for real-time tracking of drug distribution [11].

The integration of photonic and magnetic materials has given rise to "on-demand" drug release systems, in which external light or magnetic fields precisely trigger localized drug liberation, minimizing systemic side effects. However, potential long-term toxicity and biodegradability concerns necessitate careful material design and surface modification for clinical applications.

Hybrid and Biomimetic Nanocarriers

The next frontier in nanomedicine lies in hybrid and biomimetic nanocarriers, which combine the functional strengths of both organic and inorganic systems. Hybrid nanocarriers often consist of a polymeric or lipid matrix integrated with metallic or ceramic cores, enabling synergistic performance; for instance, combining the biocompatibility of polymers with the diagnostic sensitivity of inorganic materials.

Biomimetic nanocarriers, meanwhile, utilize biological components such as cell membranes, proteins, or exosomes to cloak synthetic NPs. This biomimicry confers immune evasion, prolonged circulation, and homotypic tumor targeting, as these carriers can mimic the surface markers of native cells. Cell membrane-coated NPs (e.g., red blood cell, platelet, or cancer cell membranes) have shown exceptional promise in improving tumor penetration, reducing opsonization, and minimizing immunogenicity, thereby enhancing therapeutic index [16].

2.4 *Mechanisms of Action in Tumor Targeting and Drug Delivery*

Smart nanomedicines exploit a range of biophysical, biochemical, and molecular mechanisms to achieve tumor-specific localization, internalization, and controlled drug release. These mechanisms collectively enable selective drug accumulation in malignant tissues while minimizing off-target distribution and systemic toxicity. The interplay of passive and active targeting, coupled with stimuli-triggered release and intracellular trafficking control, forms the foundation of their superior therapeutic performance.

2.4.1 Enhanced Permeability and Retention Effect

The Enhanced Permeability and Retention (EPR) effect represents the basis of passive tumor targeting in nanomedicine. Tumor tissues are characterized by abnormal angiogenesis, leaky vasculature, and poor lymphatic drainage, creating a permissive environment for NPs (typically 50–200 nm) to extravasate and accumulate selectively within the tumor interstitium [9]. Figure 1 displays the schematic depiction of drug targeting by passive targeting mode and active targeting mode. This phenomenon allows polymeric NPs, liposomes, micelles, and dendrimers to achieve passive targeting without the need for surface modification. Once accumulated, their prolonged retention results from poor lymphatic clearance in tumors, leading to higher local drug concentrations and reduced systemic exposure.

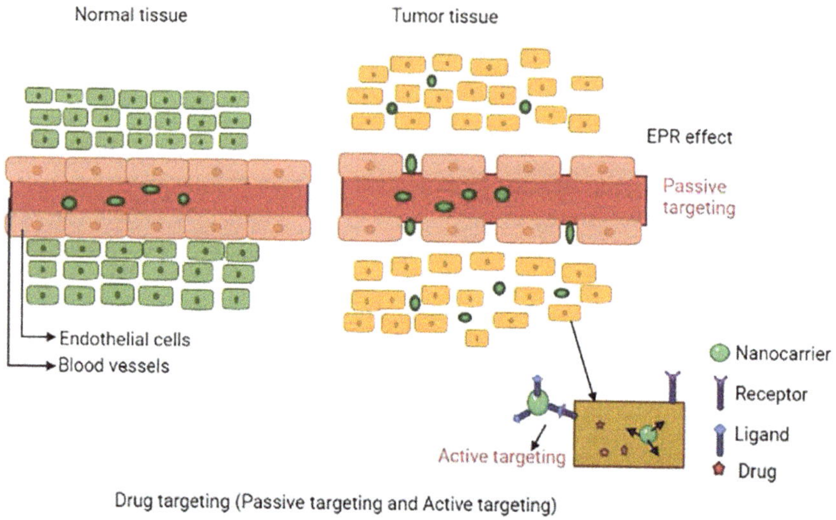

Fig. 1 Schematic representation of drug targeting by passive (EPR effect) and active targeting modes [9]

However, the heterogeneity of tumor vasculature often limits EPR uniformity among cancer types and patients. To overcome this, emerging strategies employ AI-assisted modeling and tumor vasculature mapping to predict EPR efficiency and personalize NPs size and surface properties for optimal accumulation [8].

2.4.2 Ligand-Mediated Active Targeting

While the EPR effect ensures passive accumulation, active targeting adds a layer of molecular precision by exploiting specific receptor–ligand interactions on cancer cells. Smart nanocarriers are functionalized with ligands, antibodies, peptides, or aptamers that recognize and bind to overexpressed receptors such as folate receptor, transferrin receptor, HER2, EGFR, or integrins on tumor surfaces [8].

Upon ligand binding, NPs are internalized via receptor-mediated endocytosis, enhancing intracellular delivery and reducing off-target uptake by normal tissues. For instance, folate-conjugated liposomes preferentially enter folate receptor-positive tumors, while anti-HER2 antibody-modified NPs target HER2-overexpressing breast cancer cells.

Advanced computational methods are now used to design and optimize ligand density and orientation to maximize binding affinity while maintaining stealth properties. ML and AI-guided modeling help predict ligand–receptor binding dynamics, accelerating the rational design of high-specificity nanocarriers [13].

2.4.3 Stimuli-Triggered Drug Release

Smart nanomedicines achieve precise, on-demand therapeutic action through stimuli-triggered release mechanisms. These systems remain stable during systemic circulation but are activated by specific cues within the TME or by externally applied signals.

Endogenous stimuli can be categorized as follows:

- Acidic pH: Nanocarriers containing acid-labile linkages or pH-sensitive polymers (e.g., polyhistidine, poly(β-amino esters)) disassemble in acidic tumor or endosomal environments.
- Redox Gradients: Elevated intracellular Glutathione (GSH) levels cleave disulfide bonds, releasing the drug intracellularly.
- Enzyme Overexpression: Matrix Metalloproteinases (MMPs) or cathepsins degrade peptide crosslinks, initiating drug release specifically at tumor sites.
- Hypoxia: Hypoxia-sensitive linkers (e.g., azobenzene derivatives) undergo reduction under low oxygen, liberating encapsulated therapeutics.
- Exogenous stimuli, on the other hand, provide external spatiotemporal control:
- Light-Triggered Release: Near-Infrared (NIR) light induces photothermal or photochemical effects, disrupting nanocarrier membranes.
- Magnetic Field: Alternating magnetic fields generate localized heat (magnetic hyperthermia), releasing drugs from magnetically responsive carriers.
- Ultrasound: Acoustic cavitation enhances permeability and triggers drug diffusion from nanocarriers [7].

Dual- and multi-stimuli-responsive systems are emerging to further enhance precision by integrating multiple triggers, e.g., pH/redox or magneto-thermal combinations, which improve therapeutic depth and adaptability.

2.4.4 Cellular Internalization and Endosomal Escape

Once nanomedicines reach the tumor and are internalized by cancer cells, their intracellular trafficking determines the ultimate therapeutic outcome. Most NPs enter cells through endocytosis, leading them into endosomal and lysosomal compartments, acidic organelles where drugs may be degraded or inactivated.

To counteract this, smart nanocarriers are designed to promote endosomal escape, ensuring that the drug or genetic cargo (e.g., siRNA, mRNA) reaches the cytoplasm or nucleus intact.

A key mechanism is the "proton-sponge effect," typically exhibited by polymers such as Polyethylenimine (PEI) or Polyamidoamine (PAMAM) dendrimers. These cationic polymers buffer endosomal pH, causing osmotic swelling and rupture of the vesicle membrane, thereby releasing the cargo into the cytoplasm [13].

Alternative endosomal escape strategies include:

- Photochemical Internalization (PCI): Light activation of photosensitizers disrupts endosomal membranes.
- Membrane-Disruptive Peptides: pH-sensitive fusogenic peptides facilitate direct endosomal escape.
- Redox-Triggered Disassembly: Intracellular GSH cleaves carrier linkages, destabilizing the endosomal membrane.

By ensuring efficient cytosolic delivery, these systems maximize therapeutic efficacy, particularly for gene therapies, siRNA delivery, and immunomodulatory agents.

2.4.5 Integration and Therapeutic Synergy

These mechanisms such as EPR-driven accumulation, receptor-mediated targeting, stimuli-responsive release, and intracellular trafficking act synergistically to produce a multi-stage delivery cascade. Each stage refines the localization, activation, and therapeutic precision of the nanomedicine, ultimately resulting in: (i) higher intratumoral drug concentration; (ii) reduced systemic side effects; (iii) enhanced therapeutic index and patient-specific adaptability.

Incorporating AI into this mechanistic framework further refines performance prediction, optimizing NPs design for individual tumor architectures and microenvironmental signatures. Such AI-integrated nanomedicine platforms mark the transition from passive to self-regulating, precision-guided therapy systems capable of real-time monitoring and adaptive modulation.

3 The Role of AI in Smart Nanomedicine Design

3.1 Overview of AI and ML in Nanomedicine Research

The integration of AI and ML into nanomedicine represents a paradigm shift in biomedical research and drug delivery design. These computational technologies enable researchers to model, predict, and optimize the physicochemical, pharmacokinetic, and/or biological behavior of nanocarriers, reducing the need for costly and time-consuming experimental trials. Thus, AI catalyzes precision nanomedicine, improving therapeutic performance, safety, and patient-specific customization.

Nanomedicine focuses on the design and use of nanoscale systems (typically 1–100 nm) for targeted drug delivery, diagnostics, imaging, and therapeutic modulation. These smart nanocarriers are capable of transporting therapeutic payloads directly to disease sites, thereby enhancing efficacy and minimizing systemic side effects. However, despite decades of research and promising preclinical outcomes, clinical translation of nanomedicine remains limited. The main challenges include the following:

1. Complex nano–bio interactions, including unpredictable protein corona formation and immune recognition.
2. Interpatient variability arising from differences in tumor microenvironments and physiological barriers.
3. Nonlinear pharmacokinetics and biodistribution, leading to poor correlation between in vitro and in vivo results [17].

The emergence of AI and ML offers transformative potential to overcome these challenges by introducing data-driven design and predictive analytics into the nanomedicine workflow. These computational approaches are capable of analyzing complex, high-dimensional datasets encompassing physicochemical, biological, and clinical parameters to uncover hidden relationships between NPs' structure, behavior, and therapeutic performance.

In nanomedicine research, experimental methods often generate vast and heterogeneous datasets, ranging from NPs physicochemical properties (size, charge, hydrophobicity) to biological assay results (cytotoxicity, cellular uptake) and patient-specific omics data (genomic, proteomic, metabolomic profiles). AI algorithms can synthesize and interpret these multidimensional datasets, identifying patterns that are imperceptible to human analysis. Through pattern recognition and predictive modeling, AI uncovers Quantitative Structure–Activity Relationships (QSARs) that define how material characteristics influence therapeutic efficacy, biodistribution, and safety profiles [18].

For example: (i) supervised learning algorithms (e.g., support vector machines, random forests) can predict NPs stability or toxicity based on training data from high-throughput screens; (ii) unsupervised learning techniques (e.g., clustering, principal component analysis) can identify NPs subgroups with distinct biodistribution patterns; (iii) DL models such as Convolutional Neural Networks (CNNs) can process imaging data to analyze NPs–cell interactions and tumor accumulation kinetics.

These models not only accelerate hypothesis generation and validation but also enable virtual screening of nanomaterial candidates, reducing experimental costs and timelines by orders of magnitude.

AI supports end-to-end optimization of nanomedicine, from molecular design to clinical personalization, by enhancing each stage of development:

1. Design Optimization: ML algorithms predict the influence of key formulation parameters (e.g., size, morphology, zeta potential, surface ligands, and hydrophilicity) on therapeutic performance. For instance, neural network models trained on large NPs datasets can identify optimal feature combinations that maximize drug encapsulation efficiency and controlled release kinetics [17]. This predictive capability replaces traditional trial-and-error formulation methods with rational, data-driven design.
2. Biodistribution Prediction: AI models integrate Physiologically Based Pharmacokinetic (PBPK) modeling with ML to predict NPs transport across biological barriers, including vascular endothelium and the blood–brain barrier. These hybrid models allow researchers to simulate in vivo NPs fate (circulation, tissue

penetration, clearance) before clinical trials, thus enhancing design precision and reducing failure rates [19].

3. Toxicity and Immunogenicity Assessment: AI has revolutionized toxicity prediction by modeling the formation of the protein corona, the dynamic layer of biomolecules that adsorbs onto NPs' surfaces and influences immune recognition.

   Using deep neural networks and graph-based models, AI can forecast immune compatibility, cellular uptake pathways, and clearance mechanisms based on surface composition and charge characteristics [17].

This allows early identification of potential nanotoxicity and ensures regulatory compliance during preclinical development.

4. Therapeutic Personalization: AI-driven platforms can integrate patient-specific omics data, imaging biomarkers, and tumor microenvironment characteristics to customize nanomedicine formulations for individual patients. By linking computational models with real-time clinical data, such systems can dynamically adjust NPs composition, targeting ligands, and release profiles to match the patient's unique tumor physiology and immune landscape [18]. This approach represents a shift toward AI-driven personalized nanotherapy, where the design of a nanocarrier can be algorithmically adapted for maximum efficacy and minimal toxicity in a specific patient.

3.2 *AI-Based NPs Design and Optimization*

AI has emerged as a powerful tool in the rational design, synthesis, and optimization of NPs for biomedical applications. Traditional nanomedicine development relies heavily on trial-and-error experimentation, which is time-consuming, resource-intensive, and often limited by the complexity of nano–bio interactions. AI-driven approaches such as ML, DL, and generative modeling enable predictive, data-guided NPs engineering, thereby reducing experimental uncertainty and accelerating discovery [20].

These AI methodologies integrate physicochemical data, biological response datasets, and high-throughput experimental outcomes to optimize key NPs parameters, size, shape, surface charge, composition, stability, and release kinetics, for specific therapeutic goals.

3.2.1 Predictive Modeling for NPs Synthesis and Properties

AI-powered predictive modeling serves as a cornerstone for designing nanomaterials with predefined performance characteristics. These models use large datasets from NPs libraries and experimental databases to identify Quantitative Structure–Property Relationships (QSPRs) and Quantitative Structure–Activity Relationships (QSARs) that link NPs composition with therapeutic outcomes.

ML algorithms such as Artificial Neural Networks (ANNs) and Support Vector Machines (SVMs) have been applied to predict synthesis outcomes and optimize experimental variables (temperature, surfactant ratio, and solvent type). For instance, in the fabrication of Solid-Lipid NPs (SLNs) for progesterone delivery, an AI-DoE (Design of Experiments) hybrid model accurately predicted particle size, polydispersity, zeta potential, and drug-loading efficiency—reducing the need for repetitive synthesis cycles [21].

Similarly, AI-driven Microphysiological Systems (AI-MPSs) combine ML with organ-on-chip models to optimize NPs synthesis and evaluate performance under physiologically relevant conditions, bridging the gap between in vitro and in vivo testing [22].

AI algorithms trained on large datasets have demonstrated remarkable accuracy in predicting NPs' morphology, stability, and surface chemistry. For example, an AI-assisted Physiologically Based Pharmacokinetic (PBPK) model successfully predicted the tumor delivery efficiency of polymeric and inorganic NPs in mice, achieving a strong correlation ($R^2 = 0.83$) between predicted and experimental biodistribution values [23].

In another example, Graph Neural Networks (GNNs) and Generative Adversarial Networks (GANs) were employed to explore millions of lipid combinations for RNA-loaded Lipid NPs (LNPs), identifying novel lipids with superior encapsulation efficiency and biostability for cancer vaccine delivery [24].

Additionally, de novo design frameworks employing Recurrent Neural Networks (RNNs) and Quasi-SMILES encoding have been developed to generate entirely new nanomaterials based on predicted 3D structure–function relationships. These systems can design biocompatible nanomaterials with optimized cell uptake and minimal toxicity, validated in vitro [25].

AI-assisted computational chemistry allows real-time optimization of functional NPs, such as AuNPs and polymeric micelles, by simulating nano–bio interactions. These models accurately predict binding affinities, aggregation tendencies, and structural stability, enabling fine-tuned design for targeted drug delivery [26]. The integration of AI in computational modeling ensures the rational engineering of nanostructures with enhanced therapeutic precision and reproducibility.

3.2.2 In Silico Toxicity Prediction and Biocompatibility Screening

A critical holdup in nanomedicine development is NPs-induced toxicity, which can result from undesired interactions with plasma proteins, immune cells, or organ systems. Traditional in vivo toxicity testing is costly and ethically constrained. AI-driven in silico toxicology platforms now provide a robust, non-animal alternative to predict NPs safety.

ML models, which include decision trees, random forests, and XGBoost algorithms, use physicochemical descriptors (size, charge, surface area, oxygen content) to predict NPs' toxicity and classify materials by risk level. These models can identify key contributors to cytotoxicity and oxidative stress responses, surpassing conventional QSAR methods in precision [27].

DL approaches now extend toxicity prediction to multiomics datasets, integrating gene expression, proteomics, and metabolic data to map complex nano–bio interaction networks. Such integrative modeling allows prediction of dose-response behavior and tissue-specific toxicity prior to in vivo validation [28].

The protein corona, which represents a dynamic layer of biomolecules adsorbed onto NPs surfaces, determines their biological identity and immune compatibility. AI and ML have been instrumental in predicting corona composition and immunogenic potential by learning from NPsprotein interaction datasets. These models help tailor surface chemistry and coating strategies to minimize immune activation and improve biocompatibility [29].

Furthermore, data mining techniques have been applied to correlate cytokine induction with NPs structure, guiding the design of cross-linked polymeric nanomaterials with reduced immunotoxicity [30].

Recent efforts have focused on safe-by-design nanomedicine strategies, where AI continuously evaluates toxicity risk during the design phase. For example, Singh et al. proposed an AI–ML-empowered nanotoxicology framework combining ab initio simulations and ML-based hazard ranking to assess nano–bio interactions and establish quantitative safety benchmarks for nanomaterials [31].

Such approaches enable researchers to generate biocompatibility maps and prioritize low-risk nanostructures before synthesis.

3.3 AI-Enhanced Drug Loading and Release Kinetics Prediction

The ability to accurately predict drug-loading capacity and release kinetics is critical to the design of efficient and safe smart nanomedicines. Traditional approaches rely on empirical measurements and kinetic fitting, which often fail to capture the complex interplay between NPs composition, drug–matrix interactions, and biological stimuli. AI introduces a paradigm shift by enabling data-driven modeling and computational simulation of these processes, thereby transforming drug delivery into a predictive science rather than an experimental trial-and-error endeavor [32].

By integrating ML, DL, and mechanistic modeling, AI can forecast how drugs are loaded, retained, and released from NPs under varying environmental conditions, ultimately allowing the design of stimuli-responsive, patient-specific nanocarriers.

AI-based predictive modeling utilizes Quantitative Structure–Property Relationships (QSPRs) and multi-parameter optimization algorithms to forecast the loading capacity of NPs based on their physicochemical attributes, such as polymer composition, porosity, hydrophobicity, surface area, and zeta potential.

Data mining and ML models (SVMs, neural networks, and random forests) have been used to predict drug loading in solid-lipid NPs containing hydrophobic agents like curcumin. These models accurately captured nonlinear relationships between lipid concentration, surfactant ratio, and loading efficiency, achieving > 90% predictive accuracy and reducing experimental workload by 60% [33].

DL models are increasingly used to simulate drug-carrier interactions at the molecular level. For example, Graph Neural Networks (GNNs) and Reinforcement Learning (RL) frameworks have been applied to optimize the synthesis of PEGylated doxorubicin prodrug NPs with high loading and pH-triggered release performance [34].

Molecular Dynamics (MD) simulations, coupled with neural networks, have been employed to study the adsorption, encapsulation, and release behavior of drugs like doxorubicin and paclitaxel on carbon-based nanocarriers under varying pH and temperature conditions. AI-driven analysis of these simulations revealed selective drug binding and release patterns corresponding to cancerous versus healthy tissue pH values [35].

AI offers a more robust alternative to traditional kinetic models (Higuchi, Korsmeyer–Peppas, and Hixson–Crowell) by integrating experimental data with mechanistic simulations, thereby capturing the true complexity of NPs degradation and diffusion processes.

Stiepel et al. developed a hybrid diffusion–erosion model for polymeric NPs, where ML algorithms estimated effective diffusion coefficients and predicted release curves under various pH conditions. The resulting AI-enhanced model achieved superior fit and predictive accuracy compared to conventional kinetics, demonstrating its value in rational drug-carrier optimization [36].

In mesoporous silica NPs conjugated with transferrin, AI-assisted kinetic modeling identified a Fickian diffusion mechanism and determined how transferrin surface modification affected drug–matrix interactions and release rates [37].

Reinforcement learning and genetic algorithms have been employed to optimize dual pH-and NIR-responsive release systems by learning from experimental kinetics datasets.

These AI models predict how external triggers (light, heat, magnetic fields) influence cumulative release profiles, supporting the design of adaptive, on-demand delivery systems [38].

AI is also being integrated into closed-loop experimental platforms to enable self-optimizing release systems. These systems use feedback from in vitro or in vivo drug concentration data to adjust the NPs composition in real time. Such approaches enable predictive control over: (i) release rate modulation through degradation kinetics, (ii) therapeutic window maintenance, and (iii) patient-specific dosage adjustment [32].

The convergence of AI analytics and physics-based modeling is giving rise to hybrid predictive frameworks that simulate and optimize nanomedicine performance across biological environments. By combining ML with finite element diffusion modeling, hybrid approaches can predict drug release under varying pH, redox potential, and enzyme concentrations. For example, supramolecular polymer NPs demonstrated dual redox- and light-responsive release, which AI algorithms were able to model with remarkable precision, accurately predicting cumulative release of > 50%

within 3 h [39]. AI is increasingly applied in precision oncology, where kinetic predictions help synchronize NPs drug release with tumor metabolism cycles, achieving a dynamic dosing control. In oral cancer therapy, for example, AI-enabled nanomaterials reached 18:1 tumor-to-normal tissue ratios by optimizing release timing and responsiveness [40].

3.4 AI-Assisted Molecular Docking and Target Validation

Traditionally, molecular docking, which is a computational method for predicting ligand–receptor binding affinities, relied heavily on static structural data and heuristic scoring functions. However, these conventional approaches often lacked the precision required to capture the dynamic and multiscale nature of drug–target interactions in biological environments. The incorporation of ML, DL, and generative AI models now enables more accurate prediction of binding poses, target identification, and lead compound validation, thereby accelerating the development of smart nanomedicines [41].

AI-driven docking platforms use data-trained scoring functions and pose prediction algorithms to overcome the limitations of traditional physics-based docking. Deep neural networks trained on experimental structural data can learn complex patterns of binding energy landscapes and accurately predict receptor–ligand affinities.

For instance, Xu et al. developed a generative AI framework that interfaces with molecular docking to design target-specific compound libraries. Their AI-guided model explored chemical space conditioned on the physicochemical characteristics of a receptor's active site, successfully identifying novel binding candidates for the SARS-CoV-2 main protease (Mpro). This hybrid AI-docking pipeline demonstrated remarkable efficiency in virtual screening and ligand prioritization, reducing computational cost and enhancing hit accuracy [42].

Similarly, Yang et al. applied a multi-model AI ensemble combining Support Vector Machines (SVMs), Random Forest (RF), and Artificial Neural Networks (ANNs) for the discovery of c-Jun N-terminal kinase 1 (JNK1) inhibitors from natural products. The fusion of ML with traditional docking and molecular dynamics simulations led to the identification of several potent JNK1 inhibitors, with Tricin exhibiting strong experimental validation ($IC_{50} = 17.68$ μM) (Fig. 2). This study illustrated how AI-guided virtual screening and rescoring can refine the accuracy of binding predictions and accelerate the AI-to-experiment validation loop [41].

In contrast to conventional docking, where ligands are tested against a known protein target, reverse docking screens a single compound across a wide array of proteins to identify potential molecular targets. This approach is particularly powerful for mechanism-of-action elucidation, drug repurposing, and nanocarrier targeting validation.

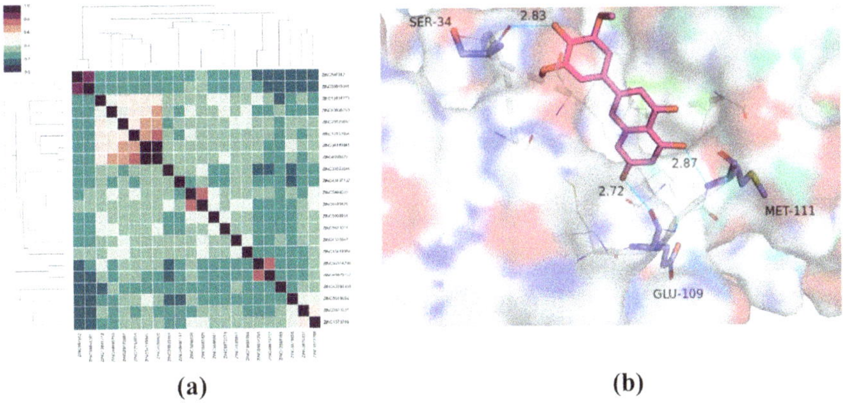

Fig. 2 **a** Correlation clustering heatmap for screening compounds; **b** the ligand–protein interaction profiles after the simulation—Tricin

Ruiz-Moreno et al. described a detailed reverse docking protocol using AI-enhanced algorithms to identify novel targets of the bioactive compound 4-hydroxycoumarin. By integrating AI-based scoring and clustering analyses, their study revealed RAC1 as a key molecular target, linking the compound's anticancer activity to specific signaling pathways. This demonstrates how AI-guided target prediction bridges the gap between computational docking and biological mechanism discovery [43].

Further advancements in AI-driven proteomics and cheminformatics have enabled large-scale target prediction for natural compounds and nanomedicine payloads. Li and Shi reported an integrated chemical proteomics and AI workflow that combined label-free proteomics data with ML-based pattern recognition to map NPs protein interactions. This approach accelerated target discovery for anticancer natural products by processing complex proteomic datasets, identifying transient or low-affinity interactions that traditional methods often overlook [44].

AI-assisted docking tools use ML models to predict binding affinity, rank docking poses, and filter false positives in large virtual screens. Algorithms such as gradient-boosted trees (XGBoost) and Convolutional Neural Networks (CNNs) learn from protein–ligand interaction data, improving both the accuracy and interpretability of docking scores.

A supercomputer-based ensemble docking pipeline used AI to cluster Molecular Dynamics (MD) trajectories and rescore docking poses, achieving remarkable predictive accuracy in screening over one billion compounds against SARS-CoV-2 targets. The study highlighted how integrating AI with High-Performance Computing (HPC) allows researchers to conduct AI-enhanced ensemble docking at unprecedented speed and scale [45].

Additionally, Malani et al. demonstrated how ML combined with molecular dynamics and metadynamics simulations could predict transporter-mediated drug toxicity by modeling substrate binding to ocular transporters. The AI model achieved

80% prediction accuracy and identified several systemic drugs as potential off-target binders, showing the application of AI-docking models for safety and off-target screening in nanomedicine [46].

Beyond virtual screening, AI systems now play a pivotal role in experimental target validation. The integration of molecular docking, AI-guided binding prediction, and bioinformatics creates a feedback loop where predicted targets are cross-validated using omics data and biological assays.

For instance, Zhang utilized AI-driven molecular docking combined with network pharmacology to identify phytochemicals (dipiperitylmagnolol, sophoranone) as potent breast cancer inhibitors. The AI-assisted docking pipeline, validated through molecular dynamics simulations, confirmed the stability and selectivity of binding, supporting the transition from in silico to in vitro verification [47].

Moreover, Deng et al. integrated molecular docking with DL models to identify galectin-3 inhibitors for neurodegenerative diseases. Their AI models, trained on physicochemical and docking descriptors, achieved correlation coefficients (R^2) up to 0.97, and molecular dynamics simulations validated the ligand–target stability, exemplifying AI-assisted target confirmation in early-stage drug discovery [48].

3.5 Case Studies of AI-Driven Smart Nanomedicine Development

The intersection between AI and nanomedicine has led to many transformative breakthroughs in cancer therapy, enabling predictive design, precise targeting, and adaptive drug delivery systems. Recent case studies highlight how AI-guided models enhance every stage of nanomedicine development, from NPs synthesis and drug formulation optimization to clinical translation and personalized therapy.

One notable example is the use of ML-driven nanoinformatics to design stimuli-responsive smart NPs capable of tumor-specific drug release. AI algorithms analyze multi-parametric data on polymer composition, drug–matrix interactions, and Tumor Microenvironment (TME) conditions to predict optimal physicochemical configurations for NPs [49]. These predictive models have enabled the creation of pH- and enzyme-responsive nanocarriers that release chemotherapeutic agents specifically in acidic tumor tissues, minimizing systemic toxicity and improving pharmacokinetics.

In another study, AI-integrated microfluidic synthesis platforms were used to produce polymeric NPs with tunable size and surface chemistry for targeted doxorubicin delivery. By combining reinforcement learning and feedback control, the platform autonomously optimized reaction conditions, achieving a 40% increase in drug loading and consistent size distribution [8].

Metabolic reprogramming is a hallmark of cancer progression, and AI has been instrumental in identifying and targeting these metabolic pathways using nanomedicine. In a 2025 study, AI-based data mining was used to design nanocarriers that disrupt glycolysis and lipid metabolism within the tumor microenvironment. The models integrated omics data and in silico simulations to predict nanocarrier accumulation efficiency and optimize drug–target pairing. These AI-optimized metabolic nanomedicines achieved selective tumor cell killing while preserving normal cell viability, demonstrating the feasibility of metabolically targeted precision nanotherapy [2].

AI has also revolutionized the design of lipid NPs systems for cancer immunotherapy. Using ML and GANs, researchers have developed predictive models to simulate RNA–lipid interactions, improving stability and targeted delivery of RNA vaccines. These models rapidly screened over 15,000 lipid structures, identifying novel lipids with superior encapsulation efficiency and tumor tropism [24]. Furthermore, digital twin simulations shortened the lyophilization optimization process from years to months, enabling personalized RNA vaccine formulations with real-time adjustment of stability and dosage.

In oncology diagnostics and therapy, AI-assisted magnetic nanostructures have emerged as multifunctional theranostic systems. A recent case study demonstrated how AI algorithms optimized the magnetic, optical, and structural properties of iron oxide–gold hybrid NPs for MRI-guided drug delivery. The resulting system achieved enhanced tumor accumulation, selective photothermal ablation, and precise image-guided release kinetics [50].

By integrating DL models to predict particle–cell interactions and optimize ligand density, the platform achieved improved targeting precision and reduced off-target toxicity. Translational studies show that AI enhances both therapeutic efficacy and clinical predictability in smart nanomedicine. For instance, AI-enabled DL models have been used to correlate NPs biodistribution data with patient-specific imaging features to optimize drug delivery schedules.

In clinical pilot trials for liver and breast cancers, AI-guided nanocarriers demonstrated superior accumulation in tumor tissues and improved patient outcomes compared to standard chemotherapy [51, 52].

A growing field, nanoinformatics, integrates AI-based data mining, predictive modeling, and biological simulation to accelerate nanomedicine translation. Through computational analysis of nanocarrier–biological interface data, AI can predict toxicological profiles, biodistribution patterns, and immune responses, drastically reducing the preclinical testing burden. This has been adopted in personalized oncology settings to customize NPs formulations according to patient-specific tumor genomics [53].

4 AI in Developing TME-Responsive Smart Nanomedicines

4.1 Overview of the TME

The TME is a highly complex and dynamic ecosystem composed of cancer cells, stromal cells, immune cells, Extracellular Matrix (ECM) components, and signaling molecules that collectively regulate tumor growth, angiogenesis, and metastasis (Fig. 3). Unlike normal tissues, the TME is characterized by a slightly acidic pH ($\approx$ 6.5–6.8), hypoxia, elevated Glutathione (GSH) levels, enzyme overexpression (such as matrix metalloproteinases and cathepsins), and an oxidative stress imbalance. These unique physicochemical features create both barriers to traditional therapies and opportunities for selective targeting by responsive nanomedicines [54, 55].

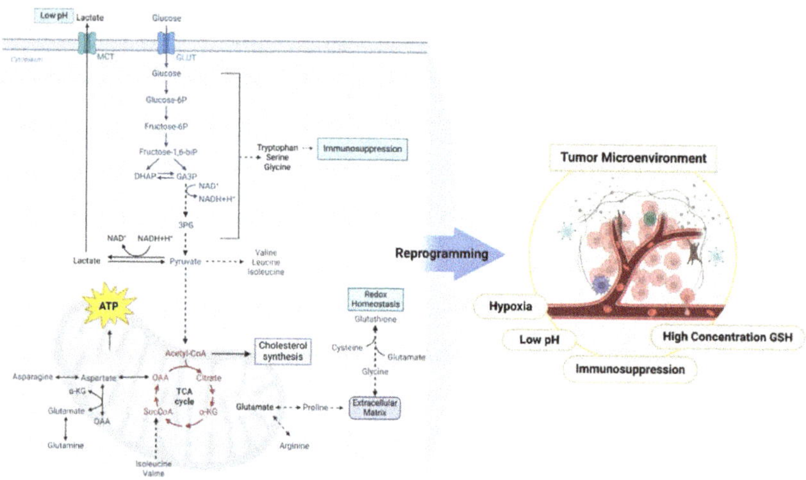

Fig. 3 Formation and characteristics of the TME. Solid tumors undergo extensive metabolic reprogramming to sustain their rapid growth and survival. These metabolic adaptations shape the Tumor Microenvironment (TME), whose unique biochemical and physiological properties can be strategically exploited for the design of nanomedicines enabling controlled and targeted drug delivery. The TME is defined by four major hallmarks: 1. Hypoxia, resulting from inadequate angiogenesis and oxygen supply; 2. Elevated lactate accumulation, a consequence of enhanced anaerobic glycolysis, which lowers the local pH; 3. Pronounced immune suppression, driven by metabolic byproducts that inhibit antitumor immune responses; and 4. Increased intracellular glutathione (GSH) levels, which support redox balance and resistance to oxidative stress in tumor cells. Key metabolic intermediates and transporters involved include MCT (Monocarboxylate Transporter), GLUT (Glucose Transporter), DHAP (Dihydroxyacetone Phosphate), GA3P (Glyceraldehyde-3-Phosphate), 3PG (3-Phosphoglycerate), NADH/NAD^+ (Reduced/Oxidized Nicotinamide Adenine Dinucleotide), OAA (Oxaloacetate), α-KG (Alpha-Ketoglutarate), and SucCoA (Succinyl-CoA) [2]

The metabolic reprogramming of cancer cells drives alterations in oxygen consumption, glucose metabolism, and redox balance, contributing to the acidic and reductive environment typical of solid tumors. Targeting these pathophysiological hallmarks forms the basis of TME-responsive nanomedicine design, where the release of therapeutic agents is triggered by intrinsic tumor cues rather than systemic factors [56].

4.2 pH-Responsive Nanocarriers

Among endogenous triggers, pH variation remains the most extensively exploited in smart nanomedicine. The extracellular environment of tumors is more acidic than normal tissue due to lactic acid accumulation from anaerobic glycolysis. pH-responsive nanocarriers are designed to exploit this gradient, releasing drugs selectively at the tumor site through mechanisms such as protonation-induced structural disassembly, acid-sensitive bond cleavage, or polymer swelling.

Polymeric nanocarriers, such as those made from poly(L-histidine), chitosan, and poly(β-amino ester), undergo conformational changes at low pH, promoting drug release and endosomal escape [57]. Inorganic pH-sensitive platforms, like mesoporous silica NPs functionalized with acid-labile linkers, demonstrate controlled degradation in acidic environments, improving delivery of chemotherapeutics and nucleic acids [58]. Recent developments include dual pH/redox-responsive nanocarriers, where polymeric micelles integrate pH-sensitive outer shells and disulfide cross-linked cores, enabling two-stage release: initial destabilization at low pH followed by intracellular release under reductive cytosolic conditions [59].

These platforms maximize drug bioavailability and minimize systemic toxicity, aligning with the goals of precision oncology.

4.3 Redox-Responsive and Enzyme-Responsive System

Redox-responsive nanomedicines harness the elevated intracellular GSH concentrations (up to 10 mM in tumor cells) to trigger drug release. Disulfide-bonded polymers and thioketal linkages are the most common chemical motifs employed, as they are stable in circulation but cleaved upon encountering reducing environments. For instance, poly(disulfide amide) NPs exhibit controlled degradation inside tumor cells, leading to site-specific release of anticancer agents [60].

Simultaneously, enzyme-responsive systems exploit the overexpression of proteases, esterases, and phosphatases in the TME. Nanocarriers incorporating peptide linkers cleavable by Matrix Metalloproteinases (MMPs) or cathepsin B can undergo structural disassembly at tumor sites, triggering selective drug release [61]. Recent hybrid systems combine enzyme sensitivity with ligand targeting, such as Arg-Gly-Asp (RGD)-functionalized NPs that recognize integrin receptors and release drugs upon enzymatic cleavage [62].

These multifunctional nanocarriers have demonstrated promising outcomes in overcoming multidrug resistance and enhancing tumor penetration.

4.4 Hypoxia-Responsive and Thermal/Light-Responsive Nanomedicines

Hypoxia, or low oxygen tension, is a defining feature of solid tumors resulting from irregular vasculature and rapid cellular proliferation. This condition can induce resistance to radiation and chemotherapy, but also provides a unique biochemical cue for targeted therapy. Hypoxia-responsive nanocarriers are typically engineered using azobenzene, nitroimidazole, or quinone moieties that undergo reduction under oxygen-deprived conditions, releasing their payload selectively within the tumor mass [63].

Polymeric nanomedicines have also been developed to alleviate hypoxia by catalyzing the decomposition of tumor-associated hydrogen peroxide (H_2O_2) into O_2. A recent study demonstrated a manganese-based nanoenzyme that performs dual catalytic and immunomodulatory functions, converting H_2O_2 into O_2 to reverse hypoxia and enhancing the tumor immune response when combined with photothermal therapy [64].

On the exogenous side, thermal-and light-responsive nanomedicines, such as gold NPs, graphene oxide, and upconversion NPs, can generate localized hyperthermia or Reactive Oxygen Species (ROS) upon exposure to near-infrared light, enabling photo-triggered drug release and synergistic photothermal/photodynamic therapy [65]. The combination of internal biochemical cues and external physical triggers represents a promising strategy for achieving both temporal and spatial precision in cancer treatment.

4.5 AI Models in the Critical Targets of TME and for Optimizing Nanocarrier Response

Traditional approaches to nanomedicine engineering often rely on empirical experimentation, which is time-consuming and limited in scope. AI-driven methodologies address these challenges by processing multidimensional datasets from omics technologies, bioimaging, and material characterization, establishing predictive relationships between nanocarrier physicochemical parameters and biological outcomes [2]. AI can also contribute to target screening based on metabolomics, enabling precision therapy (Fig. 4).

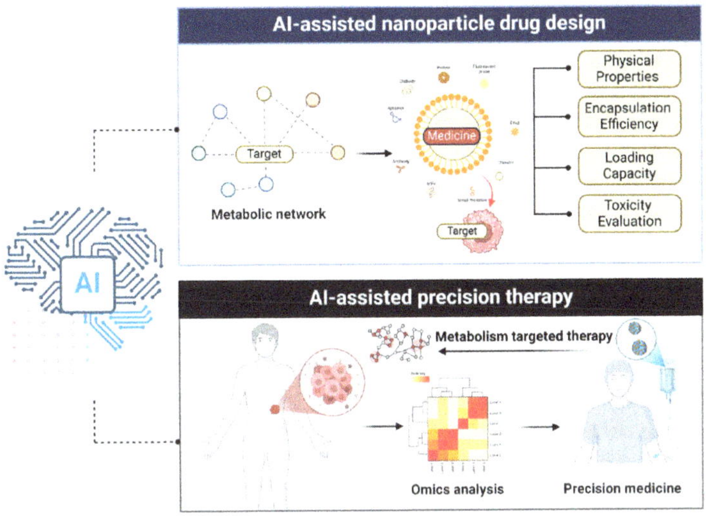

Fig. 4 AI is transforming the way NPs-based drugs are developed and personalized. By integrating metabolomics data, AI can pinpoint critical targets in tumor metabolic pathways, guiding more effective drug design and discovery. In the near future, AI-driven databases on NPs materials will enable highly accurate predictions of key factors such as physicochemical properties, encapsulation efficiency, drug-loading capacity, toxicity, and interactions within the tumor microenvironment. This approach could eliminate much of the traditional trial-and-error process, making drug development faster, more efficient, and more effective. Beyond development, AI also holds great promise for personalized medicine. By analyzing multiomics data, which include genomics, proteomics, and metabolomics, AI can recommend the most suitable treatments for each individual and provide predictive feedback on patient outcomes. This ensures that therapies are tailored to each patient's unique biology, reducing the risk of suboptimal treatment results caused by individual differences [2]

AI is already playing a key role in precision diagnosis and treatment. By assisting in patient prescreening and stratification, AI helps quickly and accurately identify target patients, enabling the creation of personalized treatment plans and drug targets. It also supports the design of nanomedicines. For example, in one study of 48 tumor samples, researchers used a random forest algorithm to classify cancer-driving mutations and built a drug–target interaction database to automatically match therapies to each patient's mutation profile, offering a new way to screen nanomedicines targeting metabolic pathways [66].

Tumor metabolism is closely tied to cancer progression and metastasis [67], and AI-driven identification of nanomedicines for metabolism-related targets shows great promise. For instance, Yangzi Chen et al. used ML to analyze gastric cancer metabolic profiles, effectively screening patients and distinguishing early cancer stages. Their model outperformed existing clinical biomarkers and provided individualized prognoses [68]. Similarly, Sunan Cui et al. applied an Actuarial Deep Learning Neural Network (ADNN) to stage III non-small cell lung cancer data, integrating radiomics and transcriptomics to improve prognostic accuracy beyond traditional models [69].

Multiomics analysis expands the diversity of data and helps reveal precise therapeutic targets for each patient. For example, ML analysis of metabolomic and transcriptomic data in triple-negative breast cancer identified N-acetyl-aspartyl-glutamate as a tumor-promoting metabolite and a potential therapeutic target [70]. With ongoing DL advancements, future research may integrate radiomics data, using imaging methods like CT and MRI combined with CNN-based analysis, to extract deeper, noninvasive insights from tumors [71, 72]. The synergy between radiomics and multiomics enables comprehensive decoding of the TME, enhancing the precision of diagnosis and treatment [73].

The TME's spatial heterogeneity poses a significant challenge to achieving uniform drug distribution and therapeutic efficacy. AI-based simulations can reconstruct spatial maps of pH, oxygen, and nutrient gradients, enabling researchers to predict how NPs behave within distinct tumor regions. CNNs applied to histological and imaging data can classify tumor subregions based on metabolic and vascular characteristics, allowing region-specific drug delivery modeling. For instance, AI-assisted diffusion models have been used to simulate NPs transport through heterogeneous tumor tissues, predicting accumulation "hotspots" and identifying zones of poor penetration due to dense extracellular matrix or low perfusion [74].

Interactions between TME metabolites are highly complex and often nonlinear. DL models capture these nonlinear patterns through activation functions between neurons, giving them an edge over traditional ML in analyzing large-scale metabolic data. This makes DL especially useful for identifying key targets in tumor metabolism [75]. For example, Fadhl M. Alakwaa et al. used a Feedforward Network (FFN) to classify ER-positive and ER-negative tumors, achieving the highest accuracy among seven tested algorithms. They identified important metabolites based on node weights in the first hidden layer and performed pathway analysis using those features. However, while effective, DL models remain difficult to interpret, and focusing only on the first layer may miss insights from deeper network layers [76].

AI-enhanced kinetic models can predict pH-dependent degradation and drug release profiles with greater precision than traditional diffusion–erosion equations. By training neural networks on large datasets of experimental release curves, researchers can identify optimal compositions and cross-linking densities for acid-labile nanocarriers that disassemble specifically in the acidic extracellular space of tumors. Similarly, Deep Reinforcement Learning (DRL) approaches have been applied to dynamically optimize synthesis parameters in real time, such as solvent ratios, reaction temperature, and polymerization time, allowing the creation of self-optimizing nanocarrier systems that autonomously adjust to achieve desired release characteristics [77]. Mingsheng Zhu et al. developed Nano-ISML, a DL-based image segmentation method for high-throughput, quantitative analysis of tumor vascular permeability. Using a U-Net CNN, the model segmented fluorescence images from two channels (vessels and NPs) to outline structures and calculate nine key metrics, including vascular permeability. This approach transformed vascular permeability studies from qualitative to quantitative, offering valuable insights for designing personalized nanomedicines [78]. Similarly, Zhen Gao et al. applied an SVM model that used drug chemical structures and clinical phenotypes, such as side effects and indications, to predict a drug's ability to cross the blood–brain barrier. Combining these features significantly improved prediction accuracy and opened new avenues for CNS drug discovery and repurposing.

When developing new nanomaterials for drug delivery, it's crucial to evaluate their potential toxicity and contamination risks. Computational approaches such as nano-QSAR, nano-QNTR (nanostructure–toxicity), and nano-QNAR (nanostructure–activity) are valuable for predicting the toxic effects of nanomaterials [79]. In inhalation nanotoxicology, direct experimental measurement of NPs absorption is difficult, so mechanical dose models like the Multi-Path Particle Dose (MPPD) model (developed by the NCRP and ICRP) are used to estimate how NPs deposit and clear within the lungs of humans and animals [80, 81]. Mei Liu et al. built an ML model that integrates chemical, physical, and phenotypic drug features to predict potential Adverse Drug Reactions (ADRs) and toxicity early in the development process. This approach enhances safety, lowers failure rates in clinical trials, and increases development efficiency [453]. Similarly, Mahnaz Ahmadi et al. used ML to predict NPs' toxicity by analyzing parameters such as particle size, surface charge, and biocompatibility. Among several algorithms tested (RF, decision tree, and SVM), the RF model performed best in classifying toxicity across cell types [82]. In addition, DNN-based QSAR models can predict Drug-Induced Liver Injury (DILI) using molecular descriptors [83]. AI systems trained on large toxicology datasets like Tox21 and ToxCast can automatically extract key features, improve prediction accuracy while reducing experimental workload [84, 85].

Digital twin systems, virtual patient models integrating AI and bioinformatics, are emerging as powerful tools for simulating how a specific nanomedicine will behave within an individual's TME. These AI-generated digital replicas account for dynamic interactions among pH, oxygenation, enzyme expression, and immune infiltration, enabling clinicians to optimize dosage, release kinetics, and timing for maximal efficacy [54].

Furthermore, AI contributes to clinical decision support by integrating patient-specific genomic, imaging, and nanomedicine performance data to identify predictive biomarkers of therapeutic success. This has accelerated the movement toward precision nanomedicine, where treatment regimens are adaptively refined based on real-time biological feedback and computational prediction.

5 Challenges and Future Perspectives

Despite remarkable advances in AI-driven nanomedicine, several scientific, technical, and ethical challenges continue to hinder clinical translation. These challenges span from biological complexities and data limitations to manufacturing constraints and regulatory uncertainties.

The transition from laboratory innovation to clinical application remains slow. While AI-designed nanomedicines have demonstrated impressive tumor targeting and controlled release in preclinical models, their performance in humans often differs due to the complexity of the TME, interpatient variability, and unpredictable immune responses. For instance, heterogeneous vascularization and immune infiltration can limit NPs accumulation and therapeutic efficacy [86]. Additionally, few AI-optimized nanomedicines have successfully reached clinical trials, largely because animal models fail to fully replicate human tumor physiology [87]. AI models require massive, high-quality datasets that integrate multiomics, imaging, and clinical data. However, inconsistencies in data collection and annotation can lead to biased or non-generalizable predictions [1]. Moreover, the "black-box" nature of DL makes it difficult to interpret how AI reaches specific conclusions about NPs behavior, which complicates regulatory approval and physician trust [3].

Producing AI-designed nanocarriers at scale while maintaining uniform physicochemical properties is another major bottleneck. Current synthesis methods lack reproducibility, and small variations in particle size, charge, or surface chemistry can drastically alter biodistribution and safety profiles [4]. Advanced microfluidic and AI-controlled fabrication platforms may help improve precision and reproducibility in NPs production [53].

Regulatory agencies currently lack standardized frameworks for evaluating AI-designed nanomedicines. Assessing algorithmic decision-making, model drift, and data privacy issues adds new layers of complexity to approval pipelines [5].

Ethical concerns also arise regarding patient data ownership, bias in training datasets, and AI's potential to amplify healthcare disparities [88].

The next decade of AI-driven nanomedicine will likely focus on integrating nanoinformatics, digital twins, and multiomics analytics to enable personalized nanotherapy design. AI-powered models will predict patient-specific responses by simulating NPs distribution, immune interactions, and metabolism in silico [49]. Furthermore, closed-loop systems coupling real-time biosensing and adaptive AI feedback could continuously adjust drug dosing and release kinetics for optimal outcomes [40].

Collaborations between clinicians, computational scientists, and regulatory bodies will be crucial to establishing data standards, ethical frameworks, and validation pipelines that accelerate safe translation from bench to bedside. With these advances, AI-driven smart nanomedicines could truly redefine precision oncology, delivering safer, more effective, and patient-tailored cancer treatments [2].

6 Conclusions

AI-driven nanomedicine represents one of the most promising frontiers in precision oncology. By combining the predictive power of artificial intelligence with the versatility of nanoscale drug delivery, researchers are moving closer to truly personalized cancer treatment. AI models can now analyze vast datasets, from genomics and metabolomics to imaging and pharmacokinetics, in order to predict how nanocarriers behave in complex tumor environments. This capability not only enhances drug targeting and minimizes toxicity but also accelerates the design of new nanotherapeutics, reducing the trial-and-error traditionally associated with drug discovery [86].

Despite this progress, real-world translation remains challenging. Biological variability, incomplete datasets, limited interpretability of DL models, and regulatory uncertainty still slow the path from bench to bedside [87]. Ethical and data privacy issues must also be addressed to build trust and ensure equitable access to AI-powered healthcare solutions [88].

Integrating nanoinformatics, digital twin modeling, and real-time AI feedback systems may revolutionize the way we design and test nanomedicines. These tools can simulate patient-specific tumor biology, predict treatment responses, and dynamically adjust drug release for optimal outcomes [49]. As multidisciplinary collaboration strengthens between data scientists, clinicians, and materials engineers, the future of nanomedicine will shift toward safer, smarter, and more personalized cancer therapies.

Acknowledgments I.N. and A.V. acknowledge the support from a grant of the Ministry of Research, Innovation and Digitalization, CNCS-UEFISCDI, project number PN-IV-PCB-RO-MD-2024-0254, within PNCDI IV and by the Romanian Ministry of Research, Innovation, and Digitalization under the Romanian National Nucleu Program LAPLAS VII—contract no. 30 N/2023.

References

1. Weerarathna IN, Kamble AR, Luharia A, Weerarathna IN, Kamble AR, Luharia A (2023) Artificial intelligence applications for biomedical cancer research: a review. Cureus 15. https://doi.org/10.7759/cureus.48307
2. Fan S, Wang W, Che W, Xu Y, Jin C, Dong L, Xia Q (2025) Nanomedicines targeting metabolic pathways in the tumor microenvironment: future perspectives and the role of AI. Metabolites 15:201. https://doi.org/10.3390/metabo15030201
3. Huhulea EN, Huang L, Eng S, Sumawi B, Huang A, Aifuwa E, Hirani R, Tiwari RK, Etienne M (2025) Artificial intelligence advancements in oncology: a review of current trends and future directions. Biomedicine 13:951. https://doi.org/10.3390/biomedicines13040951
4. Tracey SR, Smyth P, Barelle CJ, Scott CJ (2021) Development of next generation nanomedicine-based approaches for the treatment of cancer: We've barely scratched the surface. Biochem Soc Trans 49:2253–2269. https://doi.org/10.1042/BST20210343
5. Alavinejad M, Shirzad M, Javid-Naderi MJ, Rahdar A, Fathi-karkan S, Pandey S (2025) Smart Nanomedicines powered by artificial intelligence: a breakthrough in lung cancer diagnosis and treatment. Med Oncol 42:134. https://doi.org/10.1007/s12032-025-02680-x
6. Karimi M, Ghasemi A, Zangabad PS, Rahighi R, Basri SMM, Mirshekari H, Amiri M, Pishabad ZS, Aslani A, Bozorgomid M et al (2016) Smart micro/nanoparticles in stimulus-responsive drug/gene delivery systems. Chem Soc Rev 45:1457–1501. https://doi.org/10.1039/C5CS00798D
7. Rana A, Adhikary M, Singh PK, Das BC, Bhatnagar S (2023) "Smart" drug delivery: a window to future of translational medicine. Front Chem 10. https://doi.org/10.3389/fchem.2022.1095598

8. Sun L, Liu H, Ye Y, Lei Y, Islam R, Tan S, Tong R, Miao Y-B, Cai L (2023) Smart nanoparticles for cancer therapy. Signal Transduct Target Ther 8:418. https://doi.org/10.1038/s41392-023-01642-x
9. Kenchegowda M, Rahamathulla M, Hani U, Begum MY, Guruswamy S, Osmani RAM, Gowrav MP, Alshehri S, Ghoneim MM, Alshlowi A et al (2022) Smart Nanocarriers as an emerging platform for cancer therapy: a review. Molecules 27:146. https://doi.org/10.3390/molecules27010146
10. Das SS, Bharadwaj P, Bilal M, Barani M, Rahdar A, Taboada P, Bungau S, Kyzas GZ (2020) Stimuli-responsive polymeric nanocarriers for drug delivery, imaging, and theragnosis. Polymers 12:1397. https://doi.org/10.3390/polym12061397
11. Armenia I, Cuestas Ayllón C, Torres Herrero B, Bussolari F, Alfranca G, Grazú V, Martínez de la Fuente J (2022) Photonic and magnetic materials for on-demand local drug delivery. Adv Drug Deliv Rev 191:114584. https://doi.org/10.1016/j.addr.2022.114584
12. Xiao X, Teng F, Shi C, Chen J, Wu S, Wang B, Meng X, Essiet Imeh A, Li W (2022) Polymeric nanoparticles—promising carriers for cancer therapy. Front Bioeng Biotechnol 10. https://doi.org/10.3389/fbioe.2022.1024143
13. Javan Nikkhah S, Thompson D (2021) Molecular modelling guided modulation of molecular shape and charge for design of smart self-assembled polymeric drug transporters. Pharmaceutics 13:141. https://doi.org/10.3390/pharmaceutics13020141
14. Demetzos C (2016) Application of nanotechnology in drug delivery and targeting. In: Demetzos C (ed) Pharmaceutical nanotechnology: fundamentals and practical applications. Springer, Singapore, pp 77–145. ISBN:978-981-10-0791-0
15. Shukla A, Maiti P (2022) Nanomedicine and versatile therapies for cancer treatment. MedComm 3:e163. https://doi.org/10.1002/mco2.163
16. Rahman M, Afzal O, Ullah SNMN, Alshahrani MY, Alkhathami AG, Altamimi ASA, Almujri SS, Almalki WH, Shorog EM, Alossaimi MA et al (2023) Nanomedicine-based drug-targeting in breast cancer: pharmacokinetics, clinical Progress, and challenges. ACS Omega 8:48625–48649. https://doi.org/10.1021/acsomega.3c07345
17. Chou W-C, Canchola A, Zhang F, Lin Z (2025) Machine learning and artificial intelligence in nanomedicine. WIREs Nanomed Nanobiotechnol 17:e70027. https://doi.org/10.1002/wnan.70027
18. Akhtar M, Nehal N, Gull A, Parveen R, Khan S, Khan S, Ali J (2025) Explicating the transformative role of artificial intelligence in designing targeted nanomedicine. Expert Opin Drug Deliv 22:971–991. https://doi.org/10.1080/17425247.2025.2502022
19. Lin Z, Chou W-C, Cheng Y-H, He C, Monteiro-Riviere NA, Riviere JE (2022) Predicting nanoparticle delivery to Tumors using machine learning and artificial intelligence approaches. Int J Nanomedicine 17:1365–1379. https://doi.org/10.2147/IJN.S344208
20. Kapoor DU, Sharma JB, Gandhi SM, Prajapati BG, Thanawuth K, Limmatvapirat S, Sriamornsak P (2024) AI-driven design and optimization of nanoparticle-based drug delivery systems. Sci Eng Health Stud:24010003–24010003. https://doi.org/10.69598/sehs.18.24010003
21. Suriyaamporn P, Pamornpathomkul B, Wongprayoon P, Rojanarata T, Ngawhirunpat T, Opanasopit P (2024) The artificial intelligence and design of experiment assisted in the development of progesterone-loaded solid-lipid nanoparticles for transdermal drug delivery. Pharmacia 71:1–12. https://doi.org/10.3897/pharmacia.71.e123549
22. Lee Y, Min J, Kim S, Ko J (n/a) AI-driven microphysiological systems for advancing nanoparticle therapeutics. Adv Intell Syst:2500022. https://doi.org/10.1002/aisy.202500022
23. Chou W-C, Chen Q, Yuan L, Cheng Y-H, He C, Monteiro-Riviere NA, Riviere JE, Lin Z (2023) An artificial intelligence-assisted physiologically-based pharmacokinetic model to predict nanoparticle delivery to Tumors in mice. J Control Release 361:53–63. https://doi.org/10.1016/j.jconrel.2023.07.040
24. Bhujel R, Enkmann V, Burgstaller H, Maharjan R (2025) Artificial intelligence-driven strategies for targeted delivery and enhanced stability of RNA-based lipid nanoparticle cancer vaccines. Pharmaceutics 17:992. https://doi.org/10.3390/pharmaceutics17080992

25. He, Y.; Liu, F.; Min, W.; Liu, G.; Wu, Y.; Wang, Y.; Yan, X.; Yan, B. De Novo design of biocompatible nanomaterials using quasi-SMILES and recurrent neural networks. ACS Appl Mater Interfaces 2024, 16, 66367–66376, doi:https://doi.org/10.1021/acsami.4c15600
26. Alalmaie A, Alshahrani HT, Alqahtani M, Alshahrani Z, Alahmari S, Asiri A, Alqadi B, Alshahrani A, Alshahrani S, Akhter MH (2025) Integrating computational insights in gold nanoparticle-mediated drug delivery: enhancing efficacy and precision. Front Med Technol 7. https://doi.org/10.3389/fmedt.2025.1528826
27. Yousaf I (2024) AI and machine learning approaches for predicting nanoparticles toxicity the critical role of physiochemical properties
28. AE P, Selvakumar VS (2025) A review of AI-driven methods for cytotoxicity analysis and predictive modeling in the treatment of cancer using nanomaterials. In: Proceedings of the 2025 international conference on electronics, computing, communication and control technology (ICECCC), pp 1–6
29. Kopac T (2025) Leveraging artificial intelligence and machine learning for characterizing protein corona, nanobiological interactions, and advancing drug discovery. Bioengineering 12:312. https://doi.org/10.3390/bioengineering12030312
30. Elsabahy M, Wooley KL (2015) Data mining as a guide for the construction of cross-linked nanoparticles with low immunotoxicity via control of polymer chemistry and supramolecular assembly. Acc Chem Res 48:1620–1630. https://doi.org/10.1021/acs.accounts.5b00066
31. Singh AV, Rosenkranz D, Ansari MHD, Singh R, Kanase A, Singh SP, Johnston B, Tentschert J, Laux P, Luch A (2020) Artificial intelligence and machine learning empower advanced biomedical material design to toxicity prediction. Adv Intell Syst 2:2000084. https://doi.org/10.1002/aisy.202000084
32. Vidhya KS, Sultana A, Naveen Kumar M, Rangareddy H (2023) Artificial Intelligence's impact on drug discovery and development from bench to bedside. Cureus 15. https://doi.org/10.7759/cureus.47486
33. Widyati R, Ashari A (2021) Afiahayati a review of using data mining and machine learning for predicting drug loading modeling in solid lipid nanoparticles containing curcumin. J Phys Conf Ser 1918:042015. https://doi.org/10.1088/1742-6596/1918/4/042015
34. Song J, Xu B, Yao H, Lu X, Tan Y, Wang B, Wang X, Yang Z (2021) Schiff-linked PEGylated doxorubicin Prodrug forming pH-responsive nanoparticles with high drug loading and effective anticancer therapy. Front Oncol 11. https://doi.org/10.3389/fonc.2021.656717
35. Maleki, R.; Dahri, M.; Akbarialiabad, H. Loading and release of anti-cancer drugs using simultaneous pH and temperature responsive Nanohybrid 2020
36. Stiepel RT, Pena ES, Ehrenzeller SA, Gallovic MD, Lifshits LM, Genito CJ, Bachelder EM, Ainslie KM (2022) A predictive mechanistic model of drug release from surface eroding polymeric nanoparticles. J Control Release 351:883–895. https://doi.org/10.1016/j.jconrel.2022.09.067
37. Jackson N, Ortiz AC, Jerez A, Morales J, Arriagada F (2023) Kinetics and mechanism of camptothecin release from transferrin-gated Mesoporous silica nanoparticles through a pH-responsive surface linker. Pharmaceutics 15:1590. https://doi.org/10.3390/pharmaceutics15061590
38. Wang X, Yang Y, Zhang G, Tang C-Y, Law W-C, Yu C, Wu X, Li S, Liao Y (2023) NIR-cleavable and pH-responsive polymeric yolk–shell nanoparticles for controlled drug release. Biomacromolecules 24:2009–2021. https://doi.org/10.1021/acs.biomac.2c01404
39. Ma Z, Qi J (2022) Application of supramolecular polymer nanoparticles in controlled release system of anticancer drugs. J Nanomater 2022:2219602. https://doi.org/10.1155/2022/2219602
40. Tivay S, Taghipoor M, Hosseini H, Saeedzadeh M, Valanik S (2025) Next-generation drug delivery: smart nanomaterials for precision healthcare(focusing oral cancer). J Oral Dent Health Nexus 2:6–18. https://doi.org/10.61838/kman.jodhn.2.2.2
41. Yang R, Zhao G, Yan B (2022) Discovery of novel C-Jun N-terminal kinase 1 inhibitors from natural products: integrating artificial intelligence with structure-based virtual screening and biological evaluation. Molecules 27:6249. https://doi.org/10.3390/molecules27196249

42. Xu Z, Wauchope O, Frank AT (2020) Navigating chemical space by interfacing generative artificial intelligence and molecular docking. https://doi.org/10.1101/2020.06.09.143289
43. Ruiz-Moreno AJ, Dömling A, Velasco-Velázquez MA (2021) Reverse docking for the identification of molecular targets of anticancer compounds. In: Robles-Flores M (ed) Cancer cell signaling: methods and protocols. Springer US, New York, NY, pp 31–43. ISBN:978-1-0716-0759-6
44. Li G, Shi Q, Wu Q, Sui X (2025) Target identification of natural products in cancer with chemical proteomics and artificial intelligence approaches. Cancer Biol Med 22:549–597. https://doi.org/10.20892/j.issn.2095-3941.2025.0145
45. Acharya A, Agarwal R, Baker MB, Baudry J, Bhowmik D, Boehm S, Byler KG, Chen SY, Coates L, Cooper CJ et al (2020) Supercomputer-based ensemble docking drug discovery pipeline with application to Covid-19. J Chem Inf Model 60:5832–5852. https://doi.org/10.1021/acs.jcim.0c01010
46. Malani M, Hiremath MS, Sharma S, Jhunjhunwala M, Gayen S, Hota C, Nirmal J (2024) Interaction of systemic drugs causing ocular toxicity with organic Cation transporter: an artificial intelligence prediction. J Biomol Struct Dyn 42:5207–5218. https://doi.org/10.1080/07391102.2023.2226717
47. Zhang H (2024) Exploring herbal compounds as targeted therapies for breast cancer: insights from network pharmacology, molecular docking, MD simulation, ADME-toxicity and DFT profiles. Iran J Pharm Res:e153579
48. Deng L, Zhong W, Zhao L, He X, Lian Z, Jiang S, Chen CY-C (2020) Artificial intelligence-based application to explore inhibitors of neurodegenerative diseases. Front Neurorobot 14. https://doi.org/10.3389/fnbot.2020.617327
49. Soltani M, Moradi Kashkooli F, Souri M, Zare Harofte S, Harati T, Khadem A, Haeri Pour M, Raahemifar K (2021) Enhancing clinical translation of cancer using Nanoinformatics. Cancer 13:2481. https://doi.org/10.3390/cancers13102481
50. Govindan B, Sabri MA, Hai A, Banat F, Haija MA (2023) A review of advanced multifunctional magnetic nanostructures for cancer diagnosis and therapy integrated into an artificial intelligence approach. Pharmaceutics 15:868. https://doi.org/10.3390/pharmaceutics15030868
51. Bhange M, Telange D (2025) Convergence of nanotechnology and artificial intelligence in the fight against liver cancer: a comprehensive review. Discov Oncol 16:77. https://doi.org/10.1007/s12672-025-01821-y
52. Bendani H, Boumajdi N, Belyamani L, Ibrahimi A (2025) Revolutionizing breast cancer immunotherapy by integrating AI and nanotechnology approaches: review of current applications and future directions. Bioelectron Med 11:13. https://doi.org/10.1186/s42234-025-00173-w
53. Abdullah KM, Sharma G, Singh AP, Siddiqui JA (2025) Nanomedicine in cancer therapeutics: current perspectives from bench to bedside. Mol Cancer 24:169. https://doi.org/10.1186/s12943-025-02368-w
54. Sun H, Li Y, Xue M, Feng D (2025) Tumor microenvironment-responsive nanoparticles: promising cancer PTT carriers. Int J Nanomedicine 20:7987–8001. https://doi.org/10.2147/IJN.S526497
55. Wang Y, Deng T, Liu X, Fang X, Mo Y, Xie N, Nie G, Zhang B, Fan X (2024) Smart Nanoplatforms responding to the tumor microenvironment for precise drug delivery in cancer therapy. Int J Nanomedicine 19:6253–6277. https://doi.org/10.2147/IJN.S459710
56. Cui F, Liu J, Pang S, Li B (2022) Recent advance in tumor microenvironment-based stimuli-responsive nanoscale drug delivery and imaging platform. Front Pharmacol:13. https://doi.org/10.3389/fphar.2022.929854
57. Chu S, Shi X, Tian Y, Gao F (2022) pH-responsive polymer nanomaterials for tumor therapy. Front Oncol 12. https://doi.org/10.3389/fonc.2022.855019
58. Liu Y, Si L, Jiang Y, Jiang S, Zhang X, Li S, Chen J, Hu J (2025) Design of pH-responsive nanomaterials based on the tumor microenvironment. Int J Nanomedicine 20:705–721. https://doi.org/10.2147/IJN.S504629

59. Mi P (2020) Stimuli-responsive Nanocarriers for drug delivery, tumor imaging, therapy and theranostics. Theranostics 10:4557–4588. https://doi.org/10.7150/thno.38069
60. Zhou W, Jia Y, Liu Y, Chen Y, Zhao P (2022) Tumor microenvironment-based stimuli-responsive nanoparticles for controlled release of drugs in cancer therapy. Pharmaceutics 14:2346. https://doi.org/10.3390/pharmaceutics14112346
61. Li M, Zhao G, Su W-K, Shuai Q (2020) Enzyme-responsive nanoparticles for anti-tumor drug delivery. Front Chem 8. https://doi.org/10.3389/fchem.2020.00647
62. Taheri Z, Mozafari N, Moradian G, Lovison D, Dehshahri A, De Marco R (2024) Integrin-specific stimuli-responsive nanomaterials for cancer theranostics. Pharmaceutics 16:1441. https://doi.org/10.3390/pharmaceutics16111441
63. Xia Y, Duan S, Han C, Jing C, Xiao Z, Li C (2022) Hypoxia-responsive nanomaterials for tumor imaging and therapy. Front Oncol 12. https://doi.org/10.3389/fonc.2022.1089446
64. Yang Q, Wu Q, Liu H, Wu J, Ma F, Tian X (2025) Tumor microenvironment–responsive and modulatory manganese-based nanoenzyme for enhanced tumor immunotherapy. Front Pharmacol 15. https://doi.org/10.3389/fphar.2024.1518983
65. Tian L, Lu L, Qiao Y, Ravi S, Salatan F, Melancon MP (2016) Stimuli-responsive gold nanoparticles for cancer diagnosis and therapy. J Funct Biomater 7:19. https://doi.org/10.3390/jfb7030019
66. Schroeder MP, Rubio-Perez C, Tamborero D, Gonzalez-Perez A, Lopez-Bigas N (2014) OncodriveROLE classifies cancer driver genes in loss of function and activating mode of action. Bioinformatics 30:i549–i555. https://doi.org/10.1093/bioinformatics/btu467
67. Ohshima K, Morii E (2021) Metabolic reprogramming of cancer cells during tumor progression and metastasis. Metabolites 11:28. https://doi.org/10.3390/metabo11010028
68. Chen Y, Wang B, Zhao Y, Shao X, Wang M, Ma F, Yang L, Nie M, Jin P, Yao K et al (2024) Metabolomic machine learning predictor for diagnosis and prognosis of gastric cancer. Nat Commun 15:1657. https://doi.org/10.1038/s41467-024-46043-y
69. Cui S, Ten Haken RK, El Naqa I (2021) Integrating multiomics information in deep learning architectures for joint actuarial outcome prediction in non-small cell lung cancer patients after radiation therapy. Int J Radiat Oncol 110:893–904. https://doi.org/10.1016/j.ijrobp.2021.01.042
70. Xiao Y, Ma D, Yang Y-S, Yang F, Ding J-H, Gong Y, Jiang L, Ge L-P, Wu S-Y, Yu Q et al (2022) Comprehensive metabolomics expands precision medicine for triple-negative breast cancer. Cell Res 32:477–490. https://doi.org/10.1038/s41422-022-00614-0
71. Qi Y-J, Su G-H, You C, Zhang X, Xiao Y, Jiang Y-Z, Shao Z-M (2024) Radiomics in breast cancer: current advances and future directions. Cell Rep Med 5. https://doi.org/10.1016/j.xcrm.2024.101719
72. Xia T, Zhao B, Li B, Lei Y, Song Y, Wang Y, Tang T, Ju S (2024) MRI-based radiomics and deep learning in biological characteristics and prognosis of hepatocellular carcinoma: opportunities and challenges. J Magn Reson Imaging 59:767–783. https://doi.org/10.1002/jmri.28982
73. Kang W, Qiu X, Luo Y, Luo J, Liu Y, Xi J, Li X, Yang Z (2023) Application of radiomics-based multiomics combinations in the tumor microenvironment and cancer prognosis. J Transl Med 21:598. https://doi.org/10.1186/s12967-023-04437-4
74. Chamseddine IM, Rejniak KA (2020) Hybrid modeling frameworks of tumor development and treatment. WIREs Syst Biol Med 12:e1461. https://doi.org/10.1002/wsbm.1461
75. Zhang K, Yang X, Wang Y, Yu Y, Huang N, Li G, Li X, Wu JC, Yang S (2025) Artificial intelligence in drug development. Nat Med 31:45–59. https://doi.org/10.1038/s41591-024-03434-4
76. Alakwaa FM, Chaudhary K, Garmire LX (2018) Deep learning accurately predicts Estrogen receptor status in breast cancer metabolomics data. J Proteome Res 17:337–347. https://doi.org/10.1021/acs.jproteome.7b00595
77. Li H, Feng Y, Luo Q, Li Z, Li X, Gan H, Gu Z, Gong Q, Luo K (2023) Stimuli-activatable nanomedicine meets cancer theranostics. *Theranostics* 13:5386–5417. https://doi.org/10.7150/thno.87854

78. Zhu M, Zhuang J, Li Z, Liu Q, Zhao R, Gao Z, Midgley AC, Qi T, Tian J, Zhang Z et al (2023) Machine-learning-assisted single-vessel analysis of nanoparticle permeability in tumour vasculatures. Nat Nanotechnol 18:657–666. https://doi.org/10.1038/s41565-023-01323-4
79. Singh AV, Varma M, Laux P, Choudhary S, Datusalia AK, Gupta N, Luch A, Gandhi A, Kulkarni P, Nath B (2023) Artificial intelligence and machine learning disciplines with the potential to improve the nanotoxicology and nanomedicine fields: a comprehensive review. Arch Toxicol 97:963–979. https://doi.org/10.1007/s00204-023-03471-x
80. Singh AV, Ansari MHD, Rosenkranz D, Maharjan RS, Kriegel FL, Gandhi K, Kanase A, Singh R, Laux P, Luch A (2020) Artificial intelligence and machine learning in computational nanotoxicology: unlocking and empowering nanomedicine. Adv Healthc Mater 9:1901862. https://doi.org/10.1002/adhm.201901862
81. Asgharian B, Price OT, Oldham M, Chen L-C, Saunders EL, Gordon T, Mikheev VB, Minard KR, Teeguarden JG (2014) Computational modeling of nanoscale and microscale particle deposition, retention and dosimetry in the mouse respiratory tract. Inhal Toxicol 26:829–842. https://doi.org/10.3109/08958378.2014.935535
82. Ahmadi M, Ayyoubzadeh SM, Ghorbani-Bidkorpeh F (2024) Toxicity prediction of nanoparticles using machine learning approaches. Toxicology 501:153697. https://doi.org/10.1016/j.tox.2023.153697
83. Mostafa F, Chen M (2024) Computational models for predicting liver toxicity in the deep learning era. Front Toxicol 5. https://doi.org/10.3389/ftox.2023.1340860
84. Tropsha A, Isayev O, Varnek A, Schneider G, Cherkasov A (2024) Integrating QSAR modelling and deep learning in drug discovery: the emergence of deep QSAR. Nat Rev Drug Discov 23:141–155. https://doi.org/10.1038/s41573-023-00832-0
85. Yang S, Kar S (2023) Application of artificial intelligence and machine learning in early detection of adverse drug reactions (ADRs) and drug-induced toxicity. Artif Intell Chem 1:100011. https://doi.org/10.1016/j.aichem.2023.100011
86. Wang J, Zhao Y, Nie G (2023) Intelligent nanomaterials for cancer therapy: recent progresses and future possibilities. Med Rev 3:321–342. https://doi.org/10.1515/mr-2023-0028
87. Liu Y, Zhang Y, Li H, Hu TY (2025) Recent advances in the bench-to-bedside translation of cancer nanomedicines. Acta Pharm Sin B 15:97–122. https://doi.org/10.1016/j.apsb.2024.12.007
88. Adeniyi M, Ayoola VB, Samuel TE, Awosan W (2024) Artificial intelligence-driven wearable electronics and smart nanodevices for continuous cancer monitoring and enhanced diagnostic accuracy. Int J Sci Res Mod Technol 3:3–18. https://doi.org/10.38124/ijsrmt.v3i11.106

Nanorobotics: Type, Synthesis, and Medical Applications

Sanjit Kr. Roy, Kazi Asraf Ali, and Bichitra Biswas

Abstract Nanorobotics has emerged as a transformative domain in precision medicine, integrating nanotechnology, robotics, and artificial intelligence (AI) to enable diagnosis and therapy at the molecular and cellular levels. This chapter presents a comprehensive overview of the fundamentals of nanorobotics in medicine, beginning with its definition, scope, and historical evolution from early nanotechnology concepts to advanced intelligent nanosystems. The role of AI in nanorobot design, navigation, and control is emphasized, highlighting its advantages over conventional nanotechnology and traditional medical approaches. Various classifications and types of nanorobots are discussed, including biological and synthetic systems, DNA-based and protein-based nanorobots, as well as magnetically, acoustically, and light-driven nanorobots. Emerging concepts such as swarm nanorobotics and collective intelligence are also explored. The chapter examines materials and structural design considerations, focusing on biocompatible and biodegradable materials, surface functionalization, targeting ligands, and mechanical stability. Advanced synthesis and fabrication techniques, including bottom-up and top-down approaches, DNA origami, micro–nanofabrication, and AI-assisted optimization, are detailed. Navigation, control, and powering strategies using external fields, autonomous AI algorithms, propulsion mechanisms, and real-time feedback systems are reviewed. Major medical applications in drug delivery, targeted cancer therapy, diagnostics, minimally invasive surgery, and regenerative medicine are highlighted. Finally, key challenges related to toxicity, immunogenicity, ethics, regulation, scalability, and clinical translation are discussed, along with future perspectives on AI-driven nanorobotics. Overall, this chapter underscores the potential of nanorobotics to revolutionize next-generation healthcare.

Keywords Nanorobotics · Precision medicine · Artificial intelligence · Targeted drug delivery · Swarm intelligence · Nanomedicine

S. Kr. Roy (✉) · K. A. Ali · B. Biswas
Department of Pharmaceutical Technology, Maulana Abul Kalam Azad University of Technology, Haringhata, West Bengal, India

J. Chen (ed.), *AI-Driven Nanotechnology in Precision Medicine and Healthcare*, Smart Nanomaterials Technology, https://doi.org/10.1007/978-981-92-1690-1_6

1 Introduction

Nanorobotics is one of the most transformative fields in precision medicine, enabling interventions at the microscopic level. These nanoscale devices are designed to interact directly with biological components such as cells, proteins, and DNA [1]. Unlike conventional therapies, nanorobots act at the disease site, improving treatment precision and minimizing harm to healthy tissues. Nanorobotics integrates nanotechnology, robotics, biology, and medicine into a single interdisciplinary platform. The miniature size of nanorobots allows them to circulate through blood vessels and penetrate tissues that are otherwise inaccessible [2]. Their operation closely mimics natural biological processes; nanorobots can operate safely and efficiently within the body. This approach supports personalized, targeted treatments and positions nanorobotics as a foundation for next-generation medical innovation [3].

Integrating artificial intelligence (AI) with nanotechnology has accelerated the development and effectiveness of nanorobotic systems. AI enables nanorobots to analyze complex biological signals and make real-time decisions. Through machine learning, nanorobots can identify disease-specific biomarkers and differentiate healthy from diseased cells, allowing precise targeting of cancer cells, infected tissues, or abnormal molecular pathways. AI also supports autonomous navigation in dynamic environments such as the bloodstream, while predictive modeling optimizes nanorobot behavior, dosage, and treatment timing [4].

AI further enhances nanorobot adaptability by enabling learning from prior interactions. This capability improves accuracy, efficiency, and therapeutic outcomes, while AI-driven control reduces human intervention and maximizes safety [5].

This chapter covers the fundamental principles of medical nanorobotics, including design, synthesis, and functional mechanisms. It reviews various types of nanorobots, biological, synthetic, and hybrid, each tailored for specific medical applications. Synthesis strategies such as self-assembly, DNA origami, and advanced nanofabrication are discussed, highlighting their importance for creating stable, biocompatible, and efficient nanorobots. The chapter also examines clinical applications, including targeted drug delivery, diagnostics, minimally invasive surgery, and regenerative medicine, with emphasis on precise therapeutic actions and controlled release. AI improves nanorobot design, navigation, and decision-making, making treatments more accurate and efficient. The chapter also discusses safety, ethical, and scalability challenges, providing a clear and practical overview of this emerging field.

1.1 Historical Evolution and Key Milestones

The origins of nanorobotics in precision medicine and healthcare date to the mid-twentieth century, when physicist Richard Feynman introduced the concept of manipulating matter at the atomic scale. In 1959, his theoretical ideas inspired researchers to envision machines capable of operating within the human body, laying the groundwork for future nanoscale technologies [6].

In the 1970s and 1980s, the formalization of nanotechnology and advances in materials science enabled the systematic study of matter at the nanoscale. Tools such as scanning tunneling microscopy (STM) and atomic force microscopy (AFM) allowed direct visualization and manipulation of atoms and molecules, transforming nanotechnology from an abstract idea into an experimental science [6].

In the 1990s, research focused on biomedical applications, especially drug delivery and diagnostics. Scientists developed nanoparticles and nanoscale carriers [7] to target specific tissues and cells, establishing the foundation for nanorobotic systems. This era marked the shift from passive nanomaterials to active, controllable nanosystems [8, 9].

In the early 2000s, nanomedicine grew rapidly due to advances in biotechnology, molecular biology, and computational modeling. DNA-based nanorobots and bio-inspired nanosystems enabled precise interactions with biological environments. These innovations supported the goals of precision medicine, which focuses on individualized diagnosis and treatment [10].

Recently, nanorobotics has progressed toward smart, responsive systems that integrate artificial intelligence and sensor technologies. Modern nanorobots are being developed for targeted drug delivery, early disease detection, minimally invasive surgery, and real-time cellular monitoring [11]. As a result, nanorobotics has evolved from a theoretical concept into a transformative tool with significant potential for personalized healthcare and future medical innovation. The evolution of nanorobotics in precision medicine and healthcare is shown as a flowchart in Fig. 1.

1.2 Role of AI in Nanorobot Design and Decision-Making

Artificial intelligence is essential to nanorobot design and decision-making, enabling adaptive and precise functionality. In the design phase, AI optimizes size, shape, materials, and surface properties to improve biocompatibility and performance. Machine learning analyzes biological and clinical data to predict nanorobot interactions with cells and tissues. During operation, AI-equipped nanorobots detect disease-specific biomarkers and analyze molecular patterns to distinguish healthy from diseased cells, allowing targeted action only on affected cells [12]. This precision reduces side effects and enhances therapeutic outcomes. AI also supports autonomous navigation in complex environments and real-time adjustments based on physiological conditions [13]. It enables controlled drug release in response to internal stimuli, such as pH, enzymes, or temperature, and external signals like magnetic fields or light. Predictive analytics determine optimal drug dosage and timing, ensuring delivery only when and where needed. These capabilities improve treatment accuracy and minimize side effects [14].

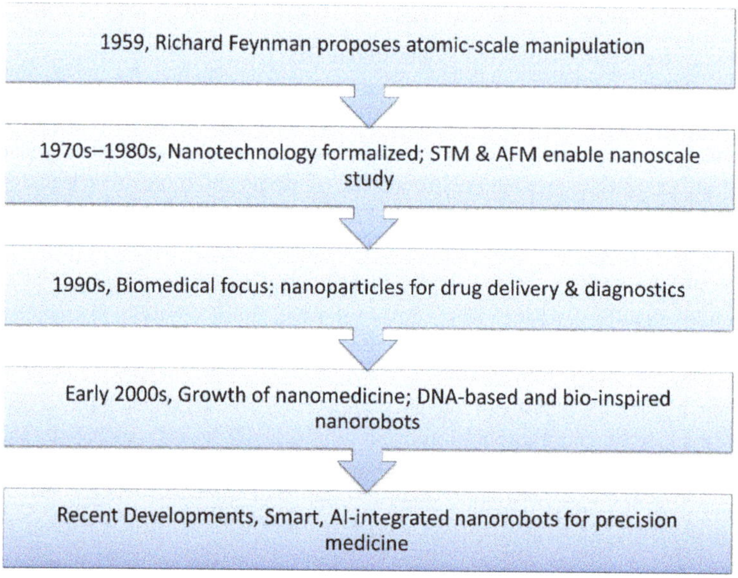

Fig. 1 Evolution of nanorobotics in precision medicine and healthcare

1.3 Advantages Over Conventional Drug Delivery and Diagnostic Systems

Nanorobotics offers significant advantages over conventional drug delivery (Table 1) and diagnostic systems by enabling targeted and precise medical interventions. Unlike traditional methods that distribute drugs throughout the body, nanorobots deliver therapeutics directly to the diseased site. This approach also reduces drug wastage and improves therapeutic efficiency [10].

2 Classification of Nanorobots

Nanorobots can be broadly classified based on their composition, driving mechanism, and functional behavior in biological systems (Fig. 2). This classification helps in understanding their design, mode of action, and suitability for specific medical applications. Some nanorobots are derived from biological components, while others are fully synthetic or hybrid in nature. They may also be categorized by how they are powered or controlled, such as magnetic, acoustic, or light-driven systems [23].

Table 1 Advantages of nanorobotics over conventional drug delivery

S. No.	Advantages	Description	References
1	Minimally invasive approach	Eliminates the need for surgical procedures for diagnosis or therapy	[15]
2	Prolonged circulation time	Designed to remain stable and functional in the bloodstream for longer durations	[16]
3	Multimodal functionality	Combines diagnosis, therapy, and monitoring in a single system	[17]
4	Enhanced bioavailability	Improves absorption and utilization of drugs at the target site	[18]
5	Ability to cross biological barriers	Can penetrate barriers such as the blood–brain barrier	[19]
6	Personalized therapy	Tailors treatment based on patient-specific biomarkers	[20]
7	Reduced dosing frequency	Sustained and controlled release lowers the need for repeated dosing	[21]
8	Improved early disease detection	Detects molecular changes before clinical symptoms appear	[22]

2.1 *Biological Nanorobots*

Biological nanorobots are tiny machines based on natural biological elements, which include cells, bacteria, viruses, and biomolecules. Since these machines consist of live biological components, they can migrate within the human body effortlessly. This is because live biological systems can propel themselves on their own and do not need any additional force [24]. Additionally, these nanorobots are biocompatible, so the human body can easily accept them without causing harm. They work just like other biological systems and are able to survive within fluids in the human body. Another aspect of these machines is their adaptability, which enables them to react to any change in their environment. They can react to temperature, pH, and chemicals, among other factors. An example of biological nanorobots is bio-modified bacteria. The bacteria can be designed to deliver drugs within the human body safely. They can migrate directly to the cancer sites in the human body and release drugs at those areas specifically. This increases the chances of success while treating a disease [25]. Additionally, these drugs will not affect any human organs negatively, and this is because the drugs target specific areas in the human body. Since these machines are biological and similar to human cells, there are reduced chances of the human immune system attacking them. This is unlike other machines, which require the production of proteins to make them acceptable to the human body, and this can be time-consuming and costly. These machines are very useful while carrying out

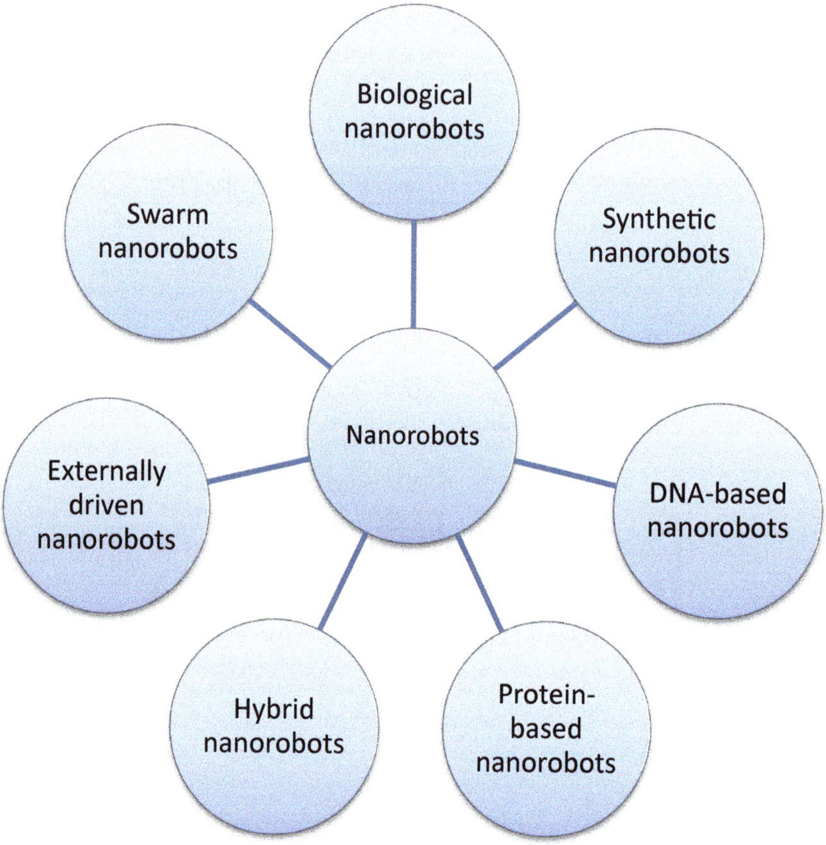

Fig. 2 Various types of nanorobots

targeted therapies. They are also widely used in biosensing to create precise diagnoses within the human body [26].

2.2 *Synthetic Nanorobots*

Synthetic nanorobots are fully designed and developed by utilizing inorganic materials or polymers. They are not made up of living elements. Therefore, their control is completely in the hands of scientists. They can be designed with particular shapes and sizes. Even the properties of their surfaces can be altered. Therefore, they can be used for particular medical purposes. The main benefit of using synthetic nanorobots is that they are fully controllable. They act as expected. Unlike living organisms, their performances are predictable. Therefore, they can easily be used in medical fields. They can also be mass-produced [27].

Materials like metals, polymers, and silica are routinely used to construct synthetic nanorobots. Such materials ensure robustness and stability. Flexibility and control over medication release are made possible through the use of polymers. Metals can be applied for imaging or magnetic control. Silica is used due to its porosity. Synthetic nanorobots are being widely explored to act as medication carriers. They are capable of releasing medication at targeted locations. They are also employed during medical imaging. They are also used during minimally invasive medical procedures. They are still effective; however, several safety issues are involved. Biocompatibility needs to be checked in the long term for synthetic nanobots. Their decomposition and removal should be safe. This ensures that there will be no toxicity. Research work has to be conducted to counter all issues [28].

2.3 DNA-Based Nanorobots

DNA-based nanorobots are microscopic devices built from DNA molecules. They are produced through processes like DNA origami and self-assembly, which involve programming DNA strands to adopt certain shapes. Such precise shaping enables researchers to assemble nanorobots with clearly defined shapes and functionalities [29].

These nanorobots can be made capable of transporting drugs or genes as therapeutic agents. One of the special things about DNA nanorobots is that they can lie dormant and be switched on, or get triggered by the presence of certain biomarkers associated with diseases that the patient has. Thus, drug delivery occurs only in the targeted region. Due to their high specificity, DNA-based nanorobots are well-suited for drug delivery and diagnostics. They are biodegradable and very compatible with the body and are therefore safe for use inside the human body. However, maintaining stability in the blood and tissues remains a challenge, and further research is needed to improve it [30].

2.4 Protein-Based Nanorobots

The main construction material for the building blocks of protein-based nanorobots is proteins or peptides, which are engineered to perform certain tasks in the body. They could mimic the activity of natural enzymes, receptors, or transport proteins. This functionality enables them to execute specialized tasks at the molecular level. Protein nanorobots may recognize and bind to specific cellular targets. Consequently, their high specificity enhances the accuracy of medical operations. They interact with cells and biological molecules efficiently. Because they strongly resemble natural proteins, they operate well within biological settings. Therefore, protein nanorobots have turned out to be very helpful in advanced medical technologies. They take their inspiration from natural biological processes [31].

The protein-based nanorobots are in wide applications, including biosensing, enzymatic therapy, and molecular recognition. They serve to detect with biosensing the markers of diseases with great accuracy, catalyze some biochemical reactions with enzymatic therapy, and perform roles in molecular recognition. Their biological origin provides good biocompatibility with body tissues, reducing toxicity or immune rejection. However, maintaining their stability in the body remains a big challenge. Proteins may degrade or lose function over time. It is also difficult to achieve large-scale production of these nanorobots. Advanced techniques are needed to perform manufacturing with good control. Further research needs to be done to achieve such improvements in stability and scalability [30].

2.5 *Hybrid Nanorobots*

The hybrid nanorobots are created by integrating biological components with artificial ones to enhance their functionality. The hybrids utilize the best qualities of both biological and artificial nanorobots, making them more efficient than both systems. As a result, they have shown increased stability when used in medicinal applications. The hybrid nanorobots are among the best developments in nanomedicine. Often, hybrid nanorobots consist of a synthetic nanoparticle coated with a biological membrane [32]. The use of a biological membrane helps the nanorobot detect targeted cells or tissues within the body. The process also helps the nanorobot evade immune detection by the body. This increases the time the nanorobot spends circulating within the bloodstream. Increased targeting ability reduces the negative side effects on the body. Hybrid nanorobots combine the capabilities of nature with those of technology. Thanks to their natural components, the nanorobots can respond as expected in the body's environment. The technology side enables the nanorobots to move, deliver medication, and turn on with a certain level of precision [33].

These nanorobots have significant applications in cancer therapy and personalized medicine. They target cancer cells specifically. Hence, the effectiveness level increases. Moreover, the hybrid nanorobots can be designed for different patients separately. Therefore, they can be used for personalized treatment.

2.6 *Externally Driven Nanorobots*

2.6.1 Magnetically Nanorobots

Nanorobots driven by magnets can be manipulated by external magnetic fields to move around in the human body. These nanorobots consist of magnets with properties capable of changing with the direction of the external magnetic field. It is also possible to control these nanorobots accurately with the help of an external magnetic field. Thus, it becomes easy to navigate through blood vessels and tissue using these nanorobots. They can be controlled with external magnetic fields without surgery. This reduces the risk of tissue damage. It also enables nanorobots to control their motions in real time. They can also be utilized effectively for targeted drug delivery. They can also serve as nanorobots in minimally invasive surgery. The main advantage is the capability to penetrate tissue effectively due to the property of magnetic fields [34, 35].

2.6.2 Acoustically Driven Nanorobots

Acoustically driven nanorobots are nanorobots that move through the use of ultrasound waves. The sound waves produce mechanical motion that propels the nanorobots in bodily fluids. These sound waves can also trigger these nanorobots at certain sites. These nanorobots do not require direct bodily contact to move. Acoustically driven nanorobots are safe, as the use of sound waves is safe and is already applied in medical practices for diagnosing and examining patients. They are ideal for use in the deep tissues of the body. These nanorobots can be remotely controlled from outside the body. They can be used for drug delivery and as therapeutic nanorobots. Their efficiency in movement depends on the sound intensities and frequencies used [36].

2.6.3 Light-Driven Nanorobots

Light-controlled nanorobots function by using particular wavelengths of light. Light sources provide power for moving and drug delivery using light energy. These nanorobots have accurate control at the target location. Light-controlled systems have high accuracy and easy actuation mechanisms. They also support non-invasive and remote control systems. These nanorobots find applications in surface modifications or near-surface processes. They also find applications in cancer treatments and diagnostics [37].

2.7 Swarm Nanorobots and Collective Intelligence

Swarm nanorobots involve the coordinated action of multiple nanorobots working together as a group. Instead of functioning individually, these nanorobots communicate and cooperate to perform complex tasks. Collective intelligence enables efficient targeting, faster response, and improved reliability. Swarm behavior is inspired by natural systems such as ants or bees. This approach is especially useful in treating large or heterogeneous disease sites. Control and coordination of large swarms remain a key research challenge [33].

3 Materials and Structural Design

Materials and structural design determine the safety, functionality, and performance of nanorobots in medical applications. The choice of biocompatible materials and optimized structural features ensures effective navigation, targeting, and stability within the body. Proper design also minimizes toxicity and enhances clinical applicability [38].

3.1 *Biocompatible and Biodegradable Materials*

Biocompatible and biodegradable materials are essential for the safe application of nanorobots in medical settings. These materials ensure that nanorobots do not cause toxicity or adverse immune responses when introduced into the body. Commonly used materials include biodegradable polymers, lipids, proteins, and specific inorganic nanoparticles. Biodegradable materials allow nanorobots to break down into non-toxic byproducts after completing their function. This reduces long-term accumulation in tissues and organs. Proper material selection enhances circulation time and therapeutic effectiveness. It also supports regulatory approval and clinical translation. Overall, these materials play a crucial role in patient safety and treatment success [39].

3.2 *Surface Functionalization and Molecular Targeting Ligands*

Surface functionalization involves modifying the outer surface of nanorobots to improve their interaction with biological systems. Molecular targeting ligands such as antibodies, peptides, or aptamers are attached to enhance specificity. These ligands enable nanorobots to recognize and bind to disease-specific biomarkers. Functionalized surfaces help avoid immune system recognition and premature clearance. They also improve cellular uptake at the target site. Controlled functionalization enhances drug loading and release efficiency. This approach is vital for targeted therapy and diagnostics. As a result, surface engineering significantly improves precision and therapeutic outcomes [40].

3.3 *Shape, Size, Flexibility, and Mechanical Stability*

The shape and size of nanorobots greatly influence their movement and distribution within the body. Smaller nanorobots can navigate narrow blood vessels and penetrate tissues more effectively. Shape affects hydrodynamics and cell interactions, influencing targeting efficiency. Flexibility allows nanorobots to adapt to dynamic biological environments. Mechanical stability ensures structural integrity during circulation and operation. Overly rigid structures may cause damage, while weak structures may collapse. Optimal design balances flexibility and strength. These factors collectively determine performance and safety [41].

3.4 Safety, Stability, and In Vivo Compatibility Considerations

Safety and stability are critical for the successful use of nanorobots in living systems. Nanorobots must remain stable under physiological conditions, including varying pH and temperature. In vivo compatibility ensures that nanorobots do not trigger immune reactions or inflammation. Long-term safety studies assess toxicity, degradation, and clearance pathways. Stability also affects shelf life and storage conditions. Proper design minimizes unwanted interactions with healthy tissues. Regulatory guidelines require a thorough evaluation of these parameters. Addressing these considerations is essential for clinical approval and patient acceptance [42].

4 Synthesis and Fabrication Techniques

Synthesis and fabrication techniques determine how nanorobots are designed, assembled, and produced at the nanoscale. Advanced methods such as self-assembly, micro–nanofabrication, and AI-assisted optimization ensure precision, efficiency, and reproducibility [11]. These techniques are essential for developing safe and functional nanorobots for medical applications.

4.1 Bottom-Up and Top-Down Fabrication Approaches

Bottom-up technology involves the fabrication of nanobots through the assembly of atoms, molecules, or nanoparticles to form functional devices. This technology makes use of various chemical interactions, along with self-assembly and molecular recognition. Bottom-up technology provides great control over composition at the nanolevel. Bottom-up technology is the most appropriate and suitable for the fabrication of biological and DNA nanobots. The top-down technology of fabricating devices involves the usage of etching, milling, and patterning. This technology provides great accuracy and uniformity. This technology has various applications and uses, especially during the synthesis of nanobots. Often, the union of both provides optimal solutions [43].

4.2 DNA Origami and Molecular Self-assembly

DNA origami is an advanced bottom-up technique that uses programmable DNA sequences to form complex nanostructures. Short DNA strands fold into predefined shapes through complementary base pairing. This enables precise control over nanorobot geometry and function. Molecular self-assembly relies on natural forces such as hydrogen bonding and electrostatic interactions. These processes allow nanorobots to form autonomously under suitable conditions. DNA-based nanorobots can respond to specific biological stimuli. They are biodegradable and biocompatible. Such systems are ideal for targeted drug delivery and biosensing. However, stability in physiological environments remains a key concern [44].

4.3 Micro–Nanofabrication and Lithographic Methods

Micro–nanofabrication involves engineering processes to assemble precisely designed nanorobots. Lithography processes like photolithography and electron-beam lithography are widely used. Such processes enable precise control over morphology, size, and topography. Microfabrication facilitates handling multiple functional components. Moreover, microfabrication technology fosters mass replication with ease. These processes are highly beneficial for micro-motors and micro-swimmers. However, high procurement costs and technical difficulties are associated with drawbacks. Specialized machinery is necessary to facilitate mass production. This technology is extremely important to the development of nanorobots [45].

4.4 AI-Assisted Modeling and Optimization of Synthesis

AI-assisted modeling plays a significant role in optimizing nanorobot synthesis and fabrication. Machine learning algorithms analyze experimental data to identify optimal synthesis parameters. AI reduces the need for extensive trial-and-error experimentation. Predictive models help improve yield, consistency, and quality. AI can simulate nanorobot behavior under biological conditions. This aids in selecting suitable materials and structures. Optimization algorithms enhance fabrication efficiency. AI also supports scalability and cost reduction. Overall, AI accelerates development and improves reliability in nanorobot synthesis [46].

5 Navigation, Control, and Powering of Nanorobots

Navigation and control of nanorobots are achieved using external fields and AI-based autonomous systems. Advanced propulsion mechanisms enable movement through complex biological environments. Efficient energy sources support sustained operation at the nanoscale. Together, these technologies ensure precise, safe, and effective nanorobot performance in medical applications.

5.1 External-Field-Guided Navigation

External-field-guided navigation helps to accurately control nanorobots using external physical fields. Magnetic fields are preferred for their safety, deep penetration, and ability to control the direction of movement. Electric fields are preferred for guiding charged nanorobots, though tissue conductivity is a limitation. Optical or light-guided navigation enables accurate control of nanorobots using specific wavelengths. These navigation mechanisms are useful for non-invasive control of nanorobot motion. External-field guidance is preferred for targeted drug delivery and therapy. External-field-guided navigation is a control mechanism that relies less on intelligence. There is improved accuracy with real-time adjustment of the external field [47].

5.2 Autonomous Navigation Using AI and Machine Learning

Autonomous navigation allows nanorobots to move independently within complex biological environments. AI and machine learning algorithms enable nanorobots to analyze local biological signals in real time. These systems help nanorobots identify obstacles such as cells and blood flow variations. Learning-based models improve navigation efficiency through adaptive behavior. Autonomous control reduces the need for continuous external guidance. Nanorobots can make decisions such as when to stop, move, or release drugs. This is particularly useful in dynamic environments like the bloodstream. AI also supports path optimization toward disease sites. Safety is improved by avoiding healthy tissues. Autonomous navigation represents a major step toward intelligent nanomedicine [48].

5.3 Energy Sources, Propulsion, and Motion Mechanisms

Powering nanorobots is a critical challenge due to their extremely small size. Various energy sources, such as chemical fuels, magnetic fields, ultrasound, and light, are used. Chemical propulsion mimics biological motors but may generate byproducts. Magnetic and acoustic propulsion offer clean and controllable motion. Motion mechanisms include rotating helices, flagella-like structures, and surface reactions. Propulsion design influences speed, direction, and stability. Efficient energy use increases operational lifetime. Motion must be compatible with physiological conditions. Excessive force can damage tissues. Optimized propulsion ensures safe and effective performance [49].

5.4 Real-Time Tracking, Imaging, and Feedback Control

Real-time tracking is essential for monitoring nanorobot location and activity inside the body. Imaging techniques such as MRI, ultrasound, and fluorescence imaging are commonly used. These methods allow visualization of nanorobot movement and accumulation. Feedback control systems use imaging data to adjust navigation and operation. This ensures accurate targeting and prevents off-target effects. Continuous monitoring improves safety and treatment effectiveness. Feedback loops allow a dynamic response to physiological changes. Integration with AI enhances decision-making. Tracking also supports regulatory validation. Overall, real-time feedback is vital for clinical translation [50].

6 Drug Delivery and Targeted Therapy Applications

Nanorobotics enables precise drug delivery directly to diseased sites, improving treatment effectiveness. Targeted therapy reduces exposure of healthy tissues to potent drugs. Controlled and stimuli-responsive release ensures optimal dosing at the right time.

6.1 Site-Specific and Personalized Drug Delivery

Site-specific and personalized drug delivery is a major advantage of nanorobotic systems in modern medicine. Nanorobots are engineered to recognize disease-specific biomarkers present at target sites. This allows drugs to be delivered precisely where they are needed. Personalized delivery considers patient-specific genetic and molecular profiles. AI assists in adjusting dosage and targeting strategies. Nanorobots navigate through blood vessels to reach affected tissues. This approach improves therapeutic efficiency significantly. Healthy tissues are largely spared from drug exposure. It reduces variability in treatment response. Personalized nanorobotic delivery supports precision medicine goals. Overall, it enhances patient outcomes and safety [51].

6.2 Stimuli-Responsive and Controlled Drug Release

Stimuli-responsive nanorobots release their drug payload in response to internal or external specific triggers. The internal stimuli include changes in pH, enzymes, or temperature at the disease site. Other external triggers that could affect the release include magnetic fields, ultrasound, or light. The controlled mechanism ensures drugs are released at the appropriate time and place. This is where AI algorithms further optimize both the release time and the dosage. The controlled release prevents early drug loss. It maintains adequate therapeutic drug levels over time. This reduces the frequency of dosing and improves compliance. Side effects are minimized because the action of drugs is localized. Such systems reinforce reliability and efficacy in treatment [52].

6.3 Cancer Therapy and Tumor Microenvironment Targeting

Nanorobots have significant applications in the treatment of cancers because they specifically target the tumor environment. Cancers have a special environment with conditions like acidic pH, poor oxygen, and blood vessels that are not normal. Nanorobots have the ability to detect and respond to the special conditions present in cancers. It helps the nanorobots selectively accumulate in the tumor tissues. They can easily reach the deep interiors of the tumor because of the nature of existing treatments, compared to other treatments that cannot reach the interiors of the tumor because of its density [53].

Nanorobots improve the effectiveness of treatment by directly delivering drugs to the site of the tumor. This targeted delivery helps to enhance the concentration of the drug at the required site. This results in increased treatment effectiveness at reduced dosages. Artificial intelligence assists in the identification of tumor-specific markers for the precise targeting of the tumor. Nanorobots are able to target multiple drugs simultaneously. This helps in targeting cancer at various interfaces. This results in the prevention of the emergence of resistance. Nanorobots are able to interfere with the process of tumor development by modifying the tumor microenvironment. They are able to cut the supply of nutrients or modify the chemical environment [54].

6.4 Minimization of Systemic Toxicity and Side Effects

Minimizing systemic toxicity is a key benefit of nanorobotic drug delivery systems. Conventional therapies often expose the entire body to drugs. Nanorobots limit drug action to diseased tissues only. This significantly reduces harmful side effects. Controlled release further enhances safety. Lower drug doses are often sufficient due to high targeting efficiency. AI assists in predicting toxicity profiles. Reduced systemic exposure improves patient quality of life. It allows safer long-term treatments. This approach is especially beneficial in chronic and cancer therapies [55].

7 Diagnostic, Surgical, and Regenerative Applications

Nanorobotics enables early disease diagnosis through in vivo biosensing, improving detection accuracy and timeliness. It supports minimally invasive surgical interventions with high precision and reduced tissue damage. Nanorobots also enhance tissue regeneration and wound healing, advancing regenerative medicine outcomes.

7.1 In Vivo Biosensing and Early Disease Diagnosis

In vivo biosensing is one of the most promising diagnostic applications of nanorobotics. They can continuously monitor physiological and biochemical parameters inside the body. They are capable of detecting disease-specific biomarkers at very early stages. This enables diagnosis before clinical symptoms become apparent. Nanorobots interact directly with cells, proteins, and metabolites. AI helps interpret complex biosignals in real time. Early detection improves treatment success rates significantly. They can transmit diagnostic data to external devices. This allows real-time health monitoring. Such systems support preventive and personalized healthcare. Overall, nanorobotic biosensing enhances diagnostic accuracy [56, 57].

7.2 Minimally Invasive Nanorobotic Surgical Interventions

Nanorobotic systems enable minimally invasive surgical procedures with high precision. These nanorobots can access hard-to-reach areas within the body. They perform tasks such as targeted tissue removal or clot dissolution. Compared to conventional surgery, tissue damage is minimal. AI-assisted control improves accuracy and safety during procedures. Nanorobots reduce the need for large incisions. This leads to faster recovery and less pain. Real-time feedback allows precise surgical actions. Infection risk is also reduced. Nanorobotic surgery represents a major advancement in medical interventions [58].

7.3 Tissue Regeneration and Wound Healing

Nanorobots play a vital role in tissue regeneration and wound healing processes. They can deliver growth factors and regenerative agents directly to damaged tissues. This promotes cell proliferation and tissue repair. Nanorobots can also remove debris and harmful substances from wounds. Controlled release ensures sustained regenerative effects. AI optimizes treatment timing and dosage. These systems accelerate healing and reduce scarring. Nanorobots support regeneration in bones, skin, and nerves. They enhance natural healing mechanisms. Overall, nanorobotics improves outcomes in regenerative medicine [59].

7.4 Integration with Imaging and Theranostic Platforms

Integration with imaging and theranostic platforms enhances the functionality of nanorobots. Theranostics combines therapy and diagnostics in a single system. Nanorobots can simultaneously diagnose and treat diseases. Imaging techniques such as MRI or fluorescence track nanorobot movement. This enables real-time monitoring of treatment progress. AI processes imaging data for precise decision-making. Feedback systems adjust therapy dynamically. This integration improves treatment accuracy and efficiency. It reduces the need for multiple procedures. Overall, theranostic nanorobots represent a comprehensive healthcare solution [60–62].

8 Challenges, Safety, and Future Perspectives

Challenges in nanorobotics include safety concerns, immune responses, regulatory barriers, and large-scale manufacturing limitations. Ensuring biocompatibility, ethical use, and clinical reliability is essential for successful translation. Future advances in AI-driven nanorobotics are expected to overcome these challenges and revolutionize precision healthcare.

8.1 Toxicity, Immunogenicity, and Biodegradation Concerns

Toxicity is a major challenge in the clinical application of nanorobotics. Some nanomaterials may generate reactive species that damage cells and tissues. Immunogenicity is another concern, as the immune system may recognize nanorobots as foreign bodies. This can lead to inflammation or rapid clearance from the bloodstream. Biodegradation behavior must be carefully controlled to avoid harmful byproducts. Non-biodegradable nanorobots may accumulate in organs over time.

Long-term exposure risks are still not fully understood. Comprehensive in vivo studies are required to assess safety. Surface modification can reduce immune reactions. Controlled degradation improves clearance. Addressing these concerns is essential for patient safety and approval [63, 64].

8.2 Ethical, Regulatory, and Clinical Translation Challenges

Ethical concerns arise due to the invasive and autonomous nature of nanorobots. Issues related to patient consent and data privacy must be addressed. Regulatory frameworks for nanorobotics are still evolving. Existing guidelines may not fully cover nanoscale devices. Standardized testing protocols are lacking. Clinical translation requires extensive safety and efficacy data. Approval processes are time-consuming and complex. Integration into healthcare systems poses additional challenges. Public acceptance is also an important factor. Ethical use of AI-driven decision-making must be ensured. Clear regulations are crucial for responsible deployment [65, 66].

8.3 Manufacturing Scalability and Cost-Effectiveness

Manufacturing nanorobots at a large scale remains a significant challenge. Many fabrication techniques are complex and expensive. Maintaining consistency and quality during mass production is difficult. Specialized equipment increases production costs. Scalability issues limit commercial viability. Cost-effective materials and processes are needed. Reproducibility is critical for regulatory approval. Automation can improve manufacturing efficiency. AI-assisted optimization may reduce waste. Affordable production is essential for widespread clinical use. Overcoming scalability barriers will determine real-world adoption [11, 67].

8.4 Future Directions in AI-Driven Nanorobotics

The future of nanorobots is tied to the development of artificial intelligence. With AI, nanorobots will not require continuous human control. Intelligent nanorobots will be able to react to changing environments within the human body. Sophisticated machine learning algorithms will enable them to better detect signals associated with disease. This will enable more precise targeting. AI will enable nanorobots to make decisions instantly while undergoing treatment. Customized treatments will become easier to develop. Patient data will make this possible. Individual nanorobots will adjust to a disease profile. The technique will make treatments more effective. It will further minimize indiscriminate side effects. In summary, AI will make nanorobots more intelligent and more effective [4, 68].

Swarm intelligence is another exciting future application of nanorobotics. Artificial intelligence is expected to make it possible for a set of nanorobots to function in a coordinated way. This approach will improve the effectiveness of nanorobots in performing complex procedures. The biosensing mechanism will improve, making it possible to conduct monitoring processes in a real-time manner. Nanorobots can effectively identify symptoms of a disease from within the body. At the same time, the development of safe and intelligent materials will take place. This will assist in making safe nanorobots. Engineers, doctors, and researchers will join hands to make advancements. Artificial intelligence-based nanorobotics is expected to bring a drastic transition in the field of precision medicine [69–71].

9 Conclusion

Nanorobotics holds immense promise for revolutionizing precision medicine by enabling targeted, intelligent, and minimally invasive healthcare solutions. Advances in AI-driven design, synthesis, and control have significantly improved the functionality and reliability of nanorobots for therapeutic, diagnostic, and regenerative applications. Despite existing challenges related to safety, regulation, and large-scale production, continuous interdisciplinary research is rapidly advancing this field. As these hurdles are addressed, nanorobotics is expected to play a central role in next-generation healthcare, transforming disease management from reactive treatment to proactive and personalized medicine.

References

1. Taha BA, Addie AJ, Abbas EM, Aubaidan BH, Ahmed NM, Haider AJ, Chaudhary V, Arsad N (2024) Biophotonics and nanorobotics for biomedical imaging, biosensing, drug delivery, and therapy. J Photochem Photobiol C Photochem Rev 60–61:100678. https://doi.org/10.1016/j.jphotochemrev.2024.100678
2. Chamkouri H (2025) Intelligent micro- and nano-robots for next-generation photobiomodulation and adaptive photomedicine. Nano Trends 100177. https://doi.org/10.1016/j.nwnano.2025.100177
3. Kong X, Gao P, Wang J, Fang Y, Hwang KC (2023) Advances of medical nanorobots for future cancer treatments. J Hematol Oncol 16:74. https://doi.org/10.1186/s13045-023-01463-z
4. Sharma PK, Chen C-Y (2025) AI-integrated micro/nanorobots for biomedical applications: recent advances in design, fabrication, and functions. Biosensors (Basel) 15:793. https://doi.org/10.3390/bios15120793
5. Olawade DB, Ige AO, Olaremu AG, Ijiwade JO, Adeola AO (2024) The synergy of artificial intelligence and nanotechnology towards advancing innovation and sustainability—a mini-review. Nano Trends 8:100052. https://doi.org/10.1016/j.nwnano.2024.100052

6. Bayda S, Adeel M, Tuccinardi T, Cordani M, Rizzolio F (2019) The history of nanoscience and nanotechnology: from chemical-physical applications to nanomedicine. Molecules 25:112. https://doi.org/10.3390/molecules25010112
7. Ali KA, Choudhuri S, Roy SKr, Reza KH, Mandal D, Chakraborty R, Hossain CM (2025) AI-generated nanoparticles in immunological research: opportunities and challenges. In: Chen J-T (ed) Advances in computational intelligence and robotics. IGI Global, pp 295–322
8. Patra JK, Das G, Fraceto LF, Campos EVR, Rodriguez-Torres MDP, Acosta-Torres LS, Diaz-Torres LA, Grillo R, Swamy MK, Sharma S, Habtemariam S, Shin H-S (2018) Nano based drug delivery systems: recent developments and future prospects. J Nanobiotechnol 16:71. https://doi.org/10.1186/s12951-018-0392-8
9. Chen G, Zhu F, Gan ASJ, Mohan B, Dey KK, Xu K, Huang G, Cui J, Solovev AA, Mei Y (2023) Towards the next generation nanorobots. Next Nanotechnol 2:100019. https://doi.org/10.1016/j.nxnano.2023.100019
10. Anto M, Mukherjee K, Mukherjee K (2025) Nano bio-robots: a new frontier in targeted therapeutic delivery. Front Robot AI 12:1639445. https://doi.org/10.3389/frobt.2025.1639445
11. Weerarathna IN, Kumar P, Dzoagbe HY, Kiwanuka L (2025) Advancements in micro/nanorobots in medicine: design, actuation, and transformative application. ACS Omega 10:5214–5250. https://doi.org/10.1021/acsomega.4c09806
12. Shirzad M, Shaban M, Mohammadzadeh V, Rahdar A, Fathi-Karkan S, Hoseini ZS, Najafi M, Kharaba Z, Aboudzadeh MA (2025) Artificial intelligence-assisted design of nanomedicines for breast cancer diagnosis and therapy: advances, challenges, and future directions. BioNanoSci 15:354. https://doi.org/10.1007/s12668-025-01980-w
13. Ali KA, Mohin S, Mondal P, Goswami S, Ghosh S, Choudhuri S (2024) Influence of artificial intelligence in modern pharmaceutical formulation and drug development. Futur J Pharm Sci 10:53. https://doi.org/10.1186/s43094-024-00625-1
14. Leng Y, Sun R (2025) Multifunctional biomedical devices with closed-loop systems for precision therapy. Adv Healthc Mater 14:e2500860. https://doi.org/10.1002/adhm.202500860
15. Jeganathan JR, Jegasothy R, Sia WT (2025) Minimally invasive surgery: a historical and legal perspective on technological transformation. J Robot Surg 19:408. https://doi.org/10.1007/s11701-025-02589-7
16. Haroon HB, Hunter AC, Farhangrazi ZS, Moghimi SM (2022) A brief history of long circulating nanoparticles. Adv Drug Deliv Rev 188:114396. https://doi.org/10.1016/j.addr.2022.114396
17. Jandoubi B, Akhloufi MA (2025) Multimodal artificial intelligence in medical diagnostics. Information 16:591. https://doi.org/10.3390/info16070591
18. Bhalani DV, Nutan B, Kumar A, Singh Chandel AK (2022) Bioavailability enhancement techniques for poorly aqueous soluble drugs and therapeutics. Biomedicine 10:2055. https://doi.org/10.3390/biomedicines10092055
19. Jiao Y, Yang L, Wang R, Song G, Fu J, Wang J, Gao N, Wang H (2024) Drug delivery across the blood-brain barrier: a new strategy for the treatment of neurological diseases. Pharmaceutics 16:1611. https://doi.org/10.3390/pharmaceutics16121611
20. Ziegler A, Koch A, Krockenberger K, Grosshennig A (2012) Personalized medicine using DNA biomarkers: a review. Hum Genet 131:1627–1638. https://doi.org/10.1007/s00439-012-1188-9
21. Adepu S, Ramakrishna S (2021) Controlled drug delivery systems: current status and future directions. Molecules 26:5905. https://doi.org/10.3390/molecules26195905
22. Zhou J, Park S, Dong S, Tang X, Wei X (2025) Artificial intelligence-driven transformative applications in disease diagnosis technology. Med Rev 5:353–377. https://doi.org/10.1515/mr-2024-0097
23. Das T, Sultana S (2024) Multifaceted applications of micro/nanorobots in pharmaceutical drug delivery systems: a comprehensive review. Futur J Pharm Sci 10:2. https://doi.org/10.1186/s43094-023-00577-y
24. Avachar AG, Ingole AM, Chimote UN, Taksande JB, Umekar MJ (2021) Nanorobots is an emerging technology which is helpful in the diagnosis and treatment of various diseases. IJPSRR 70:75–84. https://doi.org/10.47583/ijpsrr.2021.v70i01.010

25. Elnaggar A, Kang S, Tian M, Han B, Keshavarz M (2024) State of the art in actuation of micro/nanorobots for biomedical applications. Small Sci 4:2300211. https://doi.org/10.1002/smsc.202300211
26. Raychaudhuri SP, Raychaudhuri SK (2009) Biologics: target-specific treatment of systemic and cutaneous autoimmune diseases. Indian J Dermatol 54:100–109. https://doi.org/10.4103/0019-5154.53175
27. Giri G, Maddahi Y, Zareinia K (2021) A brief review on challenges in design and development of nanorobots for medical applications. Appl Sci 11:10385. https://doi.org/10.3390/app112110385
28. Islam S, Ahmed MMS, Islam MA, Hossain N, Chowdhury MA (2025) Advances in nanoparticles in targeted drug delivery—a review. Results Surf Interfaces 19:100529. https://doi.org/10.1016/j.rsurfi.2025.100529
29. Nummelin S, Shen B, Piskunen P, Liu Q, Kostiainen MA, Linko V (2020) Robotic DNA nanostructures. ACS Synth Biol 9:1923–1940. https://doi.org/10.1021/acssynbio.0c00235
30. Xu M, Qin Z, Chen Z, Wang S, Peng L, Li X, Yuan Z (2024) Nanorobots mediated drug delivery for brain cancer active targeting and controllable therapeutics. Discov Nano 19:183. https://doi.org/10.1186/s11671-024-04131-4
31. de la Rica R, Matsui H (2010) Applications of peptide and protein-based materials in bionanotechnology. Chem Soc Rev 39:3499–3509. https://doi.org/10.1039/b917574c
32. Li J, Dekanovsky L, Khezri B, Wu B, Zhou H, Sofer Z (2022) Biohybrid micro- and nanorobots for intelligent drug delivery. Cyborg Bionic Syst 2022:9824057. https://doi.org/10.34133/2022/9824057
33. Ju X, Chen C, Oral CM, Sevim S, Golestanian R, Sun M, Bouzari N, Lin X, Urso M, Nam JS, Cho Y, Peng X, Landers FC, Yang S, Adibi A, Taz N, Wittkowski R, Ahmed D, Wang W, Magdanz V, Medina-Sánchez M, Guix M, Bari N, Behkam B, Kapral R, Huang Y, Tang J, Wang B, Morozov K, Leshansky A, Abbasi SA, Choi H, Ghosh S, Borges Fernandes B, Battaglia G, Fischer P, Ghosh A, Jurado Sánchez B, Escarpa A, Martinet Q, Palacci J, Lauga E, Moran J, Ramos-Docampo MA, Städler B, Herrera Restrepo RS, Yossifon G, Nicholas JD, Ignés-Mullol J, Puigmartí-Luis J, Liu Y, Zarzar LD, Shields CW, Li L, Li S, Ma X, Gracias DH, Velev O, Sánchez S, Esplandiu MJ, Simmchen J, Lobosco A, Misra S, Wu Z, Li J, Kuhn A, Nourhani A, Maric T, Xiong Z, Aghakhani A, Mei Y, Tu Y, Peng F, Diller E, Sakar MS, Sen A, Law J, Sun Y, Pena-Francesch A, Villa K, Li H, Fan DE, Liang K, Huang TJ, Chen X-Z, Tang S, Zhang X, Cui J, Wang H, Gao W, Kumar Bandari V, Schmidt OG, Wu X, Guan J, Sitti M, Nelson BJ, Pané S, Zhang L, Shahsavan H, He Q, Kim I-D, Wang J, Pumera M (2025) Technology roadmap of micro/nanorobots. ACS Nano 19:24174–24334. https://doi.org/10.1021/acsnano.5c03911
34. Zhou H, Mayorga-Martinez CC, Pané S, Zhang L, Pumera M (2021) Magnetically driven micro and nanorobots. Chem Rev 121:4999–5041. https://doi.org/10.1021/acs.chemrev.0c01234
35. Liu D, Guo R, Wang B, Hu J, Lu Y (2022) Magnetic micro/nanorobots: a new age in biomedicines. Adv Intell Syst 4:2200208. https://doi.org/10.1002/aisy.202200208
36. Cao HX, Nguyen VD, Park J-O, Choi E, Kang B (2024) Acoustic actuators for the manipulation of micro/nanorobots: state-of-the-art and future outlooks. Micromachines 15:186. https://doi.org/10.3390/mi15020186
37. Shwetha KP, Amogh PN, Shrinath Vaidya S, Hariprasad N, Krishna S, Manjunatha C (2024) Nanorobotics for advancing biomedicine: progresses in materials, design, fabrication, opportunities and applications. IEEE Access 12:91664–91677. https://doi.org/10.1109/ACCESS.2024.3416818
38. Preetam S, Pritam P, Mishra R, Rustagi S, Lata S, Malik S (2024) Empowering tomorrow's medicine: energy-driven micro/nano-robots redefining biomedical applications. Mol Syst Des Eng 9:892–911. https://doi.org/10.1039/D4ME00090K
39. Kyriakides TR, Raj A, Tseng TH, Xiao H, Nguyen R, Mohammed FS, Halder S, Xu M, Wu MJ, Bao S, Sheu WC (2021) Biocompatibility of nanomaterials and their immunological properties. Biomed Mater 16. https://doi.org/10.1088/1748-605X/abe5fa

40. Liu M, Wang L, Lo Y, Shiu SC-C, Kinghorn AB, Tanner JA (2022) Aptamer-enabled nanomaterials for therapeutics, drug targeting and imaging. Cells 11:159. https://doi.org/10.3390/cells11010159
41. Ressnerova A, Heger Z, Pumera M (2025) Translational nanorobotics breaking through biological membranes. Chem Soc Rev 54:1924–1956. https://doi.org/10.1039/D4CS00483C
42. Zhang Q, Sun D (2025) Micro/nanorobots for combating brain disorders: challenges, advances, and perspectives. Adv Sci 12:e16592. https://doi.org/10.1002/advs.202516592
43. Yan X, Huang S, Wang Y, Tang Y, Tian Y (2020) Bottom-up self-assembly based on DNA nanotechnology. Nanomaterials 10:2047. https://doi.org/10.3390/nano10102047
44. Zhan P, Peil A, Jiang Q, Wang D, Mousavi S, Xiong Q, Shen Q, Shang Y, Ding B, Lin C, Ke Y, Liu N (2023) Recent advances in DNA origami-engineered nanomaterials and applications. Chem Rev 123:3976–4050. https://doi.org/10.1021/acs.chemrev.3c00028
45. Parvin N, Joo SW, Jung JH, Mandal TK (2025) Innovative micro- and nano-architectures in biomedical engineering for therapeutic and diagnostic applications. Micromachines 16:419. https://doi.org/10.3390/mi16040419
46. Winkler DA (2020) Role of artificial intelligence and machine learning in nanosafety. Small 16:2001883. https://doi.org/10.1002/smll.202001883
47. Huang G, Yin T, Zhang B, Ma A, Kang T, He Y, Long Y, Zheng S, Pan H, Cai L (2023) Cell-based intelligent micro/nanorobots for precise regulation and active biotherapy. Matter 6:4158–4194. https://doi.org/10.1016/j.matt.2023.09.013
48. Li M, Hu X, Zhao Y, Jiao N (2023) An overview of recent progress in micro/nanorobots for biomedical applications. Adv Mater Technol 8:2201928. https://doi.org/10.1002/admt.202201928
49. Kim J, Mayorga-Burrezo P, Song S-J, Mayorga-Martinez CC, Medina-Sánchez M, Pané S, Pumera M (2024) Advanced materials for micro/nanorobotics. Chem Soc Rev 53:9190–9253. https://doi.org/10.1039/D3CS00777D
50. Liu X, Jing Y, Xu C, Wang X, Xie X, Zhu Y, Dai L, Wang H, Wang L, Yu S (2023) Medical imaging technology for micro/nanorobots. Nanomaterials 13:2872. https://doi.org/10.3390/nano13212872
51. Gupta S, Tomar A, Manohar L, Panwar P (2025) Nanobots: the current scenario. Crit Rev Oncol Hematol 208:104652. https://doi.org/10.1016/j.critrevonc.2025.104652
52. Huang W, Hsu JC, Zhou J, Kwon GS, Kang L, Hua C, Cai W (2025) Stimuli-responsive nanomaterials for targeted drug delivery in inflammatory bowel disease: advances and emerging directions. Matter 8:102249. https://doi.org/10.1016/j.matt.2025.102249
53. Wang Z, Wang C, Ji Y, Yang M, Li C, Li M, Yang J, Tang H, Luo X, Hao H, Liu Z, Chen K, Chang Y, Yuan H, Feng L, Xing G, Li J (2025) Magnetically driven bionic nanorobots enhance chemotherapeutic efficacy and the tumor immune response via precise targeting. Innovation 6:100777. https://doi.org/10.1016/j.xinn.2024.100777
54. Khan Z, Khan N, Geetha M, Veettil RP, Kasote DM, Hasan A, Sadasivuni KK (2025) Therapeutic applications of nanobots and nanocarriers in cancer treatment. Anal Sci 41:1305–1324. https://doi.org/10.1007/s44211-025-00799-5
55. George M, Boukherroub R, Sanyal A, Szunerits S (2025) Treatment of lung diseases via nanoparticles and nanorobots: are these viable alternatives to overcome current treatments? Mater Today Bio 31:101616. https://doi.org/10.1016/j.mtbio.2025.101616
56. Rong G, Corrie SR, Clark HA (2017) In vivo biosensing: progress and perspectives. ACS Sens 2:327–338. https://doi.org/10.1021/acssensors.6b00834
57. Ma T, Sun J, Xue N, Hassan JNA, Abbas A (2025) Nano-engineered sensor systems for disease diagnostics: advances in smart healthcare applications. Biosensors 15:777. https://doi.org/10.3390/bios15120777
58. Soto F, Wang J, Ahmed R, Demirci U (2020) Medical micro/nanorobots in precision medicine. Adv Sci 7:2002203. https://doi.org/10.1002/advs.202002203
59. Liu Y, Wang K, Gao F, Xu Z, Zhao X, Bai X, Li Z, Wan G, Yang J, Wang Y (2025) Micro/nanorobots in wound healing: bridging the gap from concept to clinical translation. IJN 20:15343–15362. https://doi.org/10.2147/IJN.S557042

60. Anitha K, Chenchula S, Surendran V, Shvetank B, Ravula P, Milan R, Chikatipalli R, Padmavathi R (2024) Advancing cancer theranostics through biomimetics: a comprehensive review. Heliyon 10:e27692. https://doi.org/10.1016/j.heliyon.2024.e27692
61. Chohan DP, Dey B, Tarkunde A, Vyas V, De Sarkar S, Sundara BK (2024) Advancing autonomous nanomedicine: bridging the gap from concept to potential clinical studies. J Clust Sci 35:2607–2635. https://doi.org/10.1007/s10876-024-02691-0
62. Wu Y, Sun C, Jin K, Dong C (2025) Theranostic role of advanced nanotechnological tools in early brain metastases in lung cancer: an updated review. IJN 20:7215–7232. https://doi.org/10.2147/IJN.S517928
63. Xuan L, Ju Z, Skonieczna M, Zhou P-K, Huang R (2023) Nanoparticles-induced potential toxicity on human health: applications, toxicity mechanisms, and evaluation models. MedComm 4:e327. https://doi.org/10.1002/mco2.327
64. Kumar R, Kumar A, Bhardwaj S, Sikarwar M, Sriwastaw S, Sharma G, Gupta M (2025) Nanotoxicity unveiled: evaluating exposure risks and assessing the impact of nanoparticles on human health. J Trace Elem Miner 13:100252. https://doi.org/10.1016/j.jtemin.2025.100252
65. Ebbesen M, Jensen TG (2006) Nanomedicine: techniques, potentials, and ethical implications. J Biomed Biotechnol 2006:51516. https://doi.org/10.1155/JBB/2006/51516
66. Musa SS, Ibrahim AM, Alhassan MY, Musa AH, Jibo AG, Auwal AR, Okesanya OJ, Othman ZK, Abubakar MS, Ahmed MM, Barroso CJV, Sium AF, Garcia MB, Flores JB, Maikifi AS, Kouwenhoven MBN, Lucero-Prisno DE (2025) Nanotechnology and machine learning: a promising confluence for the advancement of precision medicine. Intell-Based Med 12:100267. https://doi.org/10.1016/j.ibmed.2025.100267
67. Bi Y, Xie S, Li Z, Dong S, Teng L (2025) Precise nanoscale fabrication technologies, the “last mile” of medicinal development. Acta Pharm Sin B 15:2372–2401. https://doi.org/10.1016/j.apsb.2025.03.040
68. Kazemi M, Abasi M, Salehabadi F, Nouralishahi A, Davoudi M, Farzan P, Rostamzadeh S (2025) Nanorobots and smart nanodevices with artificial intelligence for precision orthodontic surgery and treatment monitoring. Nanomed Res J 10. https://doi.org/10.22034/nmrj.2025.02.005
69. Li J, Esteban-Fernández de Ávila B, Gao W, Zhang L, Wang J (2017) Micro/nanorobots for biomedicine: delivery, surgery, sensing, and detoxification. Sci Robot 2:eaam6431. https://doi.org/10.1126/scirobotics.aam6431
70. Law J, Yu J, Tang W, Gong Z, Wang X, Sun Y (2023) Micro/nanorobotic swarms: from fundamentals to functionalities. ACS Nano 17:12971–12999. https://doi.org/10.1021/acsnano.2c11733
71. Wang B, Lu Y (2024) Multi-dimensional micro/nanorobots with collective behaviors. SmartMat 5:e1263. https://doi.org/10.1002/smm2.1263

Self-propelled Nanomotors in Smart Healthcare Systems

Balaji Esakkimuthu, Shama Anzum Ismail, Thilagavathi Gangadharan, Manoj Mani, Alwin Robert Asirvatham, Antony Prabhu Jeyabal Philomenathan, Velayuthaprabhu Shanmugam, and Vijaya Anand Arumugam

Abstract Self-propelled nanomotors represent one emerging class of autonomous nanosystems able to convert chemical, enzymatic, optical, magnetic, or acoustic energy into mechanical motion to enable precise navigation, active cargo transport, and intelligent biomedical functionalities. This chapter gives an in-depth overview of their architectural design, propulsion principles, strategies of targeting strategies, biosensing capability, and translational potential within smart healthcare ecosystems. Advanced Janus and asymmetric nanoarchitectures, stimuli-responsive materials, and AI-driven optimization have considerably improved propulsion efficiency, biocompatibility, imaging visibility, and adaptive behavior in complex physiological environments. The integration of nanomotors with microfluidics, biosensors, and IoMT platforms provides for real-time diagnostics, targeted therapeutic delivery, microinvasive intervention, and even closed-loop medical systems. Despite substantial progress so far, the clinical translation of nanomotors remains restricted due to obstacles in scalable fabrication, biosafety, regulatory validation, and reliable operation in vivo. This chapter points out some recent breakthroughs that are going to shape

B. Esakkimuthu
Sudha Hospital, Hyderabad, Telangana, India

S. A. Ismail
Department of Biochemistry, PSG College of Arts and Science, Coimbatore, Tamil Nadu, India

T. Gangadharan
Department of Information Technology, Coimbatore Institute of Technology, Coimbatore, Tamil Nadu, India

M. Mani · A. P. Jeyabal Philomenathan · V. A. Arumugam (✉)
Department of Human Genetics and Molecular Biology, Bharathiar University, Coimbatore, Tamil Nadu, India
e-mail: avahgmb@buc.edu.in

A. R. Asirvatham
Department of Endocrinology and Diabetes, Prince Sultan Military Medical City, Ministry of Defense, Riyadh, Kingdom of Saudi Arabia

V. Shanmugam (✉)
Department of Biotechnology, Bharathiar University, Coimbatore, Tamil Nadu, India
e-mail: velayuthaprabhu@buc.edu.in

J. Chen (ed.), *AI-Driven Nanotechnology in Precision Medicine and Healthcare*, Smart Nanomaterials Technology, https://doi.org/10.1007/978-981-92-1690-1_7

the future directions of intelligent nanomotor-based next-generation biomedical technology.

Keywords Self-propelled nanomotors · Active matter · Targeted therapy · Biosensing · Micro/nanorobotics

Abbreviations

AI	Artificial intelligence
CNNs	Convolutional neural networks
GANs	Generative adversarial networks
ML	Machine learning
MNMs	Micro/nanomotors
NIR	Near-infrared
PET	Positron emission tomography
TME	Tumor microenvironment

1 Introduction: The Rise of Intelligent Nanomachines in Medicine

The development of nanomaterials and nanomotors for biomedical applications and smart healthcare has rapidly evolved from simple chemically driven micro/nanostructures to intelligently engineered systems able to accomplish highly specialized biomedical tasks. Initial efforts started with enzymatic micro/nanomotors (MNMs), which proved highly biocompatible and used endogenous biochemical gradients as fuel. Capable of self-assembling into dynamic swarms, they offered advantages related to environmental adaptability and multitasking while offering fundamental insight into nanoscale autonomous behavior [1, 6, 19, 55 59, 92, 93, 100, 108, 138, 140]. As the fabrication strategy and propulsion mechanism become finer, MNMs moved toward precision diagnosis, targeted drug delivery, and minimally invasive interventions, therefore underlining their transformative impact on emerging biomedical technologies [166].

Building on these initial breakthroughs, research over the past two decades has progressed from elucidating elemental propulsion principles to exploring collective behaviors and complex theranostic applications. Current directions now include the development of hybrid "living" nanomotor systems and environmentally responsive materials to enhance functional integration within physiological settings [23]. Active nanomotors now offer clear advantages compared to conventional passive nanocarriers, especially because they can transform various forms of energy inputs into mechanical motion. This active propulsion significantly enhances navigation through biological barriers, efficiency in drug distribution, and precision in targeted therapy—all critical elements for applications in cancer treatment, bioimaging, and intelligent drug release systems [17].

The superior performance of self-propelled nanomotors is demonstrated in precision medicine, whereby passive nanoparticles generally face poor diffusion and unsatisfactory penetration across complex physiological barriers. Nanomotors can surmount these challenges through effective navigation within a microenvironment with dense structure, enhancing tissue penetration and bioavailability of drugs [17, 50]. This propulsion—from either catalytic reactions or externally applied energy sources—facilitates site-specific delivery of higher efficiency. In oncology, for example, self-propelled Janus nanomotors have been able to achieve improved tumor accumulation and deeper infiltration in malignant tissues with near-infrared (NIR)-triggered propulsion, thereby providing a measurable advantage over passive carriers [17, 184]. Besides, there are bioinspired nanomotors endowed with self-guiding ability and multi-functionality that have cropped up as promising probes for noninvasive cancer diagnosis and treatment and thus may redefine the future clinical nanomedicine (Fig. 1) [106].

Parallel advances in nanotechnology, digital health, and artificial intelligence (AI) catalyze the introduction of autonomous nanomachines into personalized healthcare ecosystems. Nanotechnology has already transformed the field of targeted delivery and molecular diagnostics, coming up with platforms like liposomes, dendrimers, quantum dots, and nanosensors to accurately modulate therapy and monitor disease in real time [117]. AI further enhances the aforementioned possibilities of intelligent design, optimization of motion control, and adaptive decision-making in medical nanorobots. These nanorobotic systems, made from biocompatible, non-toxic materials, are beginning to assume great importance in such fields as oncology, cardiology, and minimally invasive surgery, contributing much to personalized, high-precision therapeutic strategies with reduced systemic burden [91, 116].

Looking ahead to the next decade, this roadmap of nanorobotics and nanomotor technologies predicts important advances in high-resolution characterization techniques, precise kinetic control, and advanced materials engineering that will be required to surmount existing translational bottlenecks. Ongoing innovation has continued to build on the first catalytic nanomotor demonstrations in 2004, pushing biomedical and environmental utility into its next phase of translation from laboratory prototypes to clinically deployable systems [18]. Critical challenges are foreseen at the nexus of mechanics, AI integration, and nanoscale fabrication that need to be resolved as engineers progress toward fully autonomous, adaptive nanomedical systems. In the process of the field evolving from passive observation toward active nanomanipulation, machine-learning (ML)-enhanced control architectures and biohybrid nanorobots have emerged as a core set of solutions to increase autonomy, scalability, and multifunctionality [8, 165].

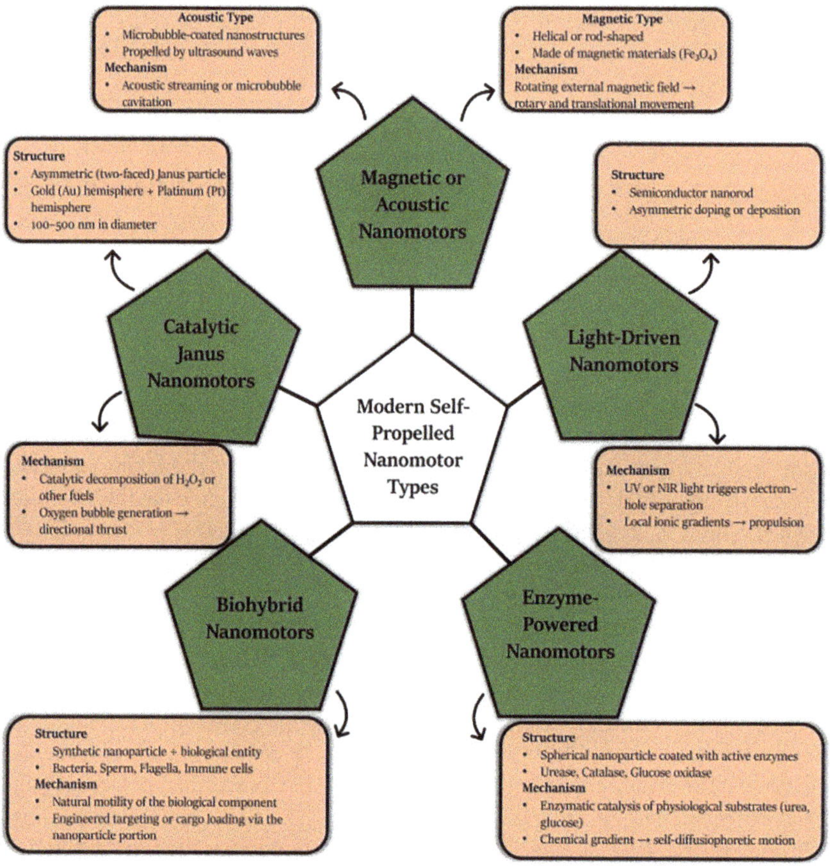

Fig. 1 Modern self-propelled nanomotor types

2 Design Principles and Propulsion Mechanisms

Self-propelled nanomotors for in vivo biomedical applications require propulsion strategies that enable autonomous motion, controlled navigation, and effective transport through complex physiological environments (Fig. 2). Among them, chemical and enzymatic propulsions constitute fundamental approaches. Chemically powered nanomotors commonly work via a redox reaction that converts chemical energy into mechanical thrust, hence allowing self-guided navigation to disease sites with a significantly enhanced active drug delivery capability [78]. Biocompatibility can be further increased by enzymatic propulsion using catalytic turnover of endogenous substrates. In enzyme-propelled vesicles, membrane-bound enzymes undergo substrate-driven activity that promotes sustained motility and is thus preferred for active delivery systems [41]. The modular assembly of nanomotors, based on enzyme–host–guest interactions, allows for the integration of imaging moieties. This, in turn, enables real-time in vivo tracking and precise monitoring of therapeutic performance [168]. Beyond steady-state propulsion, chemically powered nanomotors are also capable of dynamic interaction with oscillatory chemical environments, leading to complex spatiotemporal behaviors that can be useful for navigation within heterogeneous biological media [110]. Building on chemically and enzymatically powered systems, light-driven and magnetic-field-controlled nanomotors introduce externally tunable propulsion modes well adapted to biomedical microenvironments where chemical fuels can be scarce or undesirable. These MNMs convert external energy inputs, especially light and magnetic fields, into directed motion. This allows enhanced controllability under physiological constraints [77].

In general, light-driven MNMs offer excellent spatiotemporal precision, but their performance will be reduced due to scattering in turbid tissues. With the aim of overcoming such limitations, dual-energy-propulsion systems that combine optical

Fig. 2 Propulsion mechanisms and energy sources

and magnetic actuation have been developed. They enhance both motility and functional output by exploiting complementary energy modalities [19]. The efficiency of these systems critically depends on the structural design, actuation wavelength, and the underlying photophysical mechanisms; indeed, all three parameters will determine the achievable propulsion force and speed, which are crucial parameters for performing complex tasks in vivo [177]. Magnetic-field-controlled MNMs introduce further advantages in long operational lifetimes, robust maneuverability, and high biocompatibility. Such features render them particularly suitable for targeted therapeutic delivery in those applications where precise spatial guidance is required [78].

Other propulsion paradigms optimized for operation within physiological settings include fuel-free and biofuel-powered nanomotors. Fuel-free systems, actuated by external light, electromagnetic, electric, or acoustic fields, avoid the use of exogenous chemical fuels like hydrogen peroxide and hence have fewer toxicity problems and a higher suitability for clinical uses [77]. Controllability and long operational life make them perfect for targeted drug delivery, cellular manipulations, and nanosurgical operations [157]. For example, light-driven nanomotors can mimic phototactic and gravitactic behaviors of natural microorganisms, enabling non-invasive guiding strategies [139]. Electric and magnetic field-actuated nanowire motors also exhibit great versatility in both in vitro and in vivo conditions [143]. Biofuel-powered nanomotors, though less elaborated in those contexts, use biocompatible endogenous fuels for sustaining propulsion and thus enhance barrier penetration and therapy efficiency at the biological scale [50].

This involves synthesizing propulsion models and knowledge regarding locomotion strategies of natural microorganisms to enable efficient propulsion in viscous and structurally heterogeneous biological fluids. Examples are actuation by an external magnetic field, which, in the case of paramagnetic colloidal doublets, proved that propulsion efficiency can be optimized for tailored magnetic modulations, e.g., elliptical drive sequences, aligning experimental and computational results [135]. Similarly, biological microswimmers, including spermatozoa and flagellated algae, rely on traveling wave-like deformations to generate Purcell's scallop theorem-consistent nonreciprocal motion, maximizing propulsion for a given energetic input [69]. Extensions of classical models, such as Taylor's swimming sheet within transversely isotropic media, show that the underlying fibrous networks within fluids, like cervical mucus, can have profound influences on energetic costs and propulsion dynamics as a function of fiber orientation relative to the swimming direction [32]. The heterogeneous dual helix model found in Ray sperm combines a rigid spiral head with a soft helical tail, enabling bidirectional and adaptive locomotion in viscous media, while providing valuable design principles for engineered nanoswimmers [144].

Hybrid and multimodal propulsion strategies implement multiple energy-conversion mechanisms to achieve high performance for nanomotors. These, in turn, enhance their navigation, adaptability, and delivery of payloads under physiological conditions. These systems integrate chemical reactions, enzymatic activity, and external field-based propulsion in order to overcome biological barriers and limitations inherent in single-mode systems [78, 86]. Janus-type nanomotors, such as platinum-mesoporous silica constructs, use catalytic hydrogen peroxide decomposition to achieve rapid, directional locomotion that allows them to precisely deliver drugs to cancer cells [33]. Meanwhile, dendritic silica/carbon@Pt Janus motors use a tri-modal propulsion mode, including H_2O_2, NIR, and lipase-, which acts to promote mobility in complex physiological fluids, hence illustrating the capability of multi-input propulsion systems to be applied in advanced biomedical applications [155]. In general, all these hybrid MNMs are capable of converting different energy sources into mechanical work and, therefore, are very versatile for several applications in precision medicine, targeted therapeutics, and minimally invasive medical procedures [78, 166].

3 Advanced Nanomotor Architectures and Smart Materials

Asymmetric and Janus nanomotors are the basic design paradigms to reach enhanced self-propulsion and functional versatility in biomedical applications. The Janus architecture is identified with dual-faced asymmetric compositions that allow for the spatial separation of catalytic, magnetic, or photoreactive components of motors, while enabling them to harness diverse energy sources for propulsion. This is an important feature regarding targeted drug delivery and diagnostic functions [167]. Similarly impressive performances of asymmetric nanomotors with inherent structural, compositional, or surface-property anisotropy have been demonstrated in biological systems. For example, glutathione responsive chemotactic nanomotors selectively navigate the tumor microenvironment (TME) to facilitate localized drug delivery and induce ferroptotic cell death, hence increasing therapeutic precision [24]. Other prime examples of the sophistication of asymmetric designs are the Janus colloidal motors that couple engineered nanostructures with tailored surface chemistry. These have succeeded in realizing controlled motion under physiologically complex conditions [70]. Furthermore, asymmetric Janus microdimer swimmers, actuated by rotating magnetic fields, achieve high locomotion speed (98 $\mu m \cdot s^{-1}$), giving strong maneuverability in micro- and nanoscale biomedical environments [171]. The recent finding stated that even geometrically symmetric particles are able to display self-propulsion, further challenges long-standing assumptions in nanomotor research, and broadens the conceptual framework for future architectural designs [96].

Biodegradable organic nanomotors and polymeric microswimmers are built upon asymmetric and Janus platforms, opening up the design space toward transient, therapeutically integrated systems. Such devices have the power to mimic the natural motility of microorganisms while offering controlled degradation after fulfilling their tasks, thus minimizing toxicity and environmental accumulation in the long run [82]. Material engineering has been advanced by the introduction of helical microswimmers fabricated from enzymatically degradable gelatin methacryloyl, which featured non-cytotoxicity, full degradability by collagenases, and supported cellular adhesion and proliferation [147]. Hydrogel-based microswimmers developed to be responsive to pathological enzyme levels, such as those of matrix metalloproteinase-2, thereby allowing for environment-dependent kinetics of drug release, can be functionalized with targeting agents such as anti-ErbB2 antibody-conjugated nanoparticles for cancer cell labeling and imaging [14]. Integration of covalent organic frameworks into light-driven microswimmers further advances drug delivery in complex environments, intraocular fluids being one good example, where real-time visualization and controlled release are crucial [128].

These are combined with synthetic frameworks comprising cells, enzymes, or nanozymes for the realization of catalytic and biological functions within physiological conditions. MnO_2 nanoparticle-loaded organic stomatocytes are a class of hybrid degradable nanomotors that can efficiently penetrate tumors, maintaining sustained catalytic activity without compromising cellular compatibility [98]. Janus dendritic silica/carbon@Pt nanomotors similarly use a variety of propulsion modes, such as enzyme-mediated self-diffusiophoresis, for effective navigation through complex biological fluids and precise motion execution in biomedical tasks [155]. Polyoxometalate-nanozyme nanomotors have also been introduced,these provide robust catalytic platforms for the synergistic photothermal-catalytic therapy of tumors through reactive oxygen species formation upon peroxidase-like activity in the process of targeted tumor destruction [132]. Enzyme cascade systems on Janus colloids are evidenced to exhibit enhanced catalytic conversion and propulsion efficiency, underlining their potentialities as environmentally safer biocatalytic nanomachines [49].

In promoting these increasingly sophisticated architectures, stimuli-responsive and shape-memory smart materials have become integral to the next-generation nanomotor systems. Shape-memory alloys and polymers with reversible deformation in response to thermal, optical, or magnetic stimuli opened up new avenues for dynamic actuation and structural adaptability in nanorobotic systems [53, 134, 146]. Layer-by-layer assembly techniques allow further integration of a great variety of multifunctional components that achieve autonomous propulsion by energy conversion from the environment into mechanical motion [75]. Temperature-responsive polymer brushes have become one representative class of smart control systems because modulation of fuel accessibility, such as hydrogen peroxide, enables precise regulation of propulsion speed [136]. Stimuli-responsive nanomembranes and shape-memory nanoarchitectures also enhance the deformability, remote actuation, and multifunctionality of complex structures, thus further extending their applications in medical microbots and advanced manufacturing [53, 134].

The integration of AI into nanomotor design and materials engineering introduces a computational dimension that enhances optimization and discovery. ML algorithms explore design spaces much faster, accelerating materials development for nanomotors and technologies related to clean energy and carbon capture [16]. AI-driven modeling of optically propelled nanomotors, through the use of convolutional neural networks (CNNs) and generative adversarial networks (GANs), has enabled geometric optimizations aimed at maximizing optical torque, hence achieving fast rotation speed and precision down to sub-20 nm resolution [29]. Neural network frameworks have also succeeded in the optimal design of nanoporous metallic actuators for artificial muscle applications by solving multiscale challenges in structural and mechanical design through training with finite element datasets [129]. Various reinforcement learning-based strategies for the control of microswimmers constitute the potential of AI-enhanced adaptive locomotion on the road toward intelligent autonomous nanomotor systems [30].

4 Intelligent Navigation, Swarming Behavior, and Control

Autonomous navigation within such complex biological environments is one of the fastest-growing areas and holds immense promise for precision medicine and biotechnology. These micro/nano motors convert various energies into mechanical movements to allow them to cross through physiologically difficult territories-mucosal barriers, extracellular matrices, vascular networks, and high-precision drug payloads can thus be delivered [86]. It is critically shaped by interactions with boundary conditions, including solid–liquid and liquid–liquid interfaces, modulating hydrodynamic forces, propulsion efficiency, and trajectory stability. For the sake of optimizing MNM behavior in a boundary-rich environment characterizing biological tissues, there is a need to understand these interface-dependent dynamics [21]. In recent decades, advances have been made toward biocompatible components and biocatalytic propulsion for efficient energy conversion at physiologically mild conditions. This promotes in vivo applications along with environmentally relevant remediation strategies [44]. Deep reinforcement learning algorithms further complement these developments in improving MNM navigation and localization, thus continuously adapting their speed and orientation within dynamic microenvironments. Indeed, this forms a genuine capability in the development of applications, especially for precision surgery and targeted drug delivery [164].

Beyond individualized navigation, collective swarming and cooperative behaviors have emerged as powerful mechanisms for amplifying therapeutic performance. For example, the enzyme-powered nanomotors described by Fraire et al. show significantly enhanced barrier penetration through active mechanical disruption, leading to a number of drug delivery efficiencies up to 12-fold higher compared to those realized by passive diffusion alone [36]. In parallel, Yu et al. developed Fe_2O_3-based ultrasmall tubular nanomotors that achieve robust swarming even in high-electrolyte environments where conventional nanomotors typically lose propulsion efficiency. The ability of these nanomotors to self-organize into stable clusters at ion-rich media further extends their opportunities in deployment within realistic biological fluids [170]. In contrast, Khatri and Kapral provide mechanistic insights into these kinds of collective behaviors utilizing theoretical frameworks and simulation methods, discussing the way spatially distributed chemical activity can cause the appearance of patterns in emergent clusters and swarming configurations [61]. Consistent with these findings, Robertson et al. have shown that synthetic nanomotors can phoretically couple and self-assemble into diverse architectures that support autonomous motion and cooperative action within fluidic environments [111].

Magnetically and acoustically actuated nanomotor swarms have further expanded the toolbox for controlled navigation in vivo. External magnetic fields provide high-precision, remote manipulation capabilities, enabling targeted delivery in anatomically complex regions [83, 169]. The integration of acoustic fields increases functionality by enabling upstream motility against physiological flows, an important feature necessary to navigate bloodstreams and ductal systems [3]. Hybrid systems, such as Au-Ni-Au nanowires, use ultrasound-driven propulsion together with magnetic steering to accomplish intricate tasks that include cargo transport and localized drug delivery in biologically relevant media [39]. In addition, the development of NIR-responsive hollow magnetic nanocarriers overcomes challenges such as premature drug leakage by offering on-demand drug release and synergistic photothermal effects [20], a means of enhancing precision and further increasing therapeutic efficacy.

Biologically inspired guidance mechanisms, such as chemotaxis and phototaxis, add further levels of autonomous behavior suitable for tumor-targeted interventions. Chemotaxis-inspired nanorobots make use of chemical gradients to migrate toward pathological sites; systems mimicking bacterial chemotaxis can accumulate at tumors and release drugs based on quorum-sensing cues or biomarker concentrations [182]. Complementary control is offered by phototactic navigation, where red-light-driven nanorobots using hydrogen sulfide-based propulsion achieve deep tumor penetration via negative phototaxis, thus allowing improved intratumoral distribution of therapeutics [180]. Advanced nanosensor architectures, including nanocantilevers and biomolecular sensors, enable molecular recognition and precise site targeting [105]. Multifunctional nanorobot systems that combine magnetic navigation with light-triggered drug release further illustrate the potential of synergistic chemo-photothermal modalities for enhanced antitumor therapy [58].

ML continues to play a transformative role in the refinement of real-time trajectory control for biomedical nanomotors. Recent microrobotics research has emphasized the importance of ML in enhancing imaging, localization, and closed-loop control under various biologically relevant constraints [113]. The integration of OpenCV-based preprocessing with the Informed RRT* algorithm can advance the efficiency of path planning and the accuracy of feedback for 2D microrobotics navigation [163]. Deep reinforcement learning frameworks, including bio-inspired neural networks, have been applied to manage continuous control of MNMs with variable dynamics, enabling adaptive navigation through heterogeneous microenvironments [164]. Furthermore, CNNs are used for the optimization of the design and actuation of optically driven nanomotors to achieve high-speed rotation and precise manipulation in fluid systems [29].

5 Functionalization and Targeting Strategies

Surface functionalization is one of the most important strategies to improve nanomotor biocompatibility, prolong circulation time, and reduce immune clearance—all critical parameters related to their translational possibilities in biomedical applications. A key development in this area is represented by biomimetic integration, mainly cell membrane coating onto synthetic MNMs. Such biomimetic interfaces use the intrinsic self-recognition and immunomodulatory properties of native membranes to minimize immune activation and, at the same time, offer multifunctional benefits toward drug delivery, molecular recognition, and real-time imaging [181]. Complementary strategies illustrate that tuning of surface composition, charge, size, and shape further optimizes cellular uptake and systemic compatibility, underlining the importance of precise physicochemical tuning toward clinical translation [115]. Among surface modifications, zwitterionic ones proved to have effective protein-repelling and antifouling properties, allowing tunable functionalization of nanodiamonds as well as other nanomaterials while preserving stability in different biological environments [87]. Such a variety of different approaches is further supported by advanced surface engineering methodologies such as polymer and lipid coatings, able to reduce cytotoxicity, attenuate immunogenicity, and prolong in vivo residence time, therefore mitigating adverse immune responses, which are still one of the major challenges in nanomedicine [38].

With the development of biocompatibility optimization, ligand-based targeting has emerged as a powerful strategy to enhance the specificity of self-propelled nanomotors. MNMs functionalized with peptides and antibodies can recognize and selectively bind diseased tissues, therefore improving therapeutic precision. Antimicrobial peptides have been used to functionalize nanomotors for the direct delivery of antibiotics to infection sites, thus decreasing systemic exposure and the development of antimicrobial resistance [45]. In oncology, nanomotors functionalized with antibodies, such as anti-EpCAM-modified magnetic nanobots, achieve targeted navigation toward tumor cells, enhancing tissue penetration and allowing for spatially controlled drug release in the TME [4]. These ligand-mediated strategies also support navigation across physiological barriers, including mucosal interfaces and vascular endothelial layers, some of the challenging biological barriers for passive drug delivery systems [86].

In this direction, stimuli-responsive drug release mechanisms have been emphasized as an efficient way to realize spatiotemporally controlled therapeutic delivery. The active nanomotor carriers can be driven by exogenous stimuli like ultrasound, magnetic fields, or light to specific pathological sites where the same stimuli can trigger drug release locally [89]. Besides, different endogenous stimuli-based approaches involving pH-sensitive degradation, enzyme-responsive disassembly, and active bond cleavage exert site-specific release in acid tumor environments or glutathione-rich cancer cells [131]. In this context, redox-responsive stomatocyte nanomotors utilize disulfide bridges that cleave upon glutathione exposure, leading to nanomotor aggregation and thus payload release [137]. Hybrid nanovehicles consisting of organic and inorganic components allow even more sophisticated release modes, such as thermolabile and photolabile bond cleavage and structural disruption under external stimulation [76], broadening the therapeutic use of active nanosystems.

The utility of self-propelled nanomotors for real-time diagnostics and therapeutic monitoring is further enhanced by the integration of biosensing capabilities. These bioinspired systems make use of various propulsion mechanisms, from biochemical to optical, magnetic, or electric, thus enabling them to pass biological barriers and perform a variety of tasks, including drug delivery, molecular sensing, and localized therapeutic activation [106]. Enhanced penetration by propulsion becomes particularly important in dense tumor tissues, whereby nanomotors surmount diffusion limitations and enhance therapeutic deposition [184]. Diagnostic potential can be obtained with receptor-functionalized nanomotors that may isolate nucleic acids or proteins directly from unprocessed biological samples, coupled with electrochemically propelled motion enabling sensitive, rapid detection by means of motion-linked signal transduction [12]. Advanced multifunctionality could be achieved through NIR-powered active tumor-seeking Janus mesoporous silica nanomotors capable of selectively binding to cancer cells and then inducing thermomechanical therapy [159].

Progress in real-time imaging and tracking technologies has further enabled precise monitoring of nanomotor behavior in vivo. MNMs enhance conventional imaging modalities, including ultrasound, fluorescence, photoacoustic, and magnetic resonance imaging, through the introduction of motion-amplified contrast, thus allowing dynamic visualization of subcellular processes and microenvironmental changes [178]. Biomimetic coatings, such as cell membranes, not only improve biocompatibility but also support multifunctional imaging and tracking capabilities without immune responses [181]. Ultrafast ultrasound imaging is capable of deep-tissue tracking of MNMs at unparalleled spatiotemporal resolution, hence facilitating applications such as targeted surgery and deep organ drug delivery [162]. Advanced two-photon holographic tracking allows for high-precision real-time monitoring of fast-moving nanoscale entities, thus supporting mechanistic studies of intracellular transport. Furthermore, NIR-II fluorescence imaging-guided nanomotors have shown great promise in thrombolysis monitoring, performing with high biosafety and therapeutic efficiency by accurately navigating and visualizing thrombolytic progression [79].

6 Biomedical and Clinical Applications

6.1 Targeted Drug and Gene Delivery

Self-propelled nanomotors represent advanced platforms for the precision delivery of drugs and genes, using autonomous propulsion to achieve site-specific transport and controlled therapeutic release. Systems such as these, which convert chemical or external energy into mechanical motion, enhance intratissue navigation and improve cellular uptake while reducing off-target accumulation, especially in oncology. Widely used chemical propulsion based on catalytic reactions, exemplified by H_2O_2 decomposition, allows for fast directional transport for deep tissue penetration and efficient payload discharge. Complementing this is the use of external-field propulsion-light, magnetic fields, or ultrasound—that provides spatiotemporal control over navigation and activation of release, with refinement in therapeutic specificity [78]. Discussed herein are Janus nanomotors that mostly emerge through functional asymmetry realization, such as platinum catalytic surfaces, with mesoporous silica drug reservoirs. These usually represent highly efficient propulsion provided with glutathione-responsive release mechanisms and further improvements in the precision of delivery. Other stimuli-responsive architectures, including NIR-triggered systems, represent powerful enabling strategies for tumor-selective accumulation and enhancements in pathological tissue penetration [33].

Likewise, nanomotor and nanocarrier inventions have benefited gene delivery applications, especially for CRISPR/Cas9, siRNA, and mRNA-based therapeutics. DNA nanobots fabricated via DNA origami provide programmable frameworks with the ability to encapsulate and transport CRISPR/Cas9 complexes of high structural precision, thus enabling targeted genome editing while reducing off-target effects. These DNA-based platforms also support mRNA and siRNA delivery and hence contribute to advances in vaccine development and gene-silencing therapeutics [118]. Polymeric and lipid nanoparticles offer further versatility by accommodating larger Cas9-guide RNA complexes, enhancing endosomal escape and cytosolic transport. Gold nanoparticles confer light-activated genome-editing capabilities with enhanced biocompatibility, further expanding the functional toolbox for spatiotemporally controlled gene modification [120]. Diverse nanoarchitectures shield nucleic acid therapeutics from enzymatic degradation and facilitate stimulus-responsive release, hence improving delivery precision and stability under physiological conditions [64]. Magnetic nanoparticles are also under exploration for therapy against prostate cancer by passive or actively guided CRISPR/Cas9 and siRNA delivery, where surface functionalization strategies have been adopted to improve tumor-specific accumulation and address some critical challenges associated with targeted gene transport [97]. Nanocomposites further extend these capabilities through tunable physicochemical properties that support high-efficiency genome editing and organ-selective delivery, whereby emerging clinical trials suggest this modality is progressing toward routine clinical deployment [40].

The combination of nanomotor propulsion with therapeutic carriers is an important means to overcome TME barriers that usually impede the diffusion of drugs and greatly reduce the efficiency of therapeutic outcomes. Independent propulsion by self-sustaining motion enables nanomotors to cross hypoxic niches, acidic pH gradients, dense extracellular matrix structures, and aberrant vasculature that may impede the distribution of passive nanoparticles. Obviously, enhanced motility can act to increase the local concentration of drugs inside tumors or deeper in malignant tissues for improved therapeutic outcomes [184]. Spatial specificity is further refined in nanomotors engineered to activate under TME-associated triggers such as hypoxia or acidity, enabling on-demand payload release and reduced systemic exposure [90]. Multifunctional designs allow the integration of both diagnostic and therapeutic elements necessary for real-time monitoring and image-guided delivery. Although promising preclinical data show improved efficacy and reduced side effects, continued translational strides are hampered by reproducible problems in large-scale synthesis, validation of safety, standardization of regulation, and performance across diverse patient-specific microenvironments [160].

6.2 Theranostics and Precision Medicine

Theranostic nanomotors represent one of the key embodiments of an integrated diagnostic-therapeutic platform, squarely standing at the core of a precision-medicine approach in oncology. By taking advantage of specifically engineered physicochemical properties, nanoparticle-based systems can provide disease imaging, targeted therapy, and real-time monitoring with improved delivery accuracy and minimal systemic toxicity. Their dual functionality is all the more significant for cancer management, where early lesion detection and localized treatment improve clinical outcomes. Superparamagnetic iron oxide, gold, and bismuth nanoparticles create contrast for MRI, CT, and optical imaging, respectively, thus allowing sensitive visualization and monitoring of tumors. Their EPR properties further support selective tumor accumulation, improving therapeutic payload localization [52]. Biomarkers or active-targeting ligands increase molecular specificity and allow individualized modulation of the dose and treatment planning within a precision medicine framework [11]. Although significant progress has been made, the incomplete biocompatibility, toxicity profiling, and regulatory standardization have so far prevented their clinical translation and mandate further optimization efforts [63].

Further, the integration of nanomotors with MRI, positron emission tomography (PET), and ultrasound imaging methodologies enables spatiotemporally controlled diagnosis and therapy within one construct. Coupled to such modalities, nanomotors confer autonomous or externally guided motion, stimulus-responsive behavior, and enhanced target selectivity features that improve imaging fidelity and therapeutic precision. One of the widely used MRI contrast agents is represented by superparamagnetic iron oxide nanoparticles due to their high magnetic responsiveness and capabilities of image enhancement. Their surface can be engineered to transport therapeutic cargos, thus allowing for seamless transitions between imaging and the basic requirement in theranostic design [46]. Nanoparticle labeling of radioactive isotopes allows for high-sensitivity and real-time quantification of in vivo nanomotor biodistribution and treatment efficacy using PET imaging [174]. Ultrasound-based guidance represents a non-invasive, clinically accessible method for imaging and propulsion enhancement that may thus be used for improved in vivo navigation and targeted therapeutic release [46].

These nanomotors represent a novel approach to the theranostics of non-muscle-invasive bladder cancer through the combination of active propulsion with immune modulation. Urease-powered nanomotors that deliver STING agonists are being developed. These catalyze urea to CO_2 and NH_3, generating propulsion forces to improve intravesical mobility, enhance bladder-wall penetration, and increase drug retention despite continuous urine flow, hence overcoming the problems of rapid clearance and poor bioavailability associated with conventional intravesical therapies. Physicochemical attributes, including structural morphology and surface charge, were validated by high-resolution TEM and zeta-potential analyses to make sure of its biocompatibility [27]. Thus, activation of the cGAS–STING pathway was observed by producing type I interferon and pro-inflammatory responses, encapsulation of STING agonists within nanomotors increased their stability, reduced systemic exposure, and provoked localized immune activation in the TME. As this forms an inroad for nanotechnology in urology, precision in delivery and real-time monitoring significantly enhance the therapeutic outcomes. Besides the adaptability to other malignancies, the wider clinical application of urease-driven nanomotors is confronted with various challenges in manufacturing, regulatory, and design optimization issues [151].

6.3 Minimally Invasive Surgery and Micro-manipulation

Nanomotor-assisted microsurgery and in vivo tissue repair are the future directions of minimal invasive intervention, enabling manipulations at higher precision with fewer postoperative complications on cellular and subcellular scales. MNMs have been used to provide controlled locomotion, high maneuverability, and nanoscale working dimensions for the completion of such complex surgical tasks that may otherwise not be feasible with conventional tools. Sub-millimeter micromotors, such as piezoelectric ultrasonic resonant systems, represent an important class of such devices, with high power density values of 18.4 kW/m^3, thus efficiently navigating through narrow vasculature while supporting precise manipulations in confined anatomical regions [150]. Beyond mechanical maneuverability, MNMs contribute to regenerative medicine by enabling minimally invasive and precisely targeted cell delivery, improving cellular proliferation and differentiation in the damaged tissue sites [80]. In addition, supplementing microsurgical capability for ultra-precise incisions into very sensitive structures, such as peripheral nerves, with in vivo safety and without evident neurotoxicity, nanoknives with ~ 20 nm cutting edges [15] have been developed. All these technologies have pointed out the broad prospects of nanorobotics for operations such as cancer excision, neural repair, and organ-level treatments, while clinical readiness is further influenced by challenges in biocompatibility, navigational control, and ethical regulation [103].

Specialized applications, including biofilm disruption, thrombolysis, and targeted cardiovascular repair, further demonstrate the versatility of nanomotor-enabled manipulation within therapeutic micro-environments. Ultrasound-activated nanomotors, exemplified by hBT-Pt@Pm constructs, generate reactive oxygen species capable of degrading fibrin networks in blood clots; platelet-membrane functionalization enhances thrombus-specific targeting and reduces off-target cytotoxicity [183]. Typically, hBT-Pt@Pm constructs refer to nanomotors made from platinum nanoparticles functionalized with hyaluronic acid as a biocompatible coating and modified with platelet membranes. Such nanovehicles have demonstrated enhanced ultrasound-induced thrombolysis in fresh thrombi on account of their generation of sonoactive cavitation. Indeed, hBT-Pt@Pm constructs have been noted to demonstrate remarkably effective ultrasound-enhanced in vitro thrombolysis in fresh thrombi owing to sonoactive cavitation generated from the surface-functionalized Pt nanoparticles. By virtue of NIR-photothermal conversion coupled with controlled drug release, dual-drive polydopamine nanomotors have precisely ablated thrombi with simultaneous promotion of endothelial recovery [57]. The principles of active nanomotor propulsion thereby enhance transport to atherosclerotic plaques, overcoming hemodynamic washout and thereby improving therapeutic retention at diseased sites, strengthening plaque-targeted drug delivery and imaging precision [102].

6.4 *Biosensing and Smart Diagnostics*

Active nanomotor-based biosensors indeed represent a breakthrough toward the real-time detection of biomarkers because of their ability to enhance sensitivity, specificity, and speed of analysis through self-propulsion and functionalized interfaces. Electrochemically propelled nanomotors enable the direct isolation of biomolecules from raw samples with minimal preprocessing, which drastically reduces assay time along with operational complexity. Surface modification with ssDNA, aptamers, or antibodies enables the selective recognition of nucleic acids, proteins, and circulating tumor cells, while quantifiable analytical outputs are provided by changes in their motion parameters, such as velocity or displacement [12]. The integration of nanomaterials like gold nanoparticles and carbon nanotubes further raises the level of electrochemical and optical responsiveness of sensors, therefore ultimately enabling real-time, label-free, and multiplex biomarker detection that is essential for early diagnosis and continuous patient monitoring [25]. While there is significant promise, challenges persist in terms of nanomaterial stability, scalable manufacturing, and compliance with regulatory standards in the clinic. Integration with microfluidic architectures is likely to overcome many of these bottlenecks via improved automation, reproducibility of assays, and deployment in a variety of diagnostic settings [60].

Integrating nanomotors with lab-on-a-chip and point-of-care diagnostic platforms advances smart diagnostics by combining autonomous nanomotor motion with precision microfluid control. Within these miniaturized systems, nanomotors enable active sample transport, selective target capture, and improved mixing dynamics; together, these features enhance diagnostic accuracy and speed both in the clinic and in resource-limited settings. Microfluidic channels allow the manipulation of nano- to microscale volumes of fluids with great precision. This thus enables the sequential execution of bioassays within compact devices. Active targeting of analytes can be made directly from complex biological matrices using functionalized nanomotors, hence increasing efficiency in sample preparation and analytical fidelity. These systems have their propulsion enhanced through different ways, including catalytic nanomotors and ultrasound-driven microbullets, thus increasing the throughput of the assays while minimizing operator input [13]. LOC platforms integrate a variety of biosensing modalities, including optical, electrochemical, and nucleic acid-based systems, making possible the high-sensitivity detection of a wide range of biomarkers. Of these, FRET-based nanosensors, particularly quantum-dot constructs, offer exceptional sensitivity and multiplexing capability for real-time bioanalytical monitoring [31]. Integrated POC and LOC technologies are increasingly important for the rapid diagnosis of infectious diseases, chronic conditions, and cancers,however, their widespread translation continues to be constrained by challenges in scalable fabrication, cost reduction, and regulatory approval pathways [95].

7 Integration into Smart and Connected Healthcare Systems

The integration of Internet of Medical Things with nanomotor-based technologies is significantly remodeling by its continuous physiological monitoring, real-time data exchange, and autonomous therapeutic intervention capabilities. These wearable and implantable IoMT sensors allow for continuous monitoring of critical physiological metrics, creating subsequently a foundation for early disease diagnosis and personalized therapeutic interventions. Embedding nanomotors in these systems enables the realization of precise cell-level drug delivery and nanoscale diagnostic monitoring, boosting further personalization, accuracy of therapy, and overall clinical efficiency. This also extends the scope toward remote patient monitoring, clinical load reduction, and healthcare expenditure [124, 125]. The IoMT platforms facilitate not only remote patient monitoring but also reductions in clinical load and healthcare expenditure. When integrated with deep learning frameworks, such as deep CNNs, these systems achieve enhanced predictive analytics and accelerated decision-making. The integration of nanomotors enhances such capabilities through the generation of micro-level diagnostic data with localized therapeutic release, reducing toxicity associated with treatments and improving clinical outcomes [121].

The use of AI in combination with nanomotor-driven systems further cements connected healthcare through intelligent, adaptive control architectures. AI-enhanced biosensing platforms process real-time, highly dynamic physiological signals emanating from nanosensors and thus enable early disease diagnosis and patient-specific therapeutic modulation. Smart nanodevices integrated with AI algorithms will ensure enhanced biomarker monitoring and improved diagnostic accuracy, but perhaps most importantly, facilitate earlier detection of malignancies and metabolic disorders [123]. Enhancing these functionalities, Edge AI processing enables analytics at the device level, reduces latency, and promotes instantaneous clinical responses [142]. Examples of AI-enabled wearables include the state-of-the-art smartwatches that perform embedded ML-based processing on cardiac and metabolic signals for the prediction of conditions such as arrhythmias and cardiovascular events in view of timely interventions [2].

Incorporating nanomotors into connected healthcare networks paves the way for developing autonomous, closed-loop medical platforms with the capability to merge diagnosis, monitoring, and therapy into a single system. These nanoscale constructs conduct precise, cell-level operations such as targeted drug release and real-time physiological signal acquisition. Advances in nanotechnology, energy harvesting, and biocompatible power systems ensure continuous, energy-efficient operation of wearable and implantable closed-loop systems. In thrombosis management, nanoscale closed-loop device design has been developed to couple thrombus detection with targeted nanomotor-driven intervention, thus enabling timely modulation of platelet activity and coagulation pathways [179]. Nanobots and MNMs engineered for cancer therapy, microscale surgery, and advanced diagnostics leverage energy-driven motility to accomplish molecular or cellular-level tasks in high-precision manners (Fig. 3). Nanoparticles, nanodevices, and lab-on-chip systems further enhance drug delivery, diagnostics, and tissue engineering by facilitating continuous physiological monitoring and personalized treatment [7]. While these hold great promise for transformative changes, wider clinical adoption requires adequate resolution of prevailing digital-literacy gaps, regulatory ambiguity, and ethical concerns, particularly in under-resourced areas [161].

Other critical parts of smart, connected healthcare involve bio-cyber interfaces; at the nanoscale, these interfaces link nanomotors with wearable and implantable bioelectronic devices for continuous and bidirectional communication between biological systems and digital networks. The presence of wireless communication, bioelectronics, and nanotechnology in the bio-cyber interface enables diagnostics and targeting of therapy with long-term monitoring, reducing healthcare costs. These wearable and implantable bioelectronic devices, including even insulin pumps, microneedle patches, and transdermal drug-delivery systems, can perform therapeutic treatments automatically in a minimally invasive way, thus improving

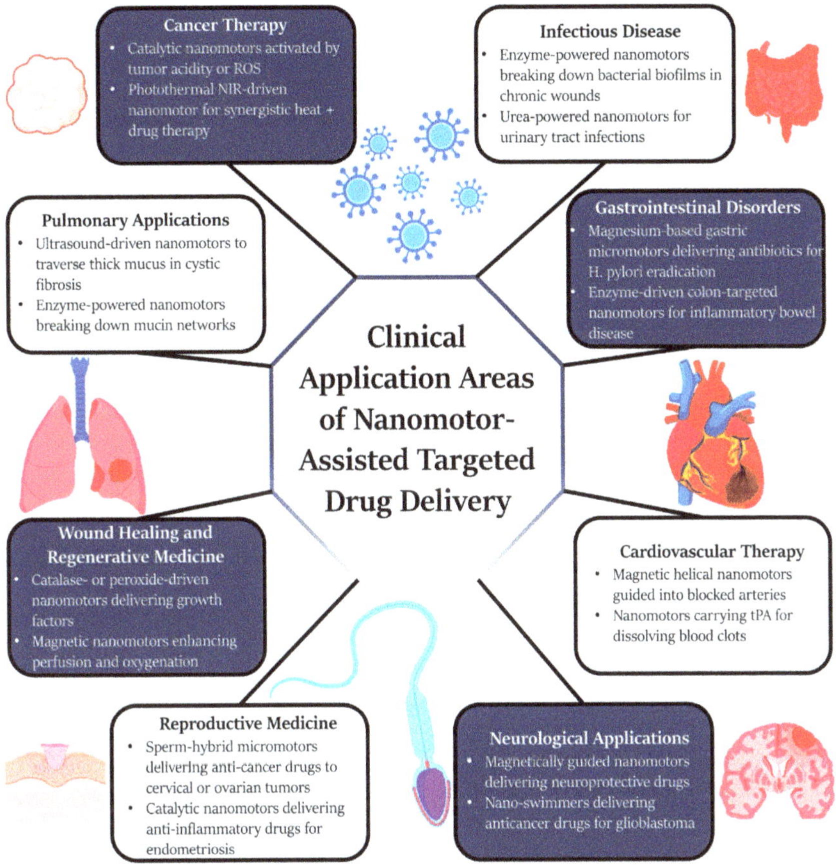

Fig. 3 Application areas of nanomotor-assisted targeted drug delivery

patient adherence [62]. Bio-cyber interfaces can convert biochemical signals to electromagnetic signals and provide intra-body communication essential for Internet of Bio-Nano Things applications such as real-time sensing and targeted drug delivery. However, such a hybrid biological-digital system introduces cybersecurity vulnerabilities, and appropriate measures are required for protection against sensitive physiological data [173]. Such functionalities are extended by flexible e-skin technologies with integrated wireless modules capable of continuous physiological monitoring and prosthetic integration with real-time bioelectronic responses while improving both patient care and mobility [94]. Intra-body molecular communication technologies and miniaturized implantable devices promoting advanced theranostic capabilities with local physiological data storage are being supported in ongoing bio-cyber interface development [67].

Convergence of AI and nanotechnology in general health systems provides a ground for high personalization, data-driven ecosystems in medicine. AI-enhanced nanosensors can track molecular biomarkers in real time by creating high-resolution diagnostic information that supports rapid clinical decision-making. AI-driven analysis of the multi-dimensional health dataset comprising clinical history, symptom profile, and lifestyle factors enhances the performance of precision diagnosis and targeted therapy strategies by improving personalized medicine frameworks [34, 51]. In spite of this fact, various challenges remain relevant: algorithmic bias, regulatory inconsistencies, infrastructure limitations, and ethical concerns regarding issues like fair equity and technological impartiality. Intensification of AI-IoMT with active nanomotor technologies may provide huge opportunities in respect to advanced, patient-centered healthcare systems. This is where healthcare professionals and nurses ought to be involved in promoting this technology-surge, setting standards for safety and efficiency, and ensuring further clinical use [161].

8 Safety, Biocompatibility, and Regulatory Considerations

Catalytic, magnetic, and enzyme-driven nanomotors have to be critically assessed regarding in-vivo application for their safety, biocompatibility, and regulatory readiness. However, although a promising preliminary biosafety profile can be seen in several classes of nanomotors, long-term stability, controlled propulsion, and clinical reliability are pending validation. On one hand, magnetic nanomotors have shown very high biocompatibility both in vitro and in vivo, without cytotoxicity, genotoxicity, or anomalies in proliferation, and remain well-tolerated at doses up to ~ 55 mg/kg in mice, with organ-specific retention predictable. Similarly, catalytic nanojet motors have little effect on human lung epithelial cells over a range of doses and exposure times. On the other hand, enzyme-driven nanomotors advantages from using endogenous enzymes and biofriendly fuels such as urea, which circumvents toxic oxidizing agents like hydrogen peroxide [104].

In addition, manufacture capable of producing uniform nanomotors both structurally and functionally in a standardized and scalable manner is required for clinical translation. Therein, predictable therapeutic activity requires consistent propulsion characteristics, reproducible surface chemistry, and precise size distribution. Complete biocompatibility of materials, nontoxic fuels, and intelligent control mechanisms remain important for clinical deployment as clinically deployable tools [126]. However, in vivo propulsion efficiency, precise targeting in complex physiological environments, and overcoming biological barriers remain bottlenecks in their performance. More importantly, active nanomotors have to show therapeutic benefits that significantly exceed passive nanoparticles in order to justify their integration into clinical theranostics [74].

Biodistribution and clearance behavior are central to nanomotor safety. Thus, different self-propelled nanomotors, including helical magnetic designs, have exhibited excellent profiles for biocompatibility, not inducing cytotoxic or genotoxic responses upon incubation at concentrations as high as 20 motors per cell or 55 mg/kg in animal models. Such nanomotors serve to effectively dissolve emboli in thrombolytic applications without secondary embolism or hemorrhagic complications, even at reduced thrombolytic drug doses [104]. Biodistribution studies reveal that systemically administered nanomotors circulate within the bloodstream, briefly adhering to the vessel walls. This is to some extent diminished by saline perfusion and hence suggests a plausible clearance mechanism. On the other hand, their extended systemic retention, as evidenced by the 7.2-h half-life of thrombolytic nanomotors, significantly enhances targeting to emboli and improves therapeutic outcomes [79].

Regulatory oversight for nanomotor-based therapies is still developing. Existing frameworks emphasize demonstrated safety, reproducible manufacturing, and validated therapeutic benefit. Precise, site-specific navigation of nanomotors offers a significant regulatory advantage due to reduced off-target burden when compared with conventional methods of drug delivery [156]. However, broader ethical and regulatory considerations include immune compatibility, informed consent, autonomous device behavior, and long-term biosurveillance. Ensuring the biocompatibility of nanomedical devices requires extensive preclinical assessment of immune interactions, toxic risk, and environmental persistence. The utilization of biocompatible materials, optimized propulsion designs, and immune-silent coatings will be important to minimize adverse biological responses [26]. Ethical considerations also make it important to protect patient privacy and place strict limits on autonomous therapeutic decisions. Harmonized globally coordinated regulatory standards are an ongoing necessity to bridge nanotechnology innovation with safe clinical integration [127].

For nanotechnology, however, the pace of evolution has still outpaced the development of the ethical and legal framework. Thus, there are critical gaps in regulation. There exists a need for specialized codes of practice and lifecycle-oriented ethical guidelines that would control the development, deployment, and long-term monitoring of nanomedical devices. Therefore, collaboration among the regulatory bodies, clinicians, technologists, and industry partners to address new emerging challenges is very much called toward the responsible translation [65].

Besides, technical and regulatory challenges impede the clinical translation of nanomotors. The SLOT mission framework spells out necessary operational criteria-survival in vivo, precise targeting, and controlled activity—which, even today, is challenging, particularly for the chemically powered nanomotors. By contrast, magnetic and ultrasound-driven systems presently show higher translational potential [145]. Other challenges include the lack of harmonized regulatory routes to nanomedicines, [84], physiological barriers due to mucosal surfaces and fluidic forces, [86],the abuse of the "nano" denomination without sufficient validation, [112], and complex interactions with the immune system which might impair or promote therapeutic performance [37].

The ability for real-time in vivo tracking and safety assessment has been extended significantly by the introduction of more sophisticated imaging modalities. Photoacoustic computed tomography allows one to visualize in real time and guide the navigation of micromotors within the gastrointestinal tract; hence, it is of great use in applications targeting drug delivery [72, 152]. High-frequency ultrasound/ photoacoustic imaging, offering both structural and functional information relevant for the optimization of therapeutic effects, allows one to precisely track magnetically actuated micromotors [9]. PET has also been successfully utilized to quantify the collective enzyme-driven nanomotor behavior in vivo, hence confirming its biocompatibility and allowing for quantitative pharmacokinetic analysis [48]. In this context, NIR-II fluorescence imaging has also been leveraged to monitor real-time self-propelled thrombolytic nanomotors with unprecedented precision and biosafety [79]. These various imaging advances are an integral part of preclinical and regulatory testing, as they allow for the transparent assessment of biodistribution, clearance, and dynamic behavior under physiologically relevant conditions.

9 Case Studies and Recent Breakthroughs

In this regard, the recent developments considerably extended the functional spectrum of nanomotors toward oncology, immunotherapy, diagnostics, and applied nanorobotics. These works underline not only advanced therapeutic functionalities but also new propulsion strategies, AI-driven engineering, and an increasing awareness of translational relevance. Biologically derived protein-based nanomotors have taken center stage in recent developments of ferroptosis-mediated cancer therapy, driven by the intrinsic biocompatibility and structural modularity of those biomolecules, which enable precise control of catalytic action. Glucose oxidase-ferritin nanomotors are an exemplary expression of this trend, eliciting superior tumor inhibition through iron-dependent lipid peroxidation while circumventing the limited penetration depth and systemic toxicity of traditional ferroptosis inducers [158]. Synergetic systems, such as protein-gold nanoclusters with the ability to stabilize and deliver Fe^{2+}, further endorsed the role of protein scaffolds by triggering strong ferroptotic killing in hepatoma cells while circumventing the off-target effects [56]. This progress has been built on an increasing understanding of the mechanistic processes of ferroptosis, centered around pathways based on GPX4 interference and ROS amplification [133]. Continuing with this momentum, the recently proposed HIFU-driven nanomotors have provided an energy-responsive platform able to enhance tumor accumulation, deep tissue penetration, and robust induction of ferroptosis while turning on antitumor immunity in models of triple-negative breast cancer [172]. Coupling these protein-based nanomotors with rationally engineered nanocarriers further reinforces their power in modulating oxidative stress and depleting glutathione for improved specificity [122].

In parallel, advances in immunologically active nanomotors have been made, especially in the development of urease-driven STING agonist nanomotors for the treatment of bladder cancer. These systems have overcome long-standing limitations of intravesical drug delivery—namely poor retention and rapid clearance—by exploiting urease-powered propulsion to achieve prolonged residence and deeper penetration into the bladder tissue [27]. They target the cGAS-STING pathway, remodel the TME effectively, and improve CD8 + T-cell recruitment. By encapsulating cyclic dinucleotide STING agonists, nanocarriers improve stability and bioavailability while reducing systemic toxicity, thereby enabling effective local and potentially systemic immunotherapy. More recently, manganese ion-containing synergistic formulations have further increased downstream signaling and achieved potent antitumor effects in preclinical models [22, 130], supporting their applicability in multi-agent immunotherapeutic regimens [149].

In parallel, the application of AI to the design of asymmetric nanomotors has accelerated the generation of highly optimized architectures for given biomedical tasks. AI-driven modeling today has the capabilities for predictive nanoscale interactions and fast screening of structural parameters that will lead to major improvements in propulsion performance and material efficiency [24]. ML-assisted optimization of optically driven nanomotors has already yielded geometries with higher torque and rapid rotation, thanks to gains in fabrication resolution [29]. Advances in this direction will have to be founded on methodologies from materials science, physics, and computational modeling, in order to bypass the inherent nonequilibrium constraints of active matter, and to develop nanomotors with adaptive and environment-responsive behavior [42].

While in parallel, significant advances have been made regarding propulsion strategies, which for chemical, magnetic, and acoustic modalities has unraveled their respective performance profiles. Chemical nanomotors—powered, indeed, essentially by redox or catalytic reactions—remain favorable for high-speed propulsion but require exogenous chemical fuels, which entail potential biocompatibility issues. In contrast, magnetic nanomotors may be propelled in a fully fuel-free fashion by the use of external fields, which increases their suitability for in vivo operation and over long-term use. Acoustic nanomotors, powered via ultrasonic fields, require no chemical fuels, too, and have reliably performed inside living cells and tissues to enable fast autonomous motion desired for drug delivery and diagnostic procedures. Hybrid magneto-acoustic systems merge advantages of these two modalities that achieve efficient propulsion with tunable directionality and speed independent of reactive fuels [71, 73, 107].

With research moving fast toward translational implementation, preclinical assessment of nanomotor-based systems has become increasingly important. Early works illustrated the possible applications of nanomotors to targeted drug delivery, microsurgery, and cell transport due to their capability to convert energy into controllable motion [148]. However, translation remains slow, hindered by biases and poor reproducibility in preclinical research practices [54]. A robust evaluation framework from in vitro, ex vivo to in vivo models is thus highly desired to characterize pharmacokinetics, pharmacodynamics, and toxicity of nanocarrier-integrated nanomotors [99]. Navigation and control remain the major challenges,fuel-dependent chemical nanomotors are inferior to externally actuated magnetic or ultrasound systems in clinical pathways [145]. Meeting an optimal balance between therapeutic efficacy and toxicity remains the key focus in preclinical optimization, underlining careful dose engineering and mechanistic validation [119].

10 Challenges, Opportunities, and Future Directions

Most of the propulsion efficiency and scalability for self-propelled nanomotors continue to be greatly hindered by such complex physicochemical landscapes. For example, one of the fundamental challenges concerns the notorious sensitivity of nanomotor dynamics to boundary conditions-most remarkably, that of solid–liquid and gas–liquid interfaces—but it substantially alters propulsion behavior and also seriously impedes predictability of performance in physiological and ecological milieus [21]. Moreover, the generally adopted propulsion mechanisms, including self-electrophoresis and self-diffusiophoresis, have a nature wherein functional quenching may occur under high-electrolyte media with the result of diminished motility or translational potentiality in vivo [170]. Further constrictions on system design come from the requirement of bioavailable fuels and materials since only a narrow fraction of all catalytic and structural components are suitable for safe biological integration [47]. For tubular nanomotors, scalability is problematic in a clinical context due to fabrication issues, the complexity of motion control, and instability of trajectories at physiologically relevant flow conditions [176].

These are being overcome through improvement in AI-driven modeling and data-centric optimization. ML frameworks, including CNN and GAN, have been employed for the design and optimization of plasmonic or optically driven nanomotors at sub-20 nm precision for high rotational velocities and enhanced optical torque under polarized light [29]. Beyond structural optimization, AI supports high-throughput characterization of nanomaterials, automated feature extraction from complex imaging datasets, and thus accelerated propulsion behavior prediction in heterogeneous environments. In nanomedicine, AI enables rational designs of nanoparticle-based therapeutics, optimizes physicochemical profiles for improved bioavailability and reduced toxicity. Predictive ML models were also adopted for propulsion velocity and contaminant-removal efficiency estimates in environmental micromotor systems, capable of identifying particle diameter and catalyst composition as critical features [175]. This is indicative of progress due to the increasing presence of AI in overcoming the design barriers toward application-specified tailoring of nanomotor systems [68].

Biohybrid and living nanomotors represent the emerging strategies to overcome the biological, mechanical, and immunological challenges of synthetic systems. In this respect, biohybrid nanorobots powered by exosomes are opening possibilities to perform targeted drug delivery with reduced cytotoxicity once the intrinsic targeting capabilities and immune stealth of extracellular vesicles have been exploited, and so far, they have been tested for the management of inflammatory and infectious diseases [101]. Biomimetic and chemotactic micro-/nanomotors further enhance the possibility of being able to navigate through the complexity of tissues and allow for programmable therapeutic functionalities [81]. Besides the mechanical functionalities of biological components, however, great promises are held in the field of hybrid interface applications. Indeed, a number of biohybrid platforms, capable of locomotion, biosensing, and modulation of the microenvironment, can be yielded so far by merging living cells or organelles with engineered micro/nanostructures [109]. Twenty years after its establishment, current research is moving toward the integration of biological intelligence, thus opening possibilities toward hybrid nanomotors that can adapt to dynamic physiological environments and be employed for advanced tasks of theranostics [18]. A paradigmatic example of such advanced performance is provided by the living micromotors optically modulated obtained from *Chlamydomonas reinhardtii*, ensuring multi-functional capabilities of manipulating microscale targets, disrupting aggregates, and pointing out perspectives for the minimally invasive performance of biomedical interventions [154].

Another avenue toward overcoming the limitations of single-motor propulsion and control comes from collective behavior and swarm robotics. Nanozyme-based swarm robots using Z-scheme heterojunctions have demonstrated enhanced photocatalytic propulsion, cooperative hydrodynamics, and depth-resolved tumor penetration with significant antitumor responses via the generation of reactive oxygen species [153]. In a related direction, enzyme-powered nanomotors using urease have shown enhanced translational motion and the capability to mechanically disrupt extracellular matrix barriers upon optical stimulation, enabling improved drug penetration and hierarchical delivery of secondary therapeutic agents [36]. Reviews on collective micro/nanorobotics highlighted emergent adaptability, robustness, and scalability, hence allowing design guidelines for clinical-grade swarms with coordinated functionalities [141].

Parallel progresses of nanomotors toward neural interfacing and smart implantable systems further illustrate opportunities and challenges. Nanomaterials engineered with finely tuned electronic properties enable the creation of stable, electroactive interfaces with neural cells, for high-fidelity communication and long-term integration [66]. Soft, stimulus-responsive biohybrid nanointerfaces expand this capability to include improved mechanical compliance with neural tissue, allowing accurate neuromodulation while minimizing chronic implantation-related inflammation [35]. Innovation in concepts such as symbiotic, minimally invasive neural connectivity through Endomycorrhizae-like interfaces, inspired by natural biological networks, provides the opportunity for long-term application of therapies to neurodegenerative diseases [114]. Integration with AI enhances these systems through real-time neural signal analyses, adaptive control of stimulation, and data-driven coordination of nanorobot activities [116]. Advances in three-dimensional, flexible, and reconfigurable nanotechnologies continue to expand the design space toward next-generation neural implants and intelligent therapeutic devices [43].

The vision for 2035 now regards the creation of autonomous, adaptive nanomedicine ecosystems based on AI-driven nanoarchitectonics and nanorobotics. Thus, AI and ML methodologies resolve longstanding problems in the translation of nanomedicine by optimizing the physicochemical parameters of nanoparticles, size, surface chemistry, and drug-loading profiles, refining therapeutic efficiency and safety [28]. AI-driven nanoarchitectonics frameworks incorporate structural programming and stimuli-responsive design into intelligent delivery platforms, while improving targeting specificity and control of drug release [10]. AI enhances risk prediction, nanotoxicity modeling, and regulatory compliance by simulation of complex nanoparticle-biological interactions and reduces preclinical evaluation pipelines by doing so. This allays old bottlenecks in personalized medicine understanding whereby AI-powered systems can analyze biomarker datasets from individual patients, design tailored nanocarriers, and orchestrate multi-agent nanorobot interventions with unparalleled precision [88]. All these together promise that in the future, nanomotors/nanorobots positioned in integrated therapeutic ecosystems will be able to make decisions autonomously, adapt, and optimize performance at the level of personalized therapy.

11 Conclusion

Self-propelled nanomotors have emerged as a powerful technological paradigm capable of reshaping diagnostics, targeted therapy, and minimally invasive biomedical intervention. Their ability to actively navigate complex physiological environments, penetrate biological barriers, and perform site-specific therapeutic or diagnostic operations provides a distinct advantage over passive nanocarriers. Advances in propulsion mechanisms, from catalytic and enzymatic to magnetic, optical, and acoustic actuation, combined with innovative Janus architectures, degradable organic frameworks, and biohybrid materials, have significantly expanded their functional versatility and biocompatibility. Furthermore, the integration of AI, microfluidics, and IoMT platforms has enabled the development of autonomous, closed-loop medical systems with real-time monitoring, adaptive control, and precision targeting. Despite remarkable progress, several key challenges must be addressed before widespread clinical adoption becomes feasible. These include scalable and reproducible manufacturing, long-term biosafety, immune compatibility, propulsion stability in vivo, and robust regulatory frameworks tailored to active nanosystems. Continued interdisciplinary collaborations such as spanning materials science, nanorobotics, computational modeling, imaging, and clinical medicine, will be essential to bridge existing gaps. As the field advances toward intelligent, adaptive nanomedical ecosystems, self-propelled nanomotors hold the potential to redefine personalized, minimally invasive, and data-driven healthcare.

Acknowledgements All the authors are thankful to their respective Universities and Institutions for their valuable supports.

Data Availability The authors confirm that the data supporting the findings of this study are available within the book/chapter.

References

1. Abidharini JD, Souparnika BR, Elizabeth J, Bavyataa V, Nivedha R, Umamaheswari A, Vignesh K, Balamuralikrishnan B, Arun M, Anand AV (2026) Silicene and germanene for

biosensing and imaging application. In: Singh RS, Deshmukh K, Hussain CM (eds) Two-dimensional nanomaterials for biosensing and imaging applications. https://doi.org/10.1002/9781394200375.ch16
2. Adeniyi M, Ayoola VB, Samuel TE, Awosan W (2024) Artificial intelligence-driven wearable electronics and smart nanodevices for continuous cancer monitoring and enhanced diagnostic accuracy. Int J Sci Res Sci Technol 3(11):3–18
3. Ahmed D, Sukhov A, Hauri D, Rodrigue D, Maranta G, Harting J, Nelson BJ (2021) Bioinspired acousto-magnetic microswarm robots with upstream motility. Nat Mach Intell 3(2):116–124
4. Andhari SS, Wavhale RD, Dhobale KD, Tawade BV, Chate GP, Patil YN, Banerjee SS (2020) Self-propelling targeted magneto-nanobots for deep tumor penetration and pH-responsive intracellular drug delivery. Sci Rep 10(1):4703
5. Arque X, Torres MD, Patino T, Boaro A, Sanchez S, de la Fuente-Nunez C (2022) Autonomous treatment of bacterial infections in vivo using antimicrobial micro-and nanomotors. ACS Nano 16(5):7547–7558
6. Arthi B, Abidharini JD, Arthi G, Girigoswami A, Gowtham P, Catharine B, Younis M, Vijaya Anand A (2024) Food microbiology and the application of nanotechnology in food processing and safety. In: Javed R, Chen JT, Khalil AT (eds) Nanomaterials for biomedical and bioengineering applications. Springer, Singapore. https://doi.org/10.1007/978-981-97-0221-3_14
7. Arya A, Yashan N (2024) The nanorevolution in medicine: advancing drug delivery systems with nanomaterials and nanodevices. https://doi.org/10.31224/4266
8. Awasthi S, Nalinan A, Khandelwal LN, Maheshwari D (2024) Nanorobitcs: miniature machines with mighty potential–exploring the building blocks, crafting techniques, and transformative real-world implementation. In: Artificial intelligence in the age of nanotechnology. IGI Global Scientific Publishing, pp 136–174
9. Aziz A, Holthof J, Meyer S, Schmidt OG, Medina-Sánchez M (2020) In vivo imaging of swimming micromotors using hybrid high-frequency ultrasound and photoacoustic imaging. bioRxiv, 2020–06.
10. Bae H, Ji H, Konstantinov K, Sluyter R, Ariga K, Kim YH, Kim JH (2025) Artificial intelligence-driven nanoarchitectonics for smart targeted drug delivery. Adv Mater 37(42):e10239
11. Bai RG, Muthoosamy K, Manickam S (2015) Nanomedicine in theranostics. In: Nanotechnology applications for tissue engineering. William Andrew Publishing, pp 195–213
12. Campuzano S, Kagan D, Orozco J, Wang J (2011) Motion-driven sensing and biosensing using electrochemically propelled nanomotors. Analyst 136(22):4621–4630
13. Castillo JIR, Sánchez AAL (2025) Biomedical devices and microfluidics: development of lab-on-a-chip systems, biosensors and diagnostic devices with applications in clinical and point-of-care settings. eVitroKhem: J Innovat Biomed Chem Eng (4), 9
14. Ceylan H, Yasa IC, Yasa O, Tabak AF, Giltinan J, Sitti M (2018) 3D-printed biodegradable microswimmer for drug delivery and targeted cell labeling. BioRxiv, 379024.
15. Chang WC, Hawkes EA, Kliot M, Sretavan DW (2007) In vivo use of a nanoknife for axon microsurgery. Neurosurgery 61(4):683–692
16. Chen H, Zheng Y, Li J, Li L, Wang X (2023) AI for nanomaterials development in clean energy and carbon capture, utilization and storage (CCUS). ACS Nano 17(11):9763–9792
17. Chen S, Chen Y, Fu M, Cao Q, Wang B, Chen W, Ma X (2022) Active nanomotors surpass passive nanomedicines: current progress and challenges. J Mater Chem B 10(37):7099–7107
18. Chen S, Fan D, Fischer P, Ghosh A, Göpfrich K, Golestanian R, Sanchez S (2025a) Nanomotors: 20 years anniversary and future roadmap. https://doi.org/10.26434/chemrxiv-2025-lczw9
19. Chen S, Prado-Morales C, Sánchez-deAlcázar D, Sánchez S (2024) Enzymatic micro/nanomotors in biomedicine: from single motors to swarms. J Mater Chem B 12(11):2711–2719

20. Chen X, Cao Q, Liang Z, Huang L, Wang J, Hu Y (2024) Hollow magnetic nanocarrier-based microrobot swarms for NIR-responsive targeted drug delivery and synergistic therapy. ACS Appl Mater Interfaces 16(44):60874–60883
21. Chen X, Ran B, Wu H, Zeng G (2025b) Micro/Nanomotors moving at the boundary: from progress to future directions. Adv Funct Mater e04234
22. Chen X, Xu Z, Li T, Thakur A, Wen Y, Zhang K, Yan Y (2024) Nanomaterial-encapsulated STING agonists for immune modulation in cancer therapy. Biomarker Res 12(1):2
23. Chen Y, Gonçalves JM, Ferrer Campos R, Villa K (2025) Dual-energy integration in photoresponsive micro/nanomotors: from strategic design to biomedical applications. Small 21(6):2410901
24. Chen Y, Jia M, Fan H, Duan J, Fu J (2025) Advances in architectural design, propulsion mechanisms, and applications of asymmetric nanomotors. Nanomaterials 15(17):1333
25. Cho Y, Choi Y, Jang Y, Seong H (2025). Nanomaterial-enhanced biosensing: mechanisms and emerging applications. *Advanced Healthcare Materials*, 2500189.
26. Chohan DP, Dey B, Tarkunde A, Vyas V, De Sarkar S, Sundara BK (2024) Advancing autonomous nanomedicine: bridging the gap from concept to potential clinical studies. J Cluster Sci 35(8):2607–2635
27. Choi H, Jeong SH, Simó C, Bakenecker A, Liop J, Lee HS, Hahn SK (2024) Urease-powered nanomotor containing STING agonist for bladder cancer immunotherapy. Nat Commun 15(1):9934
28. Chou WC, Canchola A, Zhang F, Lin Z (2025) Machine learning and artificial intelligence in nanomedicine. Wiley Interdiscip Rev: Nanomedicine Nanobiotechnology 17(4):e70027
29. Chung M, Achouri K, Martin OJ (2022) Optically driven nano-motors designed by machine learning. In: Optical trapping and optical micromanipulation XIX (p. PC121980M). SPIE
30. Cichos F, Landin SM, Pradip R (2023). Artificial intelligence (AI) enhanced nanomotors and active matter. In: Intelligent nanotechnology. Elsevier, pp 113–144
31. Claussen J, Medintz IL (2012) Using nanotechnology to improve lab on a chip devices. Biochip Tissue Chip 2(4):1609–1614
32. Cupples G, Dyson RJ, Smith DJ (2017) Viscous propulsion in active transversely isotropic media. J Fluid Mech 812:501–524
33. Díez P, Lucena-Sánchez E, Escudero A, Llopis-Lorente A, Villalonga R, Martinez-Manez R (2021) Ultrafast directional Janus Pt–mesoporous silica nanomotors for smart drug delivery. ACS Nano 15(3):4467–4480
34. Elqassas R, Alrakhawi HA, Elsobeihi MM, Habil B, Abu-Nasser BS, Abu-Naser SS (2024) Convergence of nanotechnology and artificial intelligence: revolutionizing healthcare and beyond. Int J Acad Eng Res 8(10):25–30
35. Filippi M, Balciunaite A, Katzschmann R (2024) Biohybrid nanointerfaces for neuromodulation. Nano Today 54:102094
36. Fraire JC, Guix M, Hortelao AC, Ruiz-González N, Bakenecker AC, Ramezani P, Sánchez S (2023) Light-triggered mechanical disruption of extracellular barriers by swarms of enzyme-powered nanomotors for enhanced delivery. ACS Nano 17(8):7180–7193
37. Fu D, Wang Z, Tu Y, Peng F (2021) Interactions between biomedical micro-/nano-motors and the immune molecules, immune cells, and the immune System: challenges and opportunities. Adv Healthcare Mater 10(7):2001788
38. Gamucci O, Bertero A, Gagliardi M, Bardi G (2014) Biomedical nanoparticles: overview of their surface immune-compatibility. Coatings 4(1):139–159
39. Garcia-Gradilla V, Orozco J, Sattayasamitsathit S, Soto F, Kuralay F, Pourazary A, Wang J (2013) Functionalized ultrasound-propelled magnetically guided nanomotors: toward practical biomedical applications. ACS Nano 7(10):9232–9240
40. Ghani MW, Iqbal A, Ghani H, Bibi S, Wang Z, Pei R (2023) Recent advances in nanocomposite-based delivery systems for targeted CRISPR/Cas delivery and therapeutic genetic manipulation. J Mater Chem B 11(24):5251–5271
41. Ghosh S, Mohajerani F, Son S, Velegol D, Butler PJ, Sen A (2019) Motility of enzyme-powered vesicles. Nano Lett 19(9):6019–6026

42. Gompper G, Stone HA, Kurzthaler C, Saintillan D, Peruani F, Fedosov DA, Drescher K (2025) The 2025 motile active matter roadmap. J Phys: Condens Matter 37(14):143501
43. Gracias DH (2022) Integrated nanotechnology 2.0: 3D, smart, flexible, and dynamic [Highlights]. IEEE Nanotech Magaz 16(1):11–15
44. Guo Z, Liu J, Wang DW, Xu J, Liang K (2020) Biofriendly micro/nanomotors operating on biocatalysis: from natural to biological environments. Biophys Rep 6(5):179–192
45. Guzmán E, Maestro A (2022) Synthetic micro/nanomotors for drug delivery. Technologies 10(4):96
46. Hakeem L, Saddiqa A, Rafique A, Ullah MK, Abdullah H, Abbas S, Akaml H (2022) A systematic review: advanced magnetic nanotheranostics. Compr Res Rev Eng Technol 1(1):024–028
47. He T, Yang Y, Chen X (2024) A lifetime of catalytic micro-/nanomotors. Nanomaterials 15(1):13
48. Hortelao AC, Simó C, Guix M, Guallar-Garrido S, Julián E, Vilela D, Sánchez S (2020) Monitoring the collective behavior of enzymatic nanomotors in vitro and in vivo by PET-CT. bioRxiv, 2020–06
49. Hu JQ, Zhao R, Cui RF, Kou JL, Chen JX (2024) Dynamics of nanomotors propelled by enzyme cascade reactions. Int J Mol Sci 25(23):12586
50. Huang S, Liu B, Peng F, Wang Y, Han S, Tu Y (2025) Micro/nanomotors actively penetrate physiological barriers: basic nanoarchitectonics to practical application. Adv Mater e08865
51. Ibrahim HK (2024) From nanotech to AI: the cutting-edge technologies shaping the future of medicine. Afr J Adv Pure Appl Sci (AJAPAS), 410–427.
52. Idoko DO, Adenyi M, Senejani MN, Erondu OF, Adeyeye Y (2024) Nanoparticle-assisted cancer imaging and targeted drug delivery for early-stage tumor detection and combined diagnosis-therapy systems for improved cancer management. Int J Innov Sci Res Technol 9(11):2456–2165
53. Idumah CI (2023) Multifunctional properties optimization and stimuli-responsivity of shape memory polymeric nanoarchitectures and applications. Polym Eng Sci 63(7):1857–1873
54. Ioannidis JP, Kim BY, Trounson A (2018) How to design preclinical studies in nanomedicine and cell therapy to maximize the prospects of clinical translation. Nat Biomed Eng 2(11):797–809
55. Jananisri A, Geofferina IP, Vishalini G, Sridhar S, Vimalraj P, Suruthi GSB, Sujitha S, Balamuralikrishnan B, Arun M, Anand AV (2026) 2D pnictogens for biosensing and imaging applications. In: Singh RS, Deshmukh K, Hussain CM (eds) Two-dimensional nanomaterials for biosensing and imaging applications. https://doi.org/10.1002/9781394200375.ch17
56. Jiang C, Liu X, Li Z, Zeng L, Wang R, Ding H, Zheng L (2023) Protein-gold nanoclusters loaded with ferrous ions for inducing ferroptosis of hepatoma cells. Adv Mater Technol 8(9):2201643
57. Jiang Z, Fu L, Wei C, Fu Q, Pan S (2023) Antibacterial micro/nanomotors: advancing biofilm research to support medical applications. J Nanobiotechnol 21(1):388
58. Jin Z, Nguyen KT, Go G, Kang B, Min HK, Kim SJ, Choi E (2019) Multifunctional nanorobot system for active therapeutic delivery and synergistic chemo-photothermal therapy. Nano Lett 19(12):8550–8564
59. Karpagam T, Balamuralikrishnan B, Varalakshmi B, Anand AV, Sugunabai J (2022) An insight on emerging nanomaterials for the delivery of various nutraceutical applications for the betterment of heath. In: Krishnan A, Ravindran B, Balasubramanian B, Swart HC, Panchu SJ, Prasad R (eds) Emerging nanomaterials for advanced technologies. nanotechnology in the life sciences. Springer, Cham. https://doi.org/10.1007/978-3-030-80371-1_1
60. Kaushik S, Soni V, Skotti E (eds) (2022) Nanosensors for futuristic smart and intelligent healthcare systems. CRC Press, Boca Raton, FL
61. Khatri N, Kapral R (2024) Clustering of chemically propelled nanomotors in chemically active environments. Chaos: An Interdisc J Nonlinear Sci 34(3):033103.
62. Kim S, Baek S, Sluyter R, Konstantinov K, Kim JH, Kim S, Kim YH (2023) Wearable and implantable bioelectronics as eco-friendly and patient-friendly integrated nanoarchitectonics for next-generation smart healthcare technology. EcoMat 5(8):e12356

63. Kim TH, Lee S, Chen X (2013) Nanotheranostics for personalized medicine. Expert Rev Mol Diagn 13(3):257–269
64. Komiyama M, Sumaoka J (2023) Nanoarchitectures to deliver nucleic acid drugs to disease sites. ChemNanomat 9(6):e202300069
65. Kong X, Gao P, Wang J, Fang Y, Hwang KC (2023) Advances of medical nanorobots for future cancer treatments. J Hematol Oncol 16(1):74
66. Kotov NA, Winter JO, Clements IP, Jan E, Timko BP, Campidelli S, Ballerini L (2009) Nanomaterials for neural interfaces. Adv Mater 21(40):3970–4004
67. Koucheryavy Y, Yastrebova A, Martins DP, Balasubramaniam S (2021) A review on bio-cyber interfaces for intrabody molecular communications systems. arXiv preprint arXiv:2104.14944.
68. Kumar B, Anand S, Singh N, Singh R, Banerjee P, Kujur AGP (2024) Combining artificial intelligence with nanoscale energy systems to achieve highly efficient energy harvesting and utilization. In: 2024 international conference on intelligent systems and advanced applications (ICISAA)). IEEE, pp 1–7
69. Lauga E (2020) Traveling waves are hydrodynamically optimal for long-wavelength flagella. Phys Rev Fluids 5(12):123101
70. Li H, Li Y, Liu J, He Q, Wu Y (2022) Asymmetric colloidal motors: from dissymmetric nanoarchitectural fabrication to efficient propulsion strategy. Nanoscale 14(20):7444–7459
71. Li J, Li T, Xu T, Kiristi M, Liu W, Wu Z, Wang J (2015) Magneto–acoustic hybrid nanomotor. Nano Lett 15(7):4814–4821
72. Li L, Wu Z, Yang Y, Hu P, Gao W, Wang LV (2020) Photoacoustic computed tomography guided microrobots for targeted navigation in intestines in vivo. In: Photons plus ultrasound: imaging and sensing 2020, vol 11240. SPIE
73. Li T, Li L, Zhang G, Zhou D, Chang X (2016) Propulsion mechanisms and applications of multiphysics-driven micro-and nanomotors. Chin Sci Bull 62:122–135
74. Li WW, Yu ZL, Jia J (2025) Urease-powered micro/nanomotors: Current progress and challenges. J Pharmaceut Anal 15(3):101095
75. Lin X, Wu Z, He Q (2017) Controlled molecular assembly toward self-propelled micro-/nanomotors. In: Supramolecular chemistry of biomimetic systems. Springer Singapore, Singapore, pp 259–281
76. Liu J, Detrembleur C, Mornet S, Jérôme C, Duguet E (2015) Design of hybrid nanovehicles for remotely triggered drug release: an overview. J Mater Chem B 3(30):6117–6147
77. Liu L, Gao J, Wilson DA, Tu Y, Peng F (2019) Fuel-free micro-/nanomotors as intelligent therapeutic agents. Chem–An Asian J 14(14):2325–2335
78. Liu M, Tu B, Liu L, Chen B, Tu Y (2020) Application of self-propelled micro-/nanomotors in active targeted drug delivery. J Southern Med Univ 40(3):445–452
79. Liu M, Zhang R, Chen BB, Tian Z, Ma F, Luo Y, Zhao Z (2025) NIR-II imaging-guided self-penetrating nanomotors for millimeter-scale deep thrombolysis tracking and accelerated embolus clearance. Adv Mater e12075
80. Liu S, Gao C, Peng F (2022) Micro/nanomotors in regenerative medicine. Mater Today Adv 16:100281
81. Liu W, Liu Y, Li H, Nie H, Tian M, Long W (2023) Biomedical micro-/nanomotors: design, imaging, and disease treatment. Adv Func Mater 33(15):2212452
82. Llacer-Wintle J, Rivas-Dapena A, Chen XZ, Pellicer E, Nelson BJ, Puigmartí-Luis J, Pané S (2021) Biodegradable small-scale swimmers for biomedical applications. Adv Mater 33(42):2102049
83. Lu L, Zhao H, Lu Y, Zhang Y, Wang X, Fan C, Wu Z (2024) Design and control of the magnetically actuated micro/nanorobot swarm toward biomedical applications. Adv Healthcare Mater 13(15):2400414
84. Mangla B, Kumar P, Javed S, Pathan T, Ahsan W, Aggarwal G (2025) Regulating nanomedicines: challenges, opportunities, and the path forward. Nanomedicine 20(15):1911–1927

85. Manoj M, Arockia Doss Cible AD, Sneha R, Gopika, Vanmathi R, Rajesh M, Vaishnavi, G, Vijaya Anand A (2024) Pnictogens: bridging the gap in biomedical advancements. In: Javed R, Chen JT, Khalil AT (eds) Nanomaterials for biomedical and bioengineering applications. Springer, Singapore. https://doi.org/10.1007/978-981-97-0221-3_16
86. Mao M, Wu Y, He Q (2024) Breaking through physiological barriers: nanorobotic strategies for active drug delivery. Bioconjug Chem 36(1):1–14
87. Mayerhoefer E, Parajuli H, Cimpan MR, Costea DE, Dongre HN, Krueger A (2025) Zwitterionic dipeptide surface functionalization of detonation nanodiamond for enhanced control in biological environments. Angewandte Chemie Int Edn e202501202
88. Mazumdar H, Khondakar KR, Das S, Halder A, Kaushik A (2025) Artificial intelligence for personalized nanomedicine; from material selection to patient outcomes. Expert Opin Drug Deliv 22(1):85–108
89. Medina-Sánchez M, Xu H, Schmidt OG (2018) Micro-and nano-motors: the new generation of drug carriers. Ther Deliv 9(4):303–316
90. Meng W, Huang L, Guo J, Xin Q, Liu J, Hu Y (2024) Innovative nanomedicine delivery: targeting tumor microenvironment to defeat drug resistance. Pharmaceutics 16(12):1549
91. Murugan R (2024) Nanorobots and its advancements in medical field. i-Manager's J Augmented Virtual Real (JAVR), 2(1):31
92. Muthukrishnan S, Anand AV, Palanisamy K, Gunasangkaran G, Ravi AK, Balasubramanian B (2022a) Novel organic and inorganic nanoparticles as a targeted drug delivery vehicle in cancer treatment. In: Krishnan A, Ravindran B, Balasubramanian B, Swart HC, Panchu SJ, Prasad R (eds) Emerging nanomaterials for advanced technologies. nanotechnology in the life sciences. Springer, Cham. https://doi.org/10.1007/978-3-030-80371-1_4
93. Muthukrishnan S, Anand AV, Palanisamy K, Gunasangkaran G, Ravi AK, Balasubramanian B (2022b) Novel organic and inorganic nanoparticles as a targeted drug delivery vehicle in cancer treatment. In: Krishnan A, Ravindran B, Balasubramanian B, Swart HC, Panchu SJ, Prasad R (eds) Emerging nanomaterials for advanced technologies. nanotechnology in the life sciences. Springer, Cham. https://doi.org/10.1007/978-3-030-80371-1_4
94. Navaraj W, Smith C, Dahiya R (2020) E-skin and wearable systems for health care. In: Wearable Bioelectronics. Elsevier, pp 133–178
95. Nesakumar N, Kesavan S, Li CZ, Alwarappan S (2019) Microfluidic electrochemical devices for biosensing. J Anal Test 3(1):3–18
96. Nidhi V, Allaire A, Ait Athmane Z, Guenoun P, Testard F, Renault JP, Malloggi F (2024) Making mobile nanotechnology accessible: is the explicit preparation of janus nanoparticle necessary to achieve mobility? Nanomaterials 14(22):1796
97. Panday R, Abdalla AM, Neupane M, Khadka S, Kricha A, Yang G (2021) Advances in magnetic nanoparticle-driven delivery of gene therapies towards prostate cancer. J Nanomater 2021(1):6050795
98. Pijpers IA, Cao S, Llopis-Lorente A, Zhu J, Song S, Joosten RR, Abdelmohsen LK (2020) Hybrid biodegradable nanomotors through compartmentalized synthesis. Nano Lett 20(6):4472–4480
99. Pinki K, Agarwal J, Gupta PK, Chauhan NS, Srivastava K, Gupta A (2025) Preclinical evaluation of nanocarriers: in vitro, ex vivo, and in vivo models. Nano-Brain Revolution in Alzheimer's disease: a new era of Brain Drug Delivery, 4
100. Poornima AP, Ankur TD, Priyanga P, Preethi RA, Nihala S, amalesh, M, Elamathi S, Al-Dhabi NA, Mariadhas VA, Anand AV (2026) Polymer nanocomposites of 2D nanomaterials for biosensing and imaging applications. In: Singh RS, Deshmukh K, Hussain CM (eds) Two-dimensional nanomaterials for biosensing and imaging applications. https://doi.org/10.1002/9781394200375.ch19
101. Preetam S, Lata S, Bora J, Thapliyal S, Mehta S, Ramniwas S, Malik S (2025) Exosome-driven biohybrid nanorobots: bridging nature and nanotechnology in biomedical innovation. RSC Adv 15(40):33390–33409
102. Qin YS, Yi J, Chen YJ, Zhang W, Tang SF (2024) Recent advances in micro/nanomotor for the therapy and diagnosis of atherosclerosis. ACS Appl Mater Interfaces 17(8):11443–11468

103. Rajendran S, Sundararajan P, Awasthi A, Rajendran S (2023) Nanorobotics in medicine: a systematic review of advances, challenges, and future prospects. ArXiv preprint arXiv:2309.10881.
104. Ramachandran RV, Barman A, Modak P, Bhat R, Ghosh A, Saini DK (2022) How safe are magnetic nanomotors: from cells to animals. Biomater Adv 140:213048
105. Ranjani PS, Sangeetha S, Prakaash KS, Damodharan N (2025) Nanorobots: trailblazing the future of pharmaceuticals through targeted therapy and disease monitoring. Recent Adv Drug Delivery Formul
106. Rastmanesh A, Yaraki MT, Wu J, Wang Z, Ghoderao P, Gao Y, Tan YN (2021) Bioinspired micro/nanomotors towards a self-propelled noninvasive diagnosis and treatment of cancer. Mol Syst Des Eng 6(8):566–593
107. Ren L, Wang W, Mallouk TE (2018) Two forces are better than one: combining chemical and acoustic propulsion for enhanced micromotor functionality. Acc Chem Res 51(9):1948–1956
108. Rengarajan RL, Rathinam A, Suganthy N, Balamuralikrishnan B, Anand AV, Velayuthaprabhu S (2022) Perspectives of metals and metal oxide nanoparticles for antimicrobial consequence—an overview. In: Krishnan A, Ravindran B, Balasubramanian B, Swart HC, Panchu SJ, Prasad R (eds) Emerging nanomaterials for advanced technologies. nanotechnology in the life sciences. Springer, Cham. https://doi.org/10.1007/978-3-030-80371-1_13
109. Ricotti L, Menciassi A, Morishima K (2012) Guest editorial introduction to the Special Issue on bio-hybrid systems and living machines. Biomed Microdev 14(6):965–967
110. Robertson B, Kapral R (2015) Nanomotor dynamics in a chemically oscillating medium. J Chem Phys 142(15):154902
111. Robertson B, Huang MJ, Chen JX, Kapral R (2018) Synthetic nanomotors: working together through chemistry. Acc Chem Res 51(10):2355–2364
112. Rodriguez H (2025) Terminological rigor in nanomedicine: standardizing'nano'descriptors for reproducibility and translational success. Bionat J 2(2):1–4
113. Salehi A, Hosseinpour S, Tabatabaei N, Soltani Firouz M, Zadebana N, Nauber R, Medina-Sánchez M (2024) Advancements in machine learning for microrobotics in biomedicine. Adv Intell Syst 2400458
114. Saniotis A, Henneberg M, Sawalma AR (2018) Integration of nanobots into neural circuits as a future therapy for treating neurodegenerative disorders. Front Neurosci 12:153
115. Sanità G, Carrese B, Lamberti A (2020) Nanoparticle surface functionalization: how to improve biocompatibility and cellular internalization. Front Mol Biosci 7:587012
116. Seliverstov DP (2024) Nanorevolution in medicine: synergy of nanotechnology, artificial intelligence and digital innovation. Meditsinskaya sestra 26(7):44–48
117. Sengar A (2024) Precision in practice: nanotechnology and targeted therapies for personalized care. https://doi.org/10.20944/preprints202412.0019.v1
118. Shahhoessini M, Spitzner JR, Lucas CR, Castro CE, Halley PD (2024) Exploring the potential of targeted DNA Nanobot delivery systems for gene delivery. Cancer Res 84(6_Supplement):7238–7238
119. Shankarappa SA, Koyakutty M, Nair SV (2014) Efficacy versus toxicity-The ying and yang in translating nanomedicines. Nanomater Nanotechnol 4:23
120. Sharma B, Chauhan I, Kumar G, Bhardwaj K, Tiwari RK (2024) Nanoparticle carriers: a new era of precise CRISPR/Cas9 gene editing. MicroRNA 14(2):101–111
121. Sharma HK, Tomar R (2022) IoMT and blockchain-based intelligent and secured system for smart healthcare. In: Mathematical modeling for intelligent systems. Chapman and Hall/CRC, pp 61–73
122. Sheikh A, Kesharwani P, Almalki WH, Almujri SS, Dai L, Chen ZS, Gao F (2024) Understanding the novel approach of nanoferroptosis for cancer therapy. Nano-Micro Lett 16(1):188
123. Shrivastava P, Aanjankumar S, Dhanaraj RK, Maragatharajan M, Saravanan S (2025) AI-powered wearable biosensors using openpose: transforming personalized healthcare activity monitoring of alzheimer patients. In: Proceedings of the 5th international electronic conference on biosensors. MDPI. Basel, Switzerland

124. Singh B (2025). Cruising deep convolution neural networks (DCNN) and internet of medical things (IoMT) fostering smart healthcare: runway towards quality of life and wellbeing. In: Responsible AI for digital health and medical analytics. IGI Global Scientific Publishing, pp 211–240
125. Singh B, Kaunert C (2024) Aroma of highly smart internet of medical things (IoMT) and lightweight edgetrust expansion medical care facilities for electronic healthcare systems: Fortified-chain architecture for remote patient monitoring and privacy protection beyond imagination. In: Lightweight digital trust architectures in the internet of medical things (IoMT). IGI Global, pp. 196–212
126. Somasundar A, Sen A (2021) Chemically propelled nano and micromotors in the body: quo vadis? Small 17(5):2007102
127. Souto EB, Blanco-Llamero C, Krambeck K, Kiran NS, Yashaswini C, Postwala H, Maheshwari R (2024) Regulatory insights into nanomedicine and gene vaccine innovation: safety assessment, challenges, and regulatory perspectives. Acta Biomater 180:1–17
128. Sridhar V, Yildiz E, Rodríguez-Camargo A, Lyu X, Yao L, Wrede P, Sitti M (2023) Designing covalent organic framework-based light-driven microswimmers toward therapeutic applications. Adv Mater 35(25):2301126
129. Sun S, Wang M, Jiang H, Zhang Y, Qiao H, Zhang TY (2024) Optimizing nanoporous metallic actuators through multiscale calculations and machine learning. J Mech Phys Solids 187:105611
130. Sun X, Zhang Y, Li J, Park KS, Han K, Zhou X, Moon JJ (2021) Amplifying STING activation by cyclic dinucleotide–manganese particles for local and systemic cancer metalloimmunotherapy. Nat Nanotechnol 16(11):1260–1270
131. Sunayana SRV, Pathade P, Tambe BR (2025) Stimuli-responsive nanocarriers: revolutionizing site-specific drug release. Med Pharmaceut J 4(2):79–99
132. Tang M, Ni J, Yue Z, Sun T, Chen C, Ma X, Wang L (2024) Polyoxometalate-nanozyme-integrated nanomotors (POMotors) for self-propulsion-promoted synergistic photothermal-catalytic tumor therapy. Angew Chem Int Ed 63(6):e202315031
133. Tao J, Li C, Zheng Y, Wang F, Zhang M, Wu X, Fei W (2021) Biological protein mediated ferroptotic tumor nanotherapeutics. J Mater Chem B 9(45):9262–9284
134. Tian Z, Wang Y, Chen Y, Xu B, Di Z, Mei Y (2020) Inorganic stimuli-responsive nanomembranes for small-scale actuators and robots. Adv Intell Syst 2(2):1900092
135. Tierno P, Güell O, Sagués F, Golestanian R, Pagonabarraga I (2010) Controlled propulsion in viscous fluids of magnetically actuated colloidal doublets. Phys Rev E—Statistical, Nonlinear, Soft Matter Phys 81(1):011402
136. Tu Y, Peng F, Sui X, Men Y, White PB, van Hest JC, Wilson DA (2017) Self-propelled supramolecular nanomotors with temperature-responsive speed regulation. Nat Chem 9(5):480–486
137. Tu Y, Peng F, White PB, Wilson DA (2017) Self-destroyed redox-sensitive stomatocyte nanomotor. Angew Chem Int Ed 56:7620–7624
138. Varalakshmi B, Karpagam T, Anand AV, Balamuralikrishnan B (2022) Nanoscale smart drug delivery systems and techniques of drug loading to nanoarchitectures. In: Krishnan A, Ravindran B, Balasubramanian B, Swart HC, Panchu SJ, Prasad R (eds) Emerging nanomaterials for advanced technologies. nanotechnology in the life sciences. Springer, Cham. https://doi.org/10.1007/978-3-030-80371-1_2
139. Villa K, Pumera M (2019) Fuel-free light-driven micro/nanomachines: artificial active matter mimicking nature. Chem Soc Rev 48(19):4966–4978
140. Vishnu PG, Suruthi GS, Nivedha R, Oviya M, Sivaprakash V, Murugan M, Preethi AR, Vijaya Anand A (2024) Polymer nanocomposites of 2D nanomaterials for biosensing and imaging applications. In: Javed R, Chen JT, Khalil AT (eds) Nanomaterials for biomedical and bioengineering applications. Springer, Singapore. https://doi.org/10.1007/978-981-97-0221-3_15
141. Wang B, Lu Y (2024) Multi-dimensional micro/nanorobots with collective behaviors. SmartMat 5(5):e1263

142. Wang C, He T, Zhou H, Zhang Z, Lee C (2023) Artificial intelligence enhanced sensors-enabling technologies to next-generation healthcare and biomedical platform. Bioelectron Med 9(1):17
143. Wang J (2017) U.S. Patent No. 9,698,708. U.S. Patent and Trademark Office, Washington, DC
144. Wang P, Al Azad MAR, Yang X, Martelli PR, Cheung KY, Shi J, Shen Y (2021) Self-adaptive and efficient propulsion of Ray sperms at different viscosities enabled by heterogeneous dual helixes. Proc Natl Acad Sci 118(23):e2024329118
145. Wang W, Zhou C (2021) A journey of nanomotors for targeted cancer therapy: principles, challenges, and a critical review of the state-of-the-art. Adv Healthcare Mater 10(2):2001236
146. Wang X, Qin Q, Lu Y, Mi Y, Meng J, Zhao Z, Wang N (2023) Smart triboelectric nanogenerators based on stimulus-response materials: from intelligent applications to self-powered systems. Nanomaterials 13(8):1316
147. Wang X, Qin XH, Hu C, Terzopoulou A, Chen XZ, Huang TY, Nelson BJ (2018) 3D printed enzymatically biodegradable soft helical microswimmers. Adv Func Mater 28(45):1804107
148. Wang Z, Tu Y, Chen Y, Peng F (2020) Emerging micro/nanomotor-based platforms for biomedical therapy. Adv Intell Syst 2(5):1900081
149. Wang Z, Wang Y, He Z, Liu C (2025) Emerging cGAS-STING agonist-based nanotherapeutics: mechanistic insights and applications in cancer combination therapy. Adv Sci 12(36):e09890
150. Watson B, Friend J, Yeo L (2009) Micromotor of less than 1 mm^3 volume for in vivo medical procedures. In: 2009 third international conference on quantum, nano and micro technologies. IEEE, pp 81–85
151. Wu Y, Liu Q (2025) Nanocarrier-enabled STING agonist delivery for enhanced cancer immunotherapy. J Mater Chem B 13(39):12382–12401
152. Wu Z, Li L, Yang Y, Hu P, Li Y, Yang SY, Gao W (2019) A microrobotic system guided by photoacoustic computed tomography for targeted navigation in intestines in vivo. Sci Robot 4(32):eaax0613
153. Wu Z, Yue Z, Tang M, Wang S, Liu X, Sun T, Shang L (2025) Nanozyme-based robots swarm with exciton-engineered z-scheme heterojunctions for depth-resolved tumour penetration and therapy. https://doi.org/10.21203/rs.3.rs-6432792/v1
154. Xin H, Zhao N, Wang Y, Zhao X, Pan T, Shi Y, Li B (2020) Optically controlled living micromotors for the manipulation and disruption of biological targets. Nano Lett 20(10):7177–7185
155. Xing Y, Du X, Xu T, Zhang X (2020) Janus dendritic silica/carbon@ Pt nanomotors with multiengines for H_2O_2, near-infrared light and lipase powered propulsion. Soft Matter 16(41):9553–9558
156. Xu D, Wang Y, Liang C, You Y, Sanchez S, Ma X (2020) Self-propelled micro/nanomotors for on-demand biomedical cargo transportation. Small 16(27):1902464
157. Xu T, Gao W, Xu LP, Zhang X, Wang S (2017) Fuel-free synthetic micro-/nanomachines. Adv Mater 29(9):1603250
158. Xu W, Tian H, Song Y, Qin H, Gao J, Chen Y, Tu Y (2025) Self-propelled ferroptosis nanoinducer for enhanced cancer therapy. Int J Extrem Manuf 7(3):035501
159. Xuan M, Shao J, Gao C, Wang W, Dai L, He Q (2018) Self-propelled nanomotors for thermomechanically percolating cell membranes. Angew Chem Int Ed 57(38):12463–12467
160. Yadav SK, Nandakumar UP, Vinay BC, Dikkatar MS, Rawal KB (2025) Nanomedicine strategies to overcome multi-drug resistance in cancer: innovations in targeted delivery, tumor microenvironment modulation and synergistic therapies. https://doi.org/10.20944/preprints202507.0081.v1
161. Yahia EA, Elsharkawey AE, Elsharkawey RA (2025) Nano-AI synergy in nursing: a review of intelligent nanotechnologies for precision patient care. Nanotechnol Appl Sci J 1(2):11–22
162. Yan S, Liu J, Guo X, Ta D, Huang G, Solovev AA, Xu K (2021) Ultrafast ultrasound imaging for micro-nanomotors: a phantom study. In: 2021 IEEE international ultrasonics symposium (IUS). IEEE, pp 1–4

163. Yan X, Wang M, Zhao D (2024) Navigation of visually guided miniature magnetic microrobot: optimization of path planning and realtime feedback mechanisms. In: 2024 7th international conference on mechanical, control and computer engineering (ICMCCE). IEEE, pp 39–44
164. Yang Y, Bevan MA, Li B (2020) Micro/nano motor navigation and localization via deep reinforcement learning. Adv Theory Simuls 3(6):2000034
165. Yang Z, Liu S, Li L, Chen Q, Shi C, Dong L, Fukuda T (2025) From nano robotic manipulation to nano manipulation robot. SmartBot 1(3):e12021
166. Ye J, Fan Y, Niu G, Zhou B, Kang Y, Ji X (2024) Intelligent micro/nanomotors: fabrication, propulsion, and biomedical applications. Nano Today 55:102212
167. Ye Y, Luan J, Wang M, Chen Y, Wilson DA, Peng F, Tu Y (2019) Fabrication of self-propelled micro-and nanomotors based on Janus structures. Chem A Eur J 25(37):8663–8680
168. Ye Z, Che Y, Dai D, Jin D, Yang Y, Yan X, Ma X (2024) Supramolecular modular assembly of imaging-trackable enzymatic nanomotors. Angew Chem Int Ed 63(16):e202401209
169. Yu J, Zhang L (2019) Reconfigurable colloidal microrobotic swarm for targeted delivery. In: 2019 16th international conference on ubiquitous robots (UR). IEEE, pp 615–616
170. Yu L, Yang M, Guan J, Mou F (2023) Ultrasmall Fe2O3 tubular nanomotors: the first example of swarming photocatalytic nanomotors operating in high-electrolyte media. Nanomaterials 13(8):1370
171. Yu S, Li T, Wu S (2020) Controlled propulsion of asymmetric janus microdimer swimmers under rotating magnetic fields. In: 2020 15th IEEE conference on industrial electronics and applications (ICIEA). IEEE, pp 622–624
172. Yu X, Li X, Chen Q, Wang S, Xu R, He Y, Peng S (2024) High intensity focused ultrasound-driven nanomotor for effective ferroptosis-immunotherapy of TNBC. Advanced Science 11(15):2305546
173. Zafar S, Nazir M, Bakhshi T, Khattak HA, Khan S, Bilal M, Sabah A (2021) A systematic review of bio-cyber interface technologies and security issues for internet of bio-nano things. IEEE Access 9:93529–93566
174. Zavaleta C, Ho D, Chung EJ (2018) Theranostic nanoparticles for tracking and monitoring disease state. SLAS Technol: Transl Life Sci Innov 23(3):281–293
175. Zeng M, Yuan S, Huang D, Cheng Z (2019) Accelerated design of catalytic water-cleaning nanomotors via machine learning. ACS Appl Mater Interfaces 11(43):40099–40106
176. Zha F, Wang T, Luo M, Guan J (2018) Tubular micro/nanomotors: Propulsion mechanisms, fabrication techniques and applications. Micromachines 9(2):78
177. Zhan Z, Wei F, Zheng J, Yang W, Luo J, Yao L (2018) Recent advances of light-driven micro/nanomotors: toward powerful thrust and precise control. Nanotechnol Rev 7(6):555–581
178. Zhang D, Lin L, Deng C, Osman MS, Rodriguez PES, Han F, Wang L (2025) Advanced imaging strategies based on intelligent micro/nanomotors. Cyborg Bionic Syst 6:0384
179. Zhang G, Chen Y, Zhou W, Chen C, Liu Y (2023) Bioenergy-based closed-loop medical systems for the integration of treatment, monitoring, and feedback. Small sci 3(10):2300043
180. Zhang X, Wu L, Chen X, Liu C, Wang Y, Liu R, Wang B (2024) Light-directed self-powered metal-organic framework based nanorobots for deep tumor penetration. Adv Mater 2415121
181. Zhao K, Xu J, Song W, Cheng J, Al Mamun A, Tu Y, Zhang N (2025a) Recent advances in membrane-coated micro/nanomotors in biological applications. Int J Nanomed 9427–9446
182. Zhao Q, Lin L (2023) Modeling and simulation of a bio-inspired nanorobotic drug delivery system. In: International conference on bio-inspired information and communication technologies. Springer Nature Switzerland, Cham, pp 64–76
183. Zhao Y, He J, Liu Y, Soummane H, Ran P, Yang Q, Zhao L (2025) Ultrasound-actuated platelet mimetic nanomotors enable targeted piezocatalytic ROS storm for precision thrombolysis. J Nanobiotechnology 23(1):585
184. Zhu Y, Huang H, Zhao Q, Qin J (2024) Novel micro/nanomotors for tumor diagnosis and therapy: motion mechanisms, advantages and applications. J Sci: Adv Mater Devices 9(2):100718

Smart Nanotechnology for Precision Immunomodulation

Anita Ioana Visan and Irina Negut

Abstract Conventional immunomodulation suffers from systemic toxicity and a lack of spatiotemporal precision. Smart nanotechnology provides a paradigm-shifting solution by operating at the same scale as immune recognition itself. This chapter examines the sophisticated interplay between engineered nanomaterials—including lipid-based, polymeric, inorganic, and biomimetic platforms—and the immune system. We detail fundamental principles such as protein corona formation and cellular uptake, which dictate nanomaterial fate and function. The review then surveys cutting-edge applications, from stimulating anti-tumor and anti-pathogen immunity to inducing antigen-specific tolerance in autoimmunity and transplantation. A dedicated section highlights the transformative role of artificial intelligence and machine learning in accelerating nanomaterial design, enabling personalized immunotherapy through predictive modeling and digital twins. Finally, the chapter addresses critical translational challenges in biosafety, manufacturing, and clinical integration, while outlining future frontiers such as closed-loop autonomous systems and logic-gated nanotherapeutics. Ultimately, the convergence of nanotechnology, immunology, and AI is forging a new discipline of intelligent, precise, and personalized immune engineering.

Keywords Precision immunomodulation · Smart nanotechnology · Nanovaccines · Targeted drug delivery · Artificial intelligence (AI) in nanomedicine · Immune tolerance · Theranostic nanoparticles

A. I. Visan · I. Negut (✉)
Plasma and Radiation Physics, National Institute for Lasers, Magurele, Romania
e-mail: negut.irina@inflpr.ro

A. I. Visan
e-mail: Anita.visan@inflpr.ro

J. Chen (ed.), *AI-Driven Nanotechnology in Precision Medicine and Healthcare*, Smart Nanomaterials Technology, https://doi.org/10.1007/978-981-92-1690-1_8

1 Introduction: The New Paradigm of Immune Engineering

The human immune system functions as a remarkably intricate defense network, utilizing a web of specialized cells and signaling molecules to maintain health through molecular recognition and adaptive memory [1]. While this complexity is vital for survival, its dysregulation leads to chronic diseases that are notoriously difficult to treat. Traditional therapies, such as systemic corticosteroids or broad cytokines, often trap clinicians in a "therapeutic paradox": they may suppress the intended inflammation but do so at the cost of global immune compromise [2]. This lack of specificity highlights a critical need for interventions that can be controlled both in location and timing within the body's shifting physiological environment [3].

The rise of precision immunomodulation nanotechnology provides a way to bridge this gap by operating at the exact scale of biological signaling [4]. "Smart" nanostructures (typically 1 to 200 nm) are now engineered with biomimetic features that allow them to respond to environmental triggers [5]. By using these systems, we can achieve precision immunomodulation—delivering potent drugs directly to specific immune subsets while keeping systemic exposure to a minimum [6]. This shift moves us away from passive drug carriers toward active biological tools that can "sense" pathological cues like acidity or specific enzymes to release their cargo [7].

Integrating artificial intelligence for modern immunoengineering is currently being revolutionized by the convergence of material science and computational intelligence [8]. Machine learning (ML) and artificial intelligence (AI) are now used to sift through vast chemical libraries to predict how a nanoparticle will behave in a living system [9]. This allows for the development of personalized vaccines and therapies tailored to an individual's unique immune profile, navigating biological barriers that previously rendered many drugs ineffective [10]. This chapter explores these mechanistic foundations and the future of intelligent, clinically deployable therapies [11–15].

1.1 Fundamental Principles of Nanomaterial-Immune Interactions

The biological odyssey of an administered nanoparticle initiates the moment it encounters complex physiological fluids, a contact that triggers a series of molecular events dictating both therapeutic success and toxicological profile [16]. Central to this journey is the formation of the "protein corona," a dynamic layer of proteins, lipids, and glycans that spontaneously adsorbs onto the nanoparticle surface upon intravenous injection [17]. This biomolecular coating effectively masks the nanoparticle's synthetic surface, providing it with a "biological identity" that the immune system perceives and reacts to [18]. The specific makeup of this corona is highly sensitive to the particle's physicochemical traits, including surface charge—where cationic surfaces are notably prone to opsonin adsorption—as well as hydrophobicity, size, and curvature [19]. Furthermore, this interface is characterized by the Vroman effect, a competitive process where proteins with higher abundance are gradually displaced by those with higher affinity [20]. Because plasma proteomes vary significantly between individuals, the same nanoparticle may develop distinct coronas in different patients, complicating the reproducibility of therapeutic outcomes in clinical settings [21].

Once the biological identity is established, immune cells utilize specialized internalization pathways to process these materials. Professional phagocytes, such as macrophages and dendritic cells, are naturally adept at engulfing larger particles exceeding 500 nm, especially when they are "flagged" by opsonins like complement proteins or immunoglobulins [22]. While this is beneficial for vaccine delivery to antigen-presenting cells, it often limits the circulation time required for other therapies [23]. Conversely, smaller nanoparticles ranging from 20 to 200 nm can be functionalized with specific ligands—such as antibodies or aptamers—to exploit receptor-mediated endocytosis, allowing for precise targeting of cell subsets like CD4 + T cells [24]. Other entry points include pinocytosis and non-specific endocytosis, which are generally less efficient and used by smaller, hydrophilic particles [25]. After entry, nanoparticles typically move through endosomal compartments toward lysosomes; therefore, designing carriers with pH-sensitive polymers or fusogenic lipids is vital to facilitate endosomal escape, ensuring that cargo like mRNA or DNA can reach its functional destination in the cytosol or nucleus [26].

The ultimate fate of a nanoparticle—its biodistribution and eventual clearance—is a function of its physical dimensions, geometry, and surface chemistry [3]. Size plays a decisive role: particles smaller than 10 nm face rapid renal filtration, while those between 10 and 150 nm are better suited for prolonged circulation and can leverage the Enhanced Permeability and Retention (EPR) effect to accumulate in tumor tissues [27]. Particles larger than 200 nm are typically filtered out by the liver and spleen [28]. Beyond size, high-aspect-ratio shapes like rods or flexible disc-like structures can navigate endothelial barriers more effectively than traditional spheres [29]. To shield these particles from the immune system, polyethylene glycol (PEG) conjugation is frequently used to provide "stealth" properties by reducing opsonization, though the rising prevalence of anti-PEG antibodies has spurred interest in alternatives [30]. These emerging strategies include the use of zwitterionic polymers, " "don't eat me" signals like CD47-mimetic peptides, or cloaking the particles in natural cell membranes to ensure biocompatibility and extended half-life [2, 31–33].

1.2 Nanomaterial Platforms for Immunoengineering

The field of precision immunomodulation utilizes an expansive library of nanomaterial platforms, each offering distinct architectural advantages for targeted delivery, cellular reprogramming, and vaccine amplification [34]. Lipid-based systems, most notably liposomes and solid lipid nanoparticles (SLNs), represent the most clinically mature class, recently exemplified by the global implementation of lipid nanoparticles (LNPs) for mRNA-based immunization [35]. These platforms are highly valued for their exceptional biocompatibility, high payload capacity for both lipophilic and hydrophilic agents, and their inherent fusogenic potential, which facilitates the direct delivery of nucleic acids into the cytoplasm by merging with host cell membranes [36, 37]. Despite these successes, maintaining the structural stability of lipid assemblies during storage and ensuring precise control over their internal release kinetics remain significant areas of ongoing research [38].

In parallel, polymeric nanoparticles offer a highly versatile alternative, utilizing biodegradable polymers such as poly(lactic-co-glycolic acid) (PLGA), chitosan, and poly(β-amino esters) [39]. The primary strength of polymeric systems lies in their tunable degradation profiles, which allow for the sustained and controlled release of immunomodulators over periods ranging from days to months [40]. Furthermore, the ease of surface functionalization on these scaffolds enables the covalent attachment of targeting ligands or "stealth" polymers, making them ideal for directing therapies to lymphoid tissues or specific immune cell subsets like dendritic cells [41, 42]. However, the use of organic solvents during the synthesis of certain polymers and the potential for acidic degradation byproducts to trigger local inflammation are critical factors that must be managed during the design phase [25].

Inorganic nanoparticles, including gold, mesoporous silica, iron oxide, and quantum dots, provide a rigid structural foundation with unique physical properties that organic carriers lack [43]. For example, gold nanoparticles are widely used for structured antigen presentation due to their precise size control, while iron oxide particles enable magnetic resonance imaging and localized hyperthermia for "theranostic" applications [23, 43]. Mesoporous silica nanoparticles are particularly noted for their exceptionally high surface area and pore volume, allowing for the massive loading of small-molecule inhibitors or cytokines [44]. While these inorganic cores are highly stable, their long-term biosafety and the pathways for their eventual clearance from the liver and spleen remain primary concerns for chronic immunotherapeutic applications [28].

Finally, the emergence of biomimetic nanovesicles represents a shift toward "nature-inspired" engineering [45]. This class includes exosomes, bacterial membrane vesicles, and synthetic particles cloaked in membranes derived from red blood cells, leukocytes, or cancer cells [46]. By inheriting the complex surface chemistry of the source cells—such as " "don't eat me" signals (CD47) or specific adhesion molecules—these biomimetic systems achieve superior immune evasion and natural tissue homing capabilities that synthetic coatings cannot fully replicate [31, 47]. While these platforms offer unparalleled biocompatibility and targeting, the complexity of their fabrication and the high costs associated with large-scale, standardized production present substantial hurdles for widespread clinical translation [48]. These diverse platforms, along with their core mechanisms and translational hurdles, are summarized conceptually in Fig. 1.

This conceptual schematic illustrates the core components of nanotechnology-enabled immune engineering. The central diamond highlights four major classes

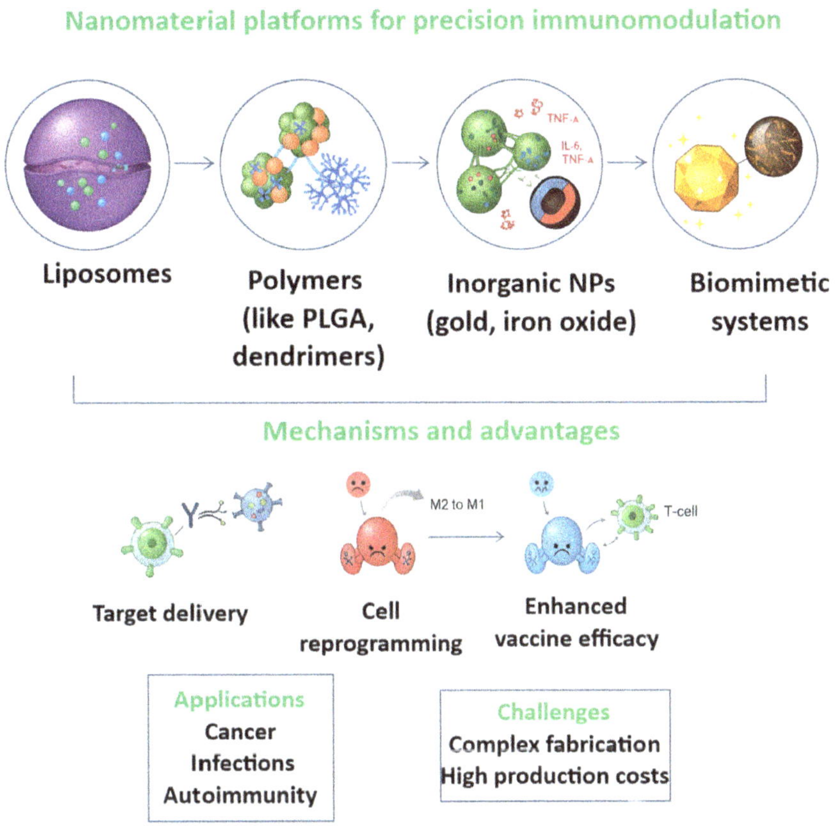

Fig. 1 Overview of nanomaterial platforms for precision immunomodulation

of engineered nanoplatforms: liposomes (phospholipid vesicles for drug encapsulation), polymers (biodegradable carriers like PLGA and dendrimers with tunable properties), inorganic nanoparticles (gold, iron oxide, and others with unique optical/magnetic functions), and biomimetic systems (exosomes or cell membrane-coated vesicles with natural targeting). These platforms enable key mechanisms and advantages, including targeted delivery to specific immune cells or tissues, cell reprogramming to alter immune cell phenotypes, and enhanced vaccine efficacy through structured antigen presentation and adjuvant co-delivery. The lower section outlines the broad applications across major disease domains—cancer immunotherapies, infectious disease vaccines, and autoimmunity tolerance induction—while acknowledging critical challenges such as the complex fabrication requirements and high production costs that must be addressed for clinical translation.

1.3 Nanotechnology-Enabled Immunostimulation Strategies

Immunostimulatory nanotechnologies serve as a critical bridge to overcome the inherent immunogenicity deficits of traditional vaccines and the systemic toxicity of global immunotherapies [15].

The multi-faceted strategy of nanotechnology-enabled immunostimulation is summarized conceptually in Fig. 2, which illustrates the hierarchy from core nanoplatforms and their enabling mechanisms to their distinct applications in cancer immunotherapy and infectious disease vaccination, ultimately leading to enhanced immune outcomes.

This diagram encapsulates the paradigm shift from conventional immunostimulation to nanotechnology-enabled precision immunomodulation. At its foundation are the core nanotechnology platforms—diverse material classes (lipids, polymers, inorganic materials, and biomimetic systems) that provide the physical scaffolding for immune engineering. These platforms are not passive carriers but active participants in immune recognition, possessing tunable properties that can be precisely engineered for specific immunological tasks.

The key enabling mechanisms represent the critical nanoscale design principles that address fundamental limitations of traditional vaccines and immunotherapies. Structured antigen presentation exploits the geometric principles of immune recognition, while co-delivery ensures synchronized immune activation. Size-based targeting leverages the body's natural lymphatic drainage, and stimuli-responsive materials provide temporal control over payload release. These mechanisms collectively

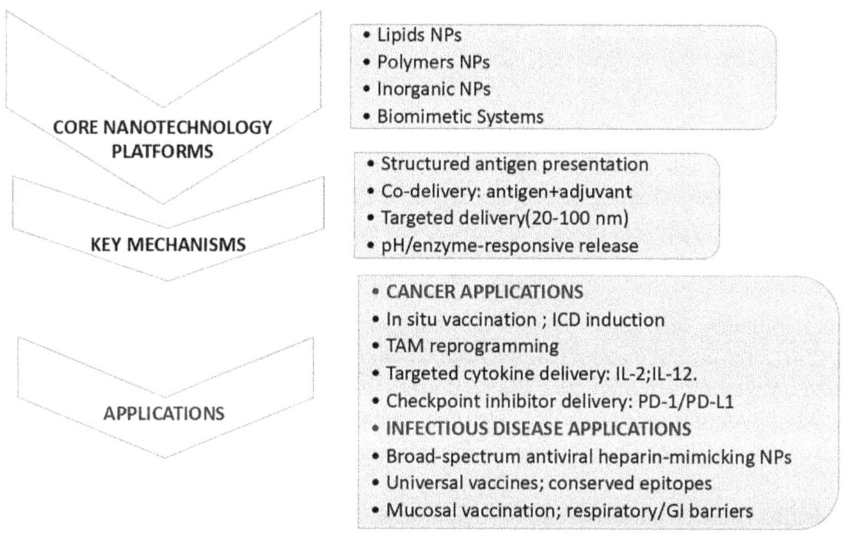

Fig. 2 Hierarchy of nanotechnology-enabled immunostimulation strategies

overcome the spatial and temporal limitations that have constrained conventional approaches.

The applications bifurcate into two major therapeutic domains. In cancer immunotherapy, the strategies focus on dismantling the immunosuppressive tumor microenvironment through multiple coordinated approaches—generating endogenous vaccines (in situ vaccination), reprogramming suppressive immune cells (TAMs, Tregs, MDSCs), localizing cytokine effects, and delivering checkpoint inhibitors with optimized pharmacokinetics. Each approach addresses a specific barrier in the cancer immunity cycle.

For infectious diseases, the applications emphasize prevention through enhanced vaccine platforms. Broad-spectrum strategies aim for pathogen-agnostic protection, universal vaccines seek to overcome antigenic drift and shift, and mucosal approaches establish frontline immunity at pathogen entry points. These applications share the common goal of creating more durable, broadly protective immune responses than traditional vaccine formulations.

The diagram's hierarchical structure emphasizes that successful immunostimulation requires integration across multiple levels: appropriate platform selection, mechanism optimization, and context-specific application. The arrows represent not just linear progression but iterative refinement, where clinical insights from applications can inform platform redesign and mechanism optimization—creating a feedback loop that drives continuous improvement in immunostimulatory nanotechnology.

Ultimately, this framework illustrates how nanotechnology transforms immunostimulation from a one-size-fits-all approach to a precision engineering discipline, where each component can be tailored to specific immune challenges, whether combating malignant cells or preventing pathogenic infections.

By engineering advanced nanovaccines, researchers can now achieve structured antigen presentation that mimics the repetitive, high-density geometry of viral surfaces [49]. This spatial arrangement is essential for effectively cross-linking B-cell receptors, which triggers a much more potent antibody response than soluble antigens alone [50]. Furthermore, nanocarriers facilitate the precise co-delivery of antigens and adjuvants to the same antigen-presenting cell (APC), a prerequisite for synchronized T-cell priming [51]. Through the use of pH-responsive or enzyme-triggered materials, these systems ensure that payloads are released within specific intracellular compartments, thereby facilitating cross-presentation on MHC class I molecules to activate cytotoxic CD8 + T cells [52]. The efficiency of these responses is further enhanced by size-controlled delivery; nanoparticles optimized within the 20–100 nm range drain rapidly from subcutaneous sites to the lymph nodes, where functionalization with mannose or anti-DC receptor antibodies concentrates the vaccine, significantly reducing the required therapeutic dose [53].

In the context of oncology, nanotechnology offers sophisticated tools to dismantle the immunosuppressive architecture of the tumor microenvironment (TME) [54]. One prominent strategy involves in situ vaccination, where nanoparticles induce immunogenic cell death (ICD) through photodynamic therapy or chemotherapy, forcing the release of endogenous tumor antigens alongside "danger signals" that activate local dendritic cells [55, 56]. Beyond direct tumor cell killing, nanomedicines are increasingly employed to reprogram immunosuppressive cell populations [57]. For instance, nanoparticles can deliver CSF-1R inhibitors or siRNA targeting the Arg-1 gene to shift tumor-associated macrophages (TAMs) from a pro-tumor M2 phenotype back to an anti-tumor M1 state [7, 58]. Similar approaches utilize targeted delivery to deplete or inhibit regulatory T cells (Tregs) and myeloid-derived suppressor cells (MDSCs), effectively lifting the "brakes" on the anti-tumor immune response [14]. Targeted cytokine delivery (e.g., IL-2, IL-12) via nanoparticles further localizes T-cell expansion within the tumor site, circumventing the life-threatening systemic toxicities traditionally associated with free cytokine therapy [59].

The therapeutic potential of these systems is maximized through combination nano-immunotherapies, which allow for the sequential or simultaneous delivery of chemotherapy and immune checkpoint inhibitors (ICIs) with optimized pharmacokinetics [11]. These multi-compartment systems can be programmed to first establish a pro-inflammatory TME and then release checkpoint blockers to sustain T-cell activity, mimicking the natural immune cascade [6]. In the realm of infectious diseases, nanotechnology expands the reach of protective immunity through broad-spectrum antiviral strategies, such as heparin-mimicking particles that act as "decoy" receptors to prevent viral entry [60, 61]. Efforts are also underway to develop universal influenza vaccines by displaying conserved epitopes on synthetic nanoparticle cores, aiming for cross-strain protection that transcends seasonal variations [62]. Finally, mucosal vaccination platforms—engineered with mucoadhesive polymers—allow for the delivery of antigens across respiratory or gastrointestinal barriers, fostering robust frontline protection at the primary sites of pathogen invasion [63].

In the management of autoimmune diseases, allergies, and organ transplantation, the therapeutic objective pivots from activation toward the selective suppression of pathological immune responses while maintaining the host's ability to fight infections [64]. Achieving this degree of precision requires targeting autoreactive or alloreactive cell populations without inducing global immunosuppression [65]. A primary strategy is the induction of antigen-specific tolerance through tolerogenic antigen presentation [66]. Nanoparticles can be engineered to carry autoantigens, such as myelin basic protein for multiple sclerosis or insulin peptides for type 1 diabetes, and targeted to steady-state dendritic cells via ligands for receptors like DEC-205 or Clec9A [67]. In the absence of pro-inflammatory co-stimulatory signals, this interaction promotes T-cell anergy, clonal deletion, or the conversion of naive T cells into regulatory T cells (Tregs) [42]. Furthermore, researchers exploit physiological clearance pathways by designing nanoparticles approximately 500 nm in size, which are naturally sequestered by marginal zone macrophages in the spleen [12]. By loading these particles with disease-relevant antigens, the immune system's natural debris-clearing mechanism is co-opted to present antigens in a natively tolerizing context [68].

To reinforce these tolerogenic signals, nanoparticles can be designed for the localized expansion of regulatory cell populations. Carriers delivering immunosuppressive cytokines like TGF-β, IL-2, or retinoic acid can preferentially boost Treg numbers at the site of inflammation [59]. More advanced "smart" systems are engineered to release these factors only in response to specific inflammatory triggers, creating a self-regulating circuit that activates only when the disease is flared [5]. This precision extends to cell-type-specific immunosuppression, where nanoparticles are functionalized to home in on pathogenic subsets [2]. In rheumatoid arthritis, for example, particles targeted to $\alpha 4\beta 1$ integrins on activated lymphocytes can deliver methotrexate directly into inflamed joints, sparing healthy tissues [41]. Similarly, nanoparticles can modulate intracellular pathways by delivering siRNA to silence master transcription factors like T-bet, RORγt, or GATA3, thereby inhibiting the specific T helper subsets (Th1, Th17, or Th2) driving Crohn's disease, psoriasis, or asthma [25, 69].

In the field of transplant immunology, nanotechnology focuses on minimizing the reliance on systemic calcineurin inhibitors, which are often associated with nephrotoxicity and infection [70]. Injectable hydrogels containing immunosuppressant-loaded nanoparticles can be placed directly at the transplant site to provide sustained, localized protection of the graft [71]. Beyond drug delivery, donor-specific tolerance can be promoted by administering nanoparticles carrying donor antigens alongside tolerogenic adjuvants prior to surgery, potentially eliminating the need for lifelong medication [72]. For type 1 diabetes, the encapsulation of pancreatic islets within nanoporous barriers allows for the free exchange of nutrients and insulin while physically shielding the cells from host immune attack [73]. Finally, in broader inflammatory disease management, such as atherosclerosis or inflammatory bowel disease (IBD), pH-responsive or enzyme-sensitive nanoparticles ensure that anti-inflammatory agents are released only within the unique microenvironments of arterial plaques or inflamed intestinal mucosa [11, 23, 34].

1.4 Artificial Intelligence in Nanomaterial Design and Personalized Immunotherapy

The structural and chemical landscape of nanomaterial engineering is defined by a combinatorial complexity that exceeds the capacity of traditional empirical research [9]. Factors such as precise material composition, particle geometry, surface charge density, and the kinetics of drug loading create a multidimensional parameter space that is virtually impossible to navigate through legacy trial-and-error methodologies [8]. To address this, artificial intelligence (AI) and machine learning (ML) have emerged as transformative engines for accelerating the discovery and personalization of smart immunotherapies [74]. These computational tools allow researchers to move from reactive observation to predictive design, utilizing large-scale datasets to identify non-obvious correlations between a nanoparticle's physical properties and its eventual biological performance in a complex immune environment [75].

At the core of this revolution is the predictive modeling of nano-bio interactions, where algorithms like Graph Neural Networks (GNNs) are trained to forecast how specific surface modifications will influence protein corona formation or cellular uptake [8, 76]. By simulating these interactions in silico, AI significantly narrows down the number of formulations that require physical laboratory testing, thereby reducing development timelines from years to months [77]. Furthermore, the integration of high-throughput automated synthesis with reinforcement learning has given rise to "self-driving laboratories" [78]. In these systems, AI agents autonomously propose new nanomaterial candidates, analyze the results of robotic experiments, and refine their models in real time to optimize for specific immunological goals, such as maximizing lymph node accumulation or minimizing systemic clearance [79, 80].

Beyond material optimization, AI is pioneering the era of personalized immunotherapy through the use of "digital twins" and advanced patient stratification [10]. By processing multi-omics data—including single-cell RNA sequencing and immune repertoire profiles—AI can match specific nanoparticle designs to a patient's unique immune phenotype [81]. Digital twin technology allows clinicians to virtually simulate a patient's immune response to a proposed nano-immunotherapy regimen, predicting efficacy and potential toxicities before the first dose is administered [8, 82]. This computational foresight is augmented by adaptive treatment planning, where real-time data from liquid biopsies or molecular imaging feed back into AI models to adjust dosage and timing dynamically [83, 84]. Ultimately, this convergence of AI and nanotechnology ensures that the manufacturing and quality control of these complex systems meet the rigorous standards required for precision medicine [6, 11].

Figure 3. illustrates a simplified AI/ML pipeline for nanomaterial design, demonstrating how data-driven approaches accelerate the development and personalization of smart immunomodulatory nanoparticles through iterative prediction, optimization, and validation cycles.

Flow arrows indicate:

- Solid arrows (→): Primary data flow and process direction
- Circular feedback (↻): Iterative learning and system improvement
- The pipeline represents a non-linear, adaptive process where each stage informs and refines the others.

The AI/ML pipeline functions as an integrated, self-optimizing system, beginning with the assimilation of comprehensive foundational knowledge. This includes curated scientific literature, established principles from prior research, high-throughput experimental datasets encompassing automated nanoparticle synthesis and characterization, and rich, patient-specific multi-omics profiles derived from advanced sequencing technologies. These diverse data streams converge to form the essential fuel for the computational engine at the heart of the workflow.

This computational core, or the AI/ML engine, then processes this information through sophisticated analytical frameworks. It employs predictive modeling to forecast the intricate relationships between a nanoparticle's physicochemical properties and its subsequent biological fate and efficacy. Building on these predictions, optimization algorithms intelligently navigate the vast design space to propose novel, high-potential nanomaterial formulations. Concurrently, the engine can generate digital twins—high-fidelity, patient-specific computational simulations that model

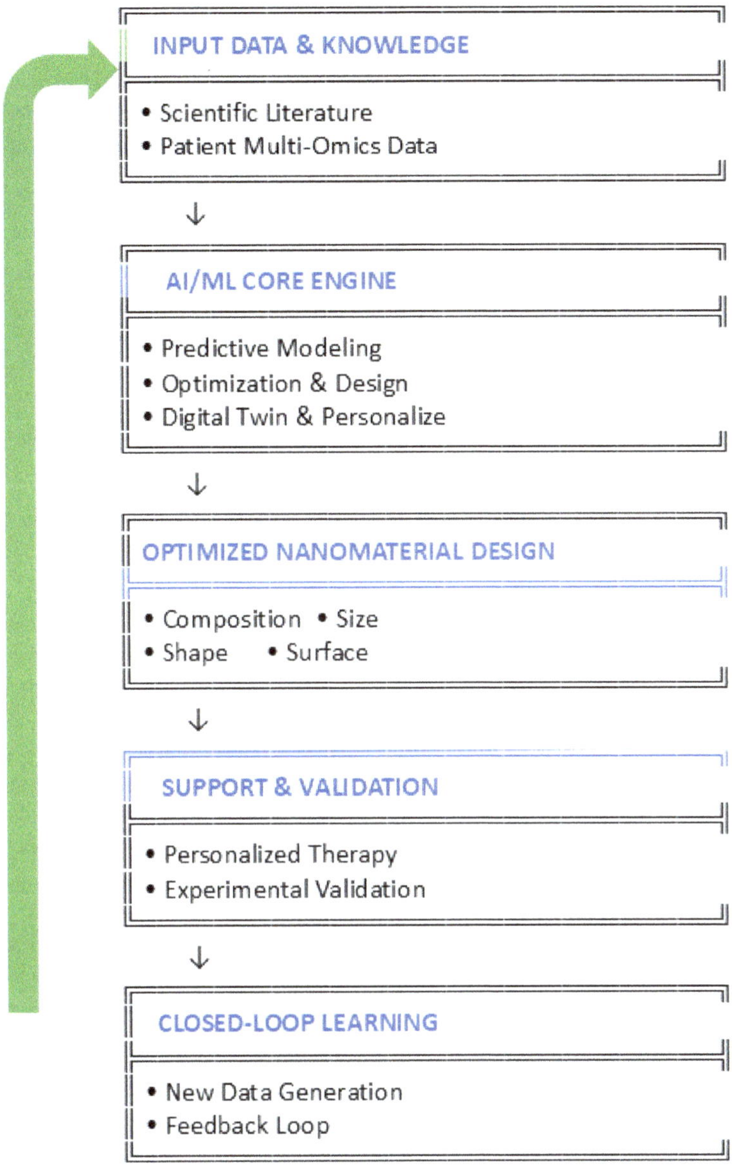

Fig. 3 AI/ML pipeline for nanomaterial design

disease progression and immune system dynamics, enabling the virtual testing and personalization of therapeutic strategies before any physical intervention.

The tangible outputs of this computational analysis are twofold: a precisely optimized nanomaterial design, ready for benchtop synthesis, and a bespoke therapeutic strategy tailored to an individual's unique biological profile. These proposals then enter a critical phase of experimental validation, where they are rigorously tested in laboratory models and, ultimately, clinical settings. Crucially, the results from this validation phase are not an endpoint but a vital new input. They are fed back into the system as fresh data, creating a closed-loop learning cycle. This continuous feedback mechanism allows the AI models to learn from both successes and failures, perpetually refining their predictions and design proposals in an iterative, self-improving process. The directional flow of data between stages and the cyclical nature of this feedback underscore that the pipeline is not a rigid, linear sequence but a dynamic, adaptive ecosystem where each component continuously informs and enhances the others

1.4.1 AI-Driven Nanomaterial Design and Optimization

The integration of computational intelligence into material science has shifted the discovery of immunotherapies from empirical observation to predictive engineering [9]. Machine learning algorithms, trained on expansive repositories of nanoparticle physicochemical data and their resulting biological signatures—such as circulation half-life, cellular internalization rates, and inherent immunogenicity—can now forecast the performance of novel formulations with high fidelity [8]. Advanced architectures, including Graph Neural Networks (GNNs), are employed to model the intricate structure–property relationships of complex polymers and lipids, while Natural Language Processing (NLP) tools scan thousands of peer-reviewed publications to extract latent insights that inform experimental design [8, 74]. These predictive models are further refined by AI-enhanced Molecular Dynamics (MD) simulations, which provide atomic-resolution views of nanoparticles interacting with lipid bilayers and organelle membranes [76]. This allows researchers to anticipate critical outcomes, such as the efficiency of endosomal escape or the specific composition of the protein corona, well before any benchtop validation occurs [75].

The acceleration of development timelines is most evident in the rise of "self-driving laboratories," where AI-powered closed-loop systems manage the entire lifecycle of nanoparticle synthesis [77]. These platforms integrate automated robotic fabrication, real-time characterization, and high-throughput biological screening into a single iterative cycle [80]. Reinforcement learning (RL) algorithms act as the central decision-makers, proposing new chemical formulations likely to outperform previous iterations based on immediate experimental feedback [78]. This autonomous optimization has been shown to reduce development schedules from several years to mere months, particularly for complex systems like multi-component lipid nanoparticles for mRNA delivery [6, 79]. By removing human bias and increasing throughput by several orders of magnitude, these systems ensure that only the most potent and stable candidates progress to clinical evaluation [11].

1.4.2 Personalized Immunotherapy and Digital Twins

As we move toward 2026, the personalization of immune engineering relies heavily on AI-driven patient stratification and the use of multi-omics data [10]. By analyzing single-cell RNA sequencing, proteomic profiles, and T-cell receptor (TCR) repertoires, AI algorithms can identify specific biomarkers that predict how a patient will respond to a given nano-immunotherapy [81]. This enables the rational matching of nanoparticle architecture to a patient's unique immune phenotype; for example, prioritizing tumor-associated macrophage (TAM) reprogramming for "cold" tumors with high myeloid infiltration, while deploying T-cell stimulating nanoparticles for "hot," lymphocytic tumors [8, 82]. This level of precision ensures that interventions are tailored to the dynamic immunological landscape of the individual rather than a population average [83].

A cornerstone of this personalized approach is the "Digital Twin" technology, which involves creating patient-specific computational models that simulate the complex dynamics of the immune system and disease progression [84]. Clinicians can virtually test various nano-immunotherapy regimens on these digital replicas to predict therapeutic efficacy and potential side effects, optimizing treatment plans before the patient receives a single dose [3]. These models are not static; they continuously ingest real-world data from liquid biopsies and molecular imaging, becoming more accurate as the treatment progresses [5, 34]. This facilitates truly adaptive treatment planning, where AI systems recommend real-time adjustments to dosage or nanoparticle composition based on the evolving biological state of the patient [2].

1.4.3 Manufacturing, Quality Control, and Regulatory Compliance

The role of AI extends beyond design into the rigorous environment of Good Manufacturing Practice (GMP). AI-driven computer vision systems now perform real-time analysis of electron microscopy and dynamic light scattering data, ensuring that every batch of nanoparticles meets stringent consistency standards for size, shape, and polydispersity [32]. Predictive maintenance algorithms further ensure the stability of the manufacturing process by monitoring equipment performance and predicting failures before they occur [33]. To meet modern regulatory demands, blockchain-integrated systems are utilized to create an unalterable audit trail of raw materials and synthesis parameters, ensuring full transparency for agencies like the FDA [28]. This computational infrastructure is essential for the transition from small-scale academic discovery to the mass production of safe, reliable, and intelligent immunotherapies [29, 30].

To illustrate the practical impact of these computational advances, Table 1 summarizes the current translational landscape of major nanomaterial platforms, highlighting the specific roles AI has played in moving these systems from the laboratory toward clinical implementation.

Table 12.1 categorizes common nanomaterial platforms and identifies the specific computational techniques—ranging from Reinforcement Learning to Graph Neural Networks—that have been instrumental in their optimization. The status updates reflect the 2026 clinical landscape, demonstrating the shift from traditional empirical formulation toward data-driven, precision-engineered therapeutics.

For example, metastatic pancreatic adenocarcinoma is characterized by a dense stroma and an abundance of immunosuppressive M2-type tumor-associated macrophages (TAMs) that render checkpoint inhibitors ineffective. A dual-action polymeric nanoparticle (PLGA-PEG) was engineered to deliver a CSF-1R inhibitor and a STING agonist. AI-driven "Digital Twin" simulations of patient biopsies predicted that sequential release—reprogramming macrophages first, followed by dendritic cell activation—would be significantly more effective than simultaneous delivery. Clinical data confirmed that this converted the TME to an M1-dominant, pro-inflammatory state, facilitating the infiltration of CD8 + T cells and sensitizing previously non-responsive patients to Pembrolizumab [7, 14, 57].

Table 1 The current state of AI-integrated immunoengineering

Platform class	Representative example	Primary AI contribution	Clinical status (2026)	References
Lipid-based (LNP)	AI-optimized CFTR mRNA-LNP	Reinforcement learning for mRNA half-life and lipid stability [35]	Phase 1/2 (Cystic Fibrosis)	[35, 78]
Polymeric NP	PLGA-encapsulated Doxorubicin	Predictive modeling of zero-order degradation kinetics [39]	Early Clinical (HCC)	[9, 39, 40]
Inorganic NP	Gold (AuNP) with AI ligands	GNN-designed ligands for 50% higher biomarker affinity [40]	Phase 1 (Liver Cancer)	[40, 43, 76]
Biomimetic	Cell membrane vesicles	ML screening of protein corona to minimize opsonization [45]	Translational R&D (Vaccines)	[17, 45, 46]
Iron Oxide (IONP)	Magnetic nanoclusters	AI-enhanced simulation of localized hyperthermia heat maps [23]	Clinical Evaluation (Combination)	[8, 11, 23]

In another example, in the case of multiple sclerosis (MS), the therapeutic goal is to stop the autoimmune attack on the myelin sheath without inducing global lymphopenia. Using the physiological clearance pathway strategy, 500 nm biodegradable PLG nanoparticles were encapsulated with myelin-derived peptides (e.g., $PLP_{139–151}$). These "debris-mimicking" particles are sequestered by the splenic marginal zone, where specialized APCs present the antigen in a non-inflammatory context. This approach successfully expanded the $CD4 + CD25 + FoxP3 + $ regulatory T cell population in patients, stabilizing disability scores (EDSS) while preserving the rest of the immune system's ability to fight opportunistic infections [66, 67, 85].

Another example related to traditional aerosolized gene therapies for cystic fibrosis (CF) frequently fails due to the dense, dehydrated mucus layer in the lungs, which traps and neutralizes standard lipid nanoparticles (LNPs) before they reach the epithelial cells. To address this, researchers utilized a reinforcement learning (RL) loop to screen over 5,000 lipid variations, optimizing for "mucus-penetrating" properties and high nebulization stability. The winning formulation, an AI-optimized LNP carrying *CFTR* mRNA, featured a specialized PEG-lipid density predicted by Graph Neural Networks to minimize adhesive interactions with mucin fibers. In 2026 trials, this platform demonstrated a 15% improvement in forced expiratory volume (FEV_1) compared to baseline, with a 30% reduction in the required dose [35, 78].

1.5 Translational Considerations and Future Frontiers

Despite the profound preclinical success of intelligent immunomodulation, the transition from laboratory prototypes to bedside applications requires overcoming several formidable biosafety and regulatory hurdles [2]. Comprehensive toxicological profiling is no longer limited to acute studies; it now demands a deep understanding of chronic immune toxicity, particularly how nanomaterials might affect long-term immune memory or inadvertently act as unintended adjuvants that trigger autoimmunity [86]. The metabolic fate of these materials is equally critical; even biodegradable constructs must be scrutinized for the toxicity of their breakdown products, while non-biodegradable inorganic cores face strict investigation regarding their accumulation in the liver and spleen over months or years [3, 28]. Furthermore, the "personalized corona" and individual genetic variability mean that a dose safe for one patient may provoke an adverse reaction in another, necessitating the integration of AI-driven safety predictions into the early stages of clinical trial design [8, 21].

The manufacturing of these complex, multifunctional systems presents its own set of challenges, particularly the need for scalable Good Manufacturing Practice (GMP) [6]. Ensuring batch-to-batch consistency in ligand density and drug loading requires advanced process engineering, such as continuous flow manufacturing equipped with real-time, in-line monitoring [84]. From a regulatory perspective, agencies like the FDA and EMA are currently refining frameworks to handle "nanosimilars," where defining bioequivalence is far more complex than for traditional small molecules [87,

88]. Additionally, the high costs associated with sophisticated nanomanufacturing must be addressed through optimized production workflows to ensure that these life-saving precision therapies do not remain inaccessible to broader patient populations [34].

Looking toward the future, the frontier of immunoengineering is defined by the integration of sensing and actuation within a single "closed-loop" nanoplatform [5]. Such autonomous systems could monitor local interferon-γ levels and release anti-inflammatory agents only when a specific pathological threshold is crossed, effectively creating a self-regulating therapeutic [11, 89]. We are also witnessing a shift toward ex vivo nano-immunoengineering, where nanoparticles replace viral vectors to deliver CRISPR/Cas9 machinery for the genetic modification of CAR-T or CAR-NK cells with enhanced safety and efficiency [7, 90]. By incorporating synthetic biology circuits, these particles can function as logic-gated systems—for instance, activating an immune response only if a specific tumor antigen is detected AND an inhibitory signal is absent [91, 92]. Finally, multi-modal theranostic platforms are merging PET or photoacoustic imaging with therapeutic delivery, while oral nanoparticles are being developed to modulate the microbiome-immune axis, opening new avenues for treating intestinal and systemic inflammatory diseases [44, 63].

1.6 Conclusion

Smart nanotechnology is fundamentally redefining the boundaries of immune modulation, evolving the field from a systemic, imprecise intervention into a highly targeted and intelligent therapeutic discipline [12]. By engineering materials that operate at the exact molecular scale of immune recognition, researchers have gained an unprecedented capacity to interface with complex biological pathways with high spatiotemporal precision [3]. This shift toward "active" engagement is further propelled by the convergence with artificial intelligence, which allows for a data-driven approach to design and personalization that finally transcends the limitations of traditional, one-size-fits-all treatments [9, 10].

The path forward necessitates a deep, sustained collaboration across diverse disciplines, including material science, immunology, computational biology, and regulatory science [2]. As the field addresses the practicalities of large-scale manufacturing and long-term biosafety, the clinical impact of these technologies becomes increasingly tangible [6]. We are moving toward a future where nanoplatforms function not merely as passive vehicles for drug delivery but as dynamic participants in immune regulation—systems capable of sensing, processing, and responding to fluctuating biological signals to restore and maintain immune homeostasis [5, 11].

Ultimately, the paradigm of precision immunomodulation through smart nanotechnology represents a transformative shift in the management of human disease [4]. Whether tackling the immunosuppressive barriers of cancer, restoring tolerance in autoimmunity, or providing frontline protection against infectious pathogens, these advanced approaches promise to harness the inherent complexity of the immune system rather than bluntly opposing it [7, 15]. By integrating molecular engineering with predictive intelligence, we are entering a new era of therapy that offers personalized hope for conditions that have historically eluded precise medical intervention [8, 34, 55].

Funding This research was funded by a grant from the Ministry of Research, Innovation and Digitalization, CCCDI—UEFISCDI, project number PN-IV-P7-7.1-PED-2024–0137, within PNCDI IV. This research was also funded by the Romanian Ministry of Research, Innovation and Digitalization under the Romanian National Nucleu Program LAPLAS VII—contract No. 30N/2023.

Competing Interests The authors declare no competing interests relevant to the content of this chapter.

Ethics Approval This chapter does not contain any primary studies with human participants or animals performed by the authors.

List of Abbreviations

AI	Artificial intelligence—Computational systems that simulate human intelligence processes, including learning, reasoning, and problem-solving.
APC	Antigen-presenting cell—Immune cells (e.g., dendritic cells, macrophages) that process and present antigens to T cells to initiate immune responses.
Arg-1	Arginase-1—An enzyme involved in the urea cycle that breaks down L-arginine; high expression in tumor-associated macrophages suppresses T-cell function.
CAR-NK	Chimeric antigen receptor natural killer cells—Natural killer cells genetically engineered to express synthetic receptors that recognize specific tumor antigens.
CAR-T	Chimeric antigen receptor T cells—T lymphocytes genetically modified to express synthetic receptors that target specific cancer cells.
CD4+	Cluster of differentiation 4 positive (T helper cells)—T lymphocytes that coordinate immune responses by activating other immune cells.
CD8+	Cluster of differentiation 8 positive (cytotoxic T cells)—T lymphocytes that directly kill infected or cancerous cells.
CRISPR	Clustered regularly interspaced short palindromic repeats—A gene-editing technology that allows precise modification of DNA sequences.
CSF-1R	Colony-stimulating factor-1 receptor—A receptor on macrophages that regulates their survival, proliferation, and differentiation; often targeted to deplete tumor-promoting macrophages.
EPR	Enhanced permeability and retention—A phenomenon where nanoparticles accumulate in tumor tissues due to leaky vasculature and impaired lymphatic drainage.

GMP	Good Manufacturing Practice—A system for ensuring that products are consistently produced and controlled according to quality standards.
ICD	Immunogenic cell death—A form of cell death that activates the immune system against dead cell-associated antigens.
IL-2	Interleukin-2—A cytokine that promotes the growth and differentiation of T cells and natural killer cells.
IL-12	Interleukin-12—A cytokine that stimulates T cells and natural killer cells, promoting Th1 immune responses.
M1/M2	Macrophage polarization states—M1 macrophages are pro-inflammatory and anti-tumor; M2 macrophages are anti-inflammatory and pro-tumor/repair.
MDSC	Myeloid-derived suppressor cells—A heterogeneous population of immature myeloid cells that suppress T-cell responses in cancer and chronic inflammation.
MHC	Major histocompatibility complex—Cell surface proteins that present antigen fragments to T cells for immune recognition.
ML	Machine learning—A subset of AI involving algorithms that improve automatically through experience and data analysis.
MRI	Magnetic resonance imaging—A medical imaging technique that uses magnetic fields and radio waves to create detailed images of organs and tissues.
PEG	Polyethylene glycol—A synthetic polymer commonly used to coat nanoparticles to increase their circulation time in blood.
PET	Positron emission tomography—A nuclear medicine imaging technique that visualizes metabolic processes in the body.
PLGA	Poly(lactic-co-glycolic acid)—A biodegradable polymer widely used in nanoparticle drug delivery systems.
RNA	Ribonucleic acid —A nucleic acid molecule essential for coding, decoding, regulation, and expression of genes.
siRNA	Small interfering RNA—Short double-stranded RNA molecules that silence gene expression by degrading complementary mRNA sequences.
TAM	Tumor-associated macrophages—Macrophages present in tumor microenvironments that often promote tumor growth and suppress anti-tumor immunity.
TGF-β	Transforming growth factor beta—A cytokine with diverse roles in cell growth, differentiation, and immune regulation; often immunosuppressive in cancer.
Treg	Regulatory T cells—A specialized subpopulation of T cells that suppress immune responses and maintain self-tolerance

References

1. Abbas AK, Lichtman AH, Pillai S (2023) Cellular and molecular immunology
2. Anselmo AC, Mitragotri S (2019) Nanoparticles in the clinic: An update. Bioeng Transl Med 4(3):e10143
3. Blanco E, Shen H, Ferrari M (2015) Principles of nanoparticle design for overcoming biological barriers to drug delivery. Nat Biotechnol 33(9):941–951
4. Germain RN (2001) The art of the probable: system control in the adaptive immune system. Science 293(5528):240–245
5. Karimi M, Ghasemi A, Zangabad PS, Rahighi R, Basri SMM, Mirshekari H, Amiri M, Pishabad ZS, Aslani A, Bozorgomid M (2016) Smart micro/nanoparticles in stimulus-responsive drug/gene delivery systems. Chem Soc Rev 45(5):1457–1501
6. Mitchell MJ, Billingsley MM, Haley RM, Wechsler ME, Peppas NA, Langer R (2021) Engineering precision nanoparticles for drug delivery. Nat Rev Drug Discovery 20(2):101–124
7. Riley RS, June CH, Langer R, Mitchell MJ (2019) Delivery technologies for cancer immunotherapy. Nat Rev Drug Discovery 18(3):175–196
8. Ho D, Wang P, Kee T (2019) Artificial intelligence in nanomedicine. Nanoscale Horizons 4(2):365–377
9. Shan X, Cai Y, Zhu B, Zhou L, Sun X, Xu X, Yin Q, Wang D, Li Y (2024) Rational strategies for improving the efficiency of design and discovery of nanomedicines. Nat Commun 15(1):9990
10. D'Orsi L, Capasso B, Lamacchia G, Pizzichini P, Ferranti S, Liverani A, Fontana C, Panunzi S, De Gaetano A, Lo Presti E (2024) Recent advances in artificial intelligence to improve immunotherapy and the use of digital twins to identify prognosis of patients with solid tumors. Int J Molecular Sci 25(21):11588
11. van der Meel R, Sulheim E, Shi Y, Kiessling F, Mulder WJ, Lammers T (2019) Smart cancer nanomedicine. Nat Nanotechnol 14(11):1007–1017
12. Smith DM, Simon JK, Baker JR Jr (2013) Applications of nanotechnology for immunology. Nat Rev Immunol 13(8):592–605
13. Farokhzad OC, Langer R (2009) Impact of nanotechnology on drug delivery. ACS Nano 3(1):16–20
14. Shi J, Kantoff PW, Wooster R, Farokhzad OC (2017) Cancer nanomedicine: progress, challenges and opportunities. Nat Rev Cancer 17(1):20–37
15. Irvine DJ, Swartz MA, Szeto GL (2013) Engineering synthetic vaccines using cues from natural immunity. Nat Mater 12(11):978–990
16. Nel AE, Mädler L, Velegol D, Xia T, Hoek EM, Somasundaran P, Klaessig F, Castranova V, Thompson M (2009) Understanding biophysicochemical interactions at the nano–bio interface. Nat Mater 8(7):543–557
17. Monopoli MP, Aberg C, Salvati A, Dawson KA (2020) Biomolecular coronas provide the biological identity of nanosized materials. Nano-enabled Med Appl 205–229
18. Mahmoudi M, Landry MP, Moore A, Coreas R (2023) The protein corona from nanomedicine to environmental science. Nat Rev Mater 8(7):422–438
19. Tenzer S, Docter D, Kuharev J, Musyanovych A, Fetz V, Hecht R, Schlenk F, Fischer D, Kiouptsi K, Reinhardt C (2020) Rapid formation of plasma protein corona critically affects nanoparticle pathophysiology. In: Nano-enabled medical applications. Jenny Stanford Publishing, pp 251–278
20. Vroman L (1962) Effect of adsorbed proteins on the wettability of hydrophilic and hydrophobic solids. Nature 196(4853):476–477
21. Corbo C, Molinaro R, Tabatabaei M, Farokhzad OC, Mahmoudi M (2017) Personalized protein corona on nanoparticles and its clinical implications. Biomater Sci 5(3):378–387
22. Gustafson HH, Holt-Casper D, Grainger DW, Ghandehari H (2015) Nanoparticle uptake: the phagocyte problem. Nano Today 10(4):487–510
23. Weissleder R, Nahrendorf M, Pittet MJ (2014) Imaging macrophages with nanoparticles. Nat Mater 13(2):125–138

24. Kou L, Sun J, Zhai Y, He Z (2013) The endocytosis and intracellular fate of nanomedicines: Implication for rational design. Asian J Pharm Sci 8(1):1–10
25. Sahay G, Alakhova DY, Kabanov AV (2010) Endocytosis of nanomedicines. J Control Release 145(3):182–195
26. Smith SA, Selby LI, Johnston AP, Such GK (2018) The endosomal escape of nanoparticles: toward more efficient cellular delivery. Bioconjug Chem 30(2):263–272
27. Maeda H (2001) The enhanced permeability and retention (EPR) effect in tumor vasculature: the key role of tumor-selective macromolecular drug targeting. Adv Enzyme Regul 41(1):189–207
28. Soo Choi H, Liu W, Misra P, Tanaka E, Zimmer JP, Itty Ipe B, Bawendi MG, Frangioni JV (2007) Renal clearance of quantum dots. Nature biotechnology 25(10):1165–1170
29. Geng Y, Dalhaimer P, Cai S, Tsai R, Tewari M, Minko T, Discher DE (2007) Shape effects of filaments versus spherical particles in flow and drug delivery. Nat Nanotechnol 2(4):249–255
30. Zhang P, Sun F, Tsao C, Liu S, Jain P, Sinclair A, Hung H-C, Bai T, Wu K, Jiang S (2015) Zwitterionic gel encapsulation promotes protein stability, enhances pharmacokinetics, and reduces immunogenicity. Proc Natl Acad Sci 112(39):12046–12051
31. Rodriguez PL, Harada T, Christian DA, Pantano DA, Tsai RK, Discher DE (2013) Minimal "Self" peptides that inhibit phagocytic clearance and enhance delivery of nanoparticles. Science 339(6122):971–975
32. Jiang W, Kim BY, Rutka JT, Chan WC (2008) Nanoparticle-mediated cellular response is size-dependent. Nat Nanotechnol 3(3):145–150
33. Wilhelm S, Tavares AJ, Dai Q, Ohta S, Audet J, Dvorak HF, Chan WC (2016) Analysis of nanoparticle delivery to tumours. Nat Rev Mater 1(5):1–12
34. Patra JK, Das G, Fraceto LF, Campos EVR, Rodriguez-Torres MDP, Acosta-Torres LS, Diaz-Torres LA, Grillo R, Swamy MK, Sharma S (2018) Nano based drug delivery systems: recent developments and future prospects. J Nanobiotechnol 16(1):71
35. Cullis PR, Hope MJ (2017) Lipid nanoparticle systems for enabling gene therapies. Mol Ther 25(7):1467–1475
36. Pantelić I, Ilić T, Nikolić I, Savić S (2022) Lipid nanoparticles employed in mRNA-based COVID-19 vaccines: an overview of materials and processes used for development and production. Arhiv za farmaciju 72(1):20–35
37. Midoux P, Pichon C (2015) Lipid-based mRNA vaccine delivery systems. Expert Rev Vaccines 14(2):221–234
38. Kulkarni JA, Witzigmann D, Leung J, Tam YYC, Cullis PR (2019) On the role of helper lipids in lipid nanoparticle formulations of siRNA. Nanoscale 11(45):21733–21739
39. Soppimath KS, Aminabhavi TM, Kulkarni AR, Rudzinski WE (2001) Biodegradable polymeric nanoparticles as drug delivery devices. J Control Release 70(1–2):1–20
40. Pridgen EM, Alexis F, Farokhzad OC (2015) Polymeric nanoparticle drug delivery technologies for oral delivery applications. Expert Opin Drug Deliv 12(9):1459–1473
41. Khandare J, Calderón M, Dagia NM, Haag R (2012) Multifunctional dendritic polymers in nanomedicine: opportunities and challenges. Chem Soc Rev 41(7):2824–2848
42. Pearson RM, Sen S, Hsu H-J, Pasko M, Gaske M, Král P, Hong S (2016) Tuning the selectivity of dendron micelles through variations of the poly (ethylene glycol) corona. ACS Nano 10(7):6905–6914
43. Dreaden EC, Alkilany AM, Huang X, Murphy CJ, El-Sayed MA (2012) The golden age: gold nanoparticles for biomedicine. Chem Soc Rev 41(7):2740–2779
44. Ghaferi M, Koohi Moftakhari Esfahani M, Raza A, Al Harthi S, Ebrahimi Shahmabadi H, Alavi SE (2021) Mesoporous silica nanoparticles: Synthesis methods and their therapeutic use-recent advances. J Drug Target 29(2):131–154
45. Rampado R, Caliceti P, Agostini M (2022) Latest advances in biomimetic cell membrane-coated and membrane-derived nanovectors for biomedical applications. Nanomaterials 12(9):1543
46. Rao L, Bu LL, Xu JH, Cai B, Yu GT, Yu X, He Z, Huang Q, Li A, Guo SS (2015) Red blood cell membrane as a biomimetic nanocoating for prolonged circulation time and reduced accelerated blood clearance. Small 11(46):6225–6236

47. Jayaseelan VP (2020) Emerging role of exosomes as promising diagnostic tool for cancer. Cancer Gene Ther 27(6):395–398
48. Burnouf T, Agrahari V, Agrahari V (2019) Extracellular vesicles as nanomedicine: hopes and hurdles in clinical translation. Int J Nanomed 8847–8859
49. Shetty S, Alvarado PC, Pettie D, Collier JH (2024) Next-generation vaccine development with nanomaterials: recent advances, possibilities, and challenges. Annu Rev Biomed Eng 26(1):273–306
50. Bagnoli F, Baudner B, Mishra RP, Bartolini E, Fiaschi L, Mariotti P, Nardi-Dei V, Boucher P, Rappuoli R (2011) Designing the next generation of vaccines for global public health. Omics: J Integrat Biol 15(9):545–566
51. Cordeiro AS, Alonso MJ (2016) Recent advances in vaccine delivery. Pharmaceutical Patent Analyst 5(1):49–73
52. Chen J, Li Z, Huang H, Yang Y, Ding Q, Mai J, Guo W, Xu Y (2011) Improved antigen cross-presentation by polyethyleneimine-based nanoparticles. Int J Nanomed 77–84
53. He A, Li X, Dai Z, Li Q, Zhang Y, Ding M, Wen Z-F, Mou Y, Dong H (2023) Nanovaccine-based strategies for lymph node targeted delivery and imaging in tumor immunotherapy. J Nanobiotechnol 21(1):236
54. Shao K, Singha S, Clemente-Casares X, Tsai S, Yang Y, Santamaria P (2015) Nanoparticle-based immunotherapy for cancer. ACS Nano 9(1):16–30
55. Luo M, Wang H, Wang Z, Cai H, Lu Z, Li Y, Du M, Huang G, Wang C, Chen X (2017) A STING-activating nanovaccine for cancer immunotherapy. Nat Nanotechnol 12(7):648–654
56. Li L, Yang Z, Chen X (2020) Recent advances in stimuli-responsive platforms for cancer immunotherapy. Acc Chem Res 53(10):2044–2054
57. Hu J, Xu X, Du Y (2022) Targeting tumor-associated macrophages for imaging. Pharmaceutics 15(1):144
58. Da Silva C, Rueda F, Löwik C, Ossendorp F, Cruz LJ (2016) Combinatorial prospects of nano-targeted chemoimmunotherapy. Biomaterials 83:308–320
59. Park J, Wrzesinski SH, Stern E, Look M, Criscione J, Ragheb R, Jay SM, Demento SL, Agawu A, Licona Limon P (2012) Combination delivery of TGF-β inhibitor and IL-2 by nanoscale liposomal polymeric gels enhances tumour immunotherapy. Nat Mater 11(10):895–905
60. Mamo T, Moseman EA, Kolishetti N, Salvador-Morales C, Shi J, Kuritzkes DR, Langer R, Andrian UV, Farokhzad OC (2010) Emerging nanotechnology approaches for HIV/AIDS treatment and prevention. Nanomedicine 5(2):269–285
61. Lin LCW, Chattopadhyay S, Lin JC, Hu CMJ (2018) Advances and opportunities in nanoparticle-and nanomaterial-based vaccines against bacterial infections. Adv Healthcare Mater 7(13):1701395
62. Kanekiyo M, Wei C-J, Yassine HM, McTamney PM, Boyington JC, Whittle JR, Rao SS, Kong W-P, Wang L, Nabel GJ (2013) Self-assembling influenza nanoparticle vaccines elicit broadly neutralizing H1N1 antibodies. Nature 499(7456):102–106
63. Zhang L, Gu F, Chan J, Wang A, Langer R, Farokhzad O (2008) Nanoparticles in medicine: therapeutic applications and developments. Clin Pharmacol Ther 83(5):761–769
64. Song S, Xia H, Guo M, Wang S, Zhang S, Ma P, Jin Y (2021) Role of macrophage in nanomedicine-based disease treatment. Drug Delivery 28(1):752–766
65. Chen Y-M, Hsiao T-H, Lin C-H, Fann YC (2025) Unlocking precision medicine: clinical applications of integrating health records, genetics, and immunology through artificial intelligence. J Biomed Sci 32(1):16
66. Rhodes K (2022) polymeric particle-based drug delivery platforms for immune modulation. Johns Hopkins University
67. Getts DR, Terry RL, Getts MT, Deffrasnes C, Müller M, van Vreden C, Ashhurst TM, Chami B, McCarthy D, Wu H (2014) Therapeutic inflammatory monocyte modulation using immune-modifying microparticles. Sci Transl Med 6(219):219ra7–219ra7
68. Shakya AK, Nandakumar KS (2013) Applications of polymeric adjuvants in studying autoimmune responses and vaccination against infectious diseases. J R Soc Interface 10(79):20120536

69. Whitehead KA, Langer R, Anderson DG (2009) Knocking down barriers: advances in siRNA delivery. Nat Rev Drug Discovery 8(2):129–138
70. Kohane DS (2007) Microparticles and nanoparticles for drug delivery. Biotechnol Bioeng 96(2):203–209
71. Alonci G, Fiorini F, Riva P, Monroy F, López-Montero I, Perretta S, De Cola L (2018) Injectable hybrid hydrogels, with cell-responsive degradation, for tumor resection. ACS Appl Bio Mater 1(5):1301–1310
72. Vancsik T (2020) Mechanism of modulated electro-hyperthermia induced tumor destruction in C26 colorectal cancer models
73. Vegas AJ, Veiseh O, Gürtler M, Millman JR, Pagliuca FW, Bader AR, Doloff JC, Li J, Chen M, Olejnik K (2016) Long-term glycemic control using polymer-encapsulated human stem cell–derived beta cells in immune-competent mice. Nat Med 22(3):306–311
74. Chou WC, Canchola A, Zhang F, Lin Z (2025) Machine learning and artificial intelligence in nanomedicine. Wiley Interdiscip Rev: Nanomedicine Nanobiotechnology 17(4):e70027
75. Wang W, Sedykh A, Sun H, Zhao L, Russo DP, Zhou H, Yan B, Zhu H (2017) Predicting nano–bio interactions by integrating nanoparticle libraries and quantitative nanostructure activity relationship modeling. ACS Nano 11(12):12641–12649
76. Kapoor DU, Sharma JB, Gandhi SM, Prajapati BG, Thanawuth K, Limmatvapirat S, Sriamornsak P (2024) AI-driven design and optimization of nanoparticle-based drug delivery systems. Science, Engineering and Health Studies, pp 24010003–24010003
77. Soldatov MA, Butova VV, Pashkov D, Butakova MA, Medvedev PV, Chernov AV, Soldatov AV (2021) Self-driving laboratories for development of new functional materials and optimizing known reactions. Nanomaterials 11(3):619
78. Elsborg J, Bhowmik A (2023) Equivariant graph-representation-based actor–critic reinforcement learning for nanoparticle design. J Chem Inf Model 63(12):3731–3741
79. Mansourinia F, Adibhosseini MS, Yazdi MK, Tabibian M (2026) Artificial intelligence-assisted biomaterials synthesis. Artificial Intelligence in Biomaterials Design and Development. Elsevier, pp 97–116
80. Tomé I, Francisco V, Fernandes H, Ferreira L (2021) High-throughput screening of nanoparticles in drug delivery. APL Bioeng 5(3)
81. Di Vizio D, Driedonks TA, Erdbrügger U, Falcon-Perez JM, Fu Q-L, Hill AF, Lenassi M, Lim SK, Mahoney MG, Mohanty S Position paper minimal information for studies of extracellular vesicles (MISEV): from basic to advanced approaches. In: Welsh JA, Goberdhan DCI, O'Driscoll L, Buzas EI, Blenkiron C, Bussolati B, Cai H
82. Garg H (2020) Digital twin technology: revolutionary to improve personalized healthcare. Sci Progress Res (SPR) 1(1)
83. De A, Jayronia S, Kuppusamy G, Park YJ (2025) AI-powered biopolymers engineering: advancements in drug delivery and efficacy. In: Biopolymers as therapeutic adjuvants: innovations and advancements. Bentham Science Publishers, pp 223–250
84. Fan Z, Hu X, Gao RX (2022) Indirect measurement methods for quality and process control in nanomanufacturing. Nanomanufacturing and Metrology 5(3):209–229
85. Kumar D, Anand P, Singh SK (2025) Mechanisms of immunotherapeutic biopolymers in autoimmune disease. In: Biopolymers as therapeutic adjuvants: innovations and advancements. Bentham Science Publishers, pp 350–378
86. Rajput D, Dwivedi A, Derashri A, Bhandari DD, Kolhe N (2024) Assessment of nanoparticle induced cytotoxicity and safety profile. Nanoparticles in Cancer Therapy. CRC Press, pp 275–300
87. Tinkle S, McNeil SE, Mühlebach S, Bawa R, Borchard G, Barenholz Y, Tamarkin L, Desai N (2014) Nanomedicines: addressing the scientific and regulatory gap. Ann N Y Acad Sci 1313(1):35–56
88. Halamoda-Kenzaoui B, Holzwarth U, Roebben G, Bogni A, Bremer-Hoffmann S (2019) Mapping of the available standards against the regulatory needs for nanomedicines. Wiley Interdisciplinary Rev: Nanomedicine and Nanobiotechnology 11(1):e1531

89. Li Y, Zhang X, Liu X, Pan W, Li N, Tang B (2021) Intelligent stimuli-responsive nano immunomodulators for cancer immunotherapy. Chem Sci 12(9):3130–3145
90. Olden BR, Cheng Y, Jonathan LY, Pun SH (2018) Cationic polymers for non-viral gene delivery to human T cells. J Control Release 282:140–147
91. Xie Z, Wroblewska L, Prochazka L, Weiss R, Benenson Y (2011) Multi-input RNAi-based logic circuit for identification of specific cancer cells. Science 333(6047):1307–1311
92. Roybal KT, Williams JZ, Morsut L, Rupp LJ, Kolinko I, Choe JH, Walker WJ, McNally KA, Lim WA (2016) Engineering T cells with customized therapeutic response programs using synthetic notch receptors. Cell, 167(2):419–432. e16

AI-Powered Nanotechnology for Combating Infectious Diseases

Bandita Datta, Sneha Thakur, and Joyeta Ghosh

Abstract Infectious diseases remain a critical global health challenge, disproportionately affecting developing nations where healthcare infrastructure and diagnostic capabilities are limited. Conventional diagnostic methods, including culture-based techniques and polymerase chain reaction, face significant limitations such as extended turnaround times, insufficient sensitivity, and vulnerability to genetic mutations in emerging pathogen variants. The convergence of artificial intelligence and nanotechnology has introduced transformative solutions that address these fundamental shortcomings through AI-enhanced biosensors, smart nanodevices, and personalized treatment platforms. This comprehensive review examines the synergistic integration of AI and nanotechnology in combating infectious diseases, focusing on diagnostic innovations, therapeutic applications, and clinical implementations. Nanomaterials, including metallic nanoparticles, carbon-based structures, polymeric systems, lipid-based carriers, and quantum dots, demonstrate unique physicochemical properties enabling ultra-sensitive pathogen detection and targeted drug delivery. Machine learning algorithms enhance these capabilities by processing complex biological data, optimizing nanomaterial properties, and enabling predictive analytics for early disease detection. Deep learning models, particularly convolutional neural networks, have revolutionized pathogen identification through analysis of microscopic images and genomic sequences, achieving diagnostic accuracy exceeding 95% in multiple applications. The review also addresses critical challenges, including data privacy concerns, algorithmic bias, regulatory hurdles, and infrastructure barriers that must be overcome for widespread clinical translation. The integration of AI and nanotechnology represents a paradigm shift from reactive diagnostic approaches towards proactive, precision-based methodologies, offering unprecedented opportunities to transform infectious disease management and advance global health equity.

B. Datta
Department of Chemistry, Amity Institute of Applied Sciences, Amity University, Kolkata, India

S. Thakur · J. Ghosh (✉)
Department of Dietetics and Applied Nutrition, Amity Institute of Applied Sciences, Amity University, Kolkata, India
e-mail: joyetaghosh01@gmail.com

J. Chen (ed.), *AI-Driven Nanotechnology in Precision Medicine and Healthcare*, Smart Nanomaterials Technology, https://doi.org/10.1007/978-981-92-1690-1_9

Keywords Nanotechnology · Infectious diseases · Machine learning · Biosensors · Point-of-care testing · Nanomedicine · Deep learning · Pathogen detection

1 Introduction

Infectious diseases continue to pose substantial challenges to global healthcare systems due to the remarkable diversity and rapid evolutionary adaptations of pathogenic organisms. Infections caused by bacteria, viruses, fungi, and parasites remain leading contributors to illness and death globally, with disproportionate impacts observed in developing nations where healthcare infrastructure and diagnostic capabilities are often limited [1]. Prompt and precise identification of infectious agents forms the cornerstone of effective clinical management, disease surveillance, and prevention of further transmission within communities. Conventional diagnostic approaches in clinical microbiology, particularly culture-based methodologies and molecular testing platforms, represent the current standard of practice. However, these established techniques face inherent limitations that compromise their effectiveness. Culture methods, despite being considered the reference standard, require extended incubation periods and frequently demonstrate insufficient sensitivity, potentially missing critical diagnostic information [2, 3]. Molecular diagnostic techniques such as polymerase chain reaction (PCR) offer accelerated turnaround times but are constrained by their targeted nature, susceptibility to environmental contamination leading to spurious positive results, and vulnerability to genetic mutations in emerging pathogen variants that may escape detection [4].

Revolutionary advances in artificial intelligence and nanotechnology have introduced innovative diagnostic solutions that address these fundamental shortcomings. Machine learning algorithms capable of integrating diverse data sources—including patient symptoms, laboratory parameters, radiological findings, and genomic information—generate sophisticated predictive models that enhance diagnostic precision, inform prognostic assessments, and facilitate individualized therapeutic strategies [5]. Simultaneously, the emergence of engineered nanomaterials, including quantum dots, colloidal gold particles, and carbon-based nanostructures, has revolutionized biosensing technology. These advanced materials enable ultra-sensitive detection of multiple pathogens simultaneously in point-of-care settings with unprecedented accuracy [6, 7].

The convergence of artificial intelligence analytics with nanosensor technology provides real-time interpretation of complex biological signals, minimizes diagnostic errors, and incorporates continuous learning capabilities to recognize novel pathogen mutations [8].

These technological innovations represent a paradigm shift from traditional reactive, one-size-fits-all diagnostic approaches towards proactive, precision-based methodologies capable of rapid pathogen identification, customized treatment selection, and continuous monitoring of therapeutic responses across diverse clinical environments [9]. This investigation examines the synergistic integration of artificial intelligence and nanotechnology, highlighting their combined potential to advance predictive diagnostics and personalized treatment paradigms for infectious disease management within the AI-Powered Nanotechnology for Combating Infectious Diseases project framework.

1.1 Types of Nanomaterials Used in Biomedicine

Nanotechnology has modernized biomedicine by introducing materials of nanoscale (1–100 nm). These materials, depending on their structure, possess unique physical, chemical, and biological properties. Based on these properties, they have exhibited excellent applications in the fields of therapeutics, diagnostics, and regenerative medicine. The major types of nanomaterials used in biomedicine are polymeric, metallic, carbon-based, lipid-based, and semiconductor nanoparticles, having unique therapeutic benefits.

1.1.1 Metallic Nanoparticles

Several metallic nanoparticles exhibit bioactivity, especially gold and silver nanoparticles. Gold nanoparticles (AuNPs) show surface plasmon resonance properties, which makes it highly valuable for use in photothermal treatment and imaging. Moreover, they can be further functionalized using various antibodies for cancer cell ablation or targeted drug delivery applications [10]. Similarly, silver nanoparticles (AgNPs) also showcase strong antimicrobial activities, which enhance their applications not only for wound dressings but also for developing medical devices for infection control. They are carefully engineered with desired size and shape and high surface area-to-volume ratio properties, which make them highly biocompatible [11].

1.1.2 Carbon-Based Nanomaterials

Graphene, fullerenes and carbon nanotubes (CNTs) are the most common carbon-based nanomaterials that possess remarkable mechanical strength and conductivity [12]. It is reported that single-walled and multi-walled CNTs can penetrate the permeable cell membrane and are used for the transportation of genes or drug molecules [13]. Graphene oxide has a large surface area, and with further functionalization, it can be employed in biosensing and tissue engineering scaffolds [14]. Scientists have proved that fullerenes, a spherical carbon molecule, exhibit excellent antioxidant properties. However, they are partially soluble and form aggregates, causing potential cytotoxicity. Surface modification of fullerenes helps in reducing aggregation with enhanced solubility [15].

1.1.3 Polymeric Nanomaterials

Micelles, hydrogels, and dendrimers fall under the category of polymeric nanomaterials. Dendrimers possess branched-like structures due to their precise molecular weights and multiple functional groups, which enable them to have a high drug-loading capacity required for targeted therapies [16]. Polymeric micelles at optimum concentration assemble themselves to form core–shell structures essential for prolonged release of hydrophobic drugs at the target cells [17]. On the contrary, nanoscale hydrogels imitate extracellular environments. These matrices are suitable for cell encapsulation in regenerative medicines [18].

1.1.4 Lipid-Based Nanomaterials

Several reports are available in the literature for lipid-based nanoparticles like solid lipid nanoparticles and liposomes, which imitate biological membranes [19]. Liposomes consist of phospholipids, which form vesicular structures enabling encapsulation of both hydrophilic and hydrophobic medications, which do not undergo any structural changes after encapsulation [20]. Such examples are available in the market, for example, a drug, Doxil, a liposome, is prescribed for cancer therapy [21]. Solid lipid nanoparticles offer combined properties of liposomes and polymeric systems. Such medications offer stability and controlled release of drugs for oral or intravenous administration [22].

1.1.5 Semiconductor Nanomaterials

Semiconductor nanomaterials like quantum dots (QDs) are nanocrystals with tunable fluorescence characteristics. QDs have performed better than the conventional dye in terms of brightness and photostability [23]. QDs, when doped with cadmium selenide, for example, have been used in multiplexed bioimaging processes at the cellular level. Furthermore, they are used in molecular diagnostics and real-time tracking of disease progression [24].

The nanomaterials mentioned above have exhibited numerous applications because of their ability to be engineered for targeted biomedical needs. In 2025, it has seen many reports on the development of hybrid nanomaterials by combining metallic cores with polymeric shells. These carefully engineered nanomaterials have expanded their utility, addressing constraints in specific classes [25].

1.2 Nano-Drug Delivery Systems

Nano-drug delivery systems (NDDS) offer a paradigm shift in pharmacotherapy by addressing bottlenecks faced by conventional medicines, for example, low solubility, off-target effects, and early metabolism [26]. NDDS are comprised of encapsulation of therapeutic agents in nanocarriers, which helps in developing drugs with controlled release at precision target and improved bioavailability. NDDS find wide applications in cancer therapy, chronic ailments, and infectious diseases [27].

Liposomal delivery is a well-known NDDS available in the market. Drugs are incorporated into the bilayer vesicles of lipid membranes containing an aqueous core, shielding them from any enzymatic degradation. Polyethylene glycol (PEG) coated PEGylated liposomes result in prolonged bioavailability and circulation of the encapsulated drugs by evading immune recognition [28].

One of the examples available in the literature is liposomal doxorubicin, which is used for its enhanced permeability and retention (EPR) effect in oncology. This drug has decreased cardiotoxicity while concentrating at the tumour locations [29]. Moreover, recent developments have been reported on stimuli-responsive liposomes that are aimed to release drugs in response to a change in pH, temperature, or light, advancing their specificity [30].

Polymeric nanoparticles, for example, those that consist of poly(lactic-co-glycolic acid) (PLGA), showcase prolonged drug release through engineered degradation rates [31]. For active targeting of the overexpressed receptors present on sick cells, they can be coupled with ligands like folic acid [32]. A combined therapy and diagnostics known as theranostics is facilitated by dendrimer-based systems. With their multivalent surfaces, they enable co-delivery of several medications or imaging agents [33].

For hydrophobic drugs, micelles and nanoemulsions have been applied. Polymeric micelles are formed through the self-assembly of amphiphilic block copolymers, resulting in a hydrophobic core surrounded by a hydrophilic shell, which makes them particularly suitable for intravenous administration. Whereas nanoemulsions, which are surfactant-stabilized oil-in water dispersions, are utilized for enhancing the oral bioavailability of poorly soluble substances [34].

Mesoporous silica nanoparticles (MSNs) and other inorganic NDDS contain both pH-sensitive caps for spontaneous release as well as high porosity for drug loading. A recent report is present on development of drug-functionalized metallic nanoparticles, which shows hyperthermia-induced release of drugs used in photothermal therapies [35].

By 2025, several smart NDDS have been reported, which integrate nanotechnology with biologics, such as CRISPR-Cas9 delivery with the usage of viral-mimicking nanoparticles. This revolutionized gene editing for genetic disorders [36]. For many such nano-formulations, FDA approvals have been received, highlighting their clinical promise; however, their use is downregulated due to the challenges faced during scaling production and guaranteed consistent particle size.

1.3 Nanosensors for Disease Detection

Nanosensors apply the sensitivity of its sizes and properties to develop them as biomarkers at extremely low concentration up to nano-level for its application as real-time monitors and early-stage diagnosis of a disease. These devices show excellent performance compared to standard assays in terms of speed, mobility, and multiplexing by converting biological signals into quantifiable outputs [37].

Optical nanosensors have surface-enhanced Raman scattering (SERS) or fluorescence signals that are used for the detection of analytes, for instance, in the case of quantum dots (QDs) or gold nanoparticles. The binding of QD-based sensors to targets like proteins or DNA produces detectable signals, allowing selective analysis of multiple biomarkers in a single sample. SERS nanosensors are used for the identification of tumour DNA for cancer screening by amplifying Raman signals to a greater extent [38].

Several reports are available on electrochemical nanosensors that integrate carbon nanotubes or graphene electrodes, presenting excellent conductivity and large surface areas. Recently, enzyme-nanoparticle-based glucose nanosensors have been developed by integrating with wearable technology for monitoring diabetes in patients [39]. Nanowire field-effect transistors provide real-time monitoring by detection of ionic changes resulting from biomarker binding. Superparamagnetic iron oxide nanoparticles (SPIONs) are employed by magnetic nanosensors to enhance MRI contrast and detect biomarkers. They facilitate rapid detection and isolation of infections or exosomes in point-of-care devices [40].

Modality integrated hybrid nanosensors have been reported, such as optoelectronic systems for identifying volatile organic compounds in breath for the diagnosis of lung cancer.

Nowadays, AI-integrated nanosensors are developed for predicting analytics and detection of viral RNA at attomolar levels. However, standardization and validation against false results need to be monitored for integration into clinical operations, regardless of the benefits of sensitivity [41] (Table 1).

1.4 Biocompatibility and Safety Considerations

Several nanomaterials can interact with the biological systems without eliciting adverse effects, which is the result of their biocompatibility. Biodistribution, immunological activation, toxicity, and long-term consequences are critical factors for translating these materials for clinical use. Biocompatible nanoparticles need to be carefully engineered with optimum size, shape, surface charge, and composition. Smaller size of the nanoparticles may cause oxidative stress or inflammation since they can cross the Blood–Brain barrier (BBB). Moreover, metallic nanoparticles like carbon-based materials may aggregate, leading to pulmonary fibrosis upon inhalation. Silver-based nanoparticles may release ions, inducing cytotoxicity [42].

Table 1 Recent advances in AI-integrated nanotechnology and biomedical applications (2020–2026)

Reference	Year	Key findings
Cheah et al. [5]	2025	Advancements in machine learning and AI for surveillance, diagnosis, and prognosis of infectious diseases
Shriwas et al. [7]	2025	The role of artificial intelligence in the specific diagnosis of viral infections and medical virology
Yuan et al. [8]	2025	New developments in infusion-based drug delivery systems targeting the brain
Badir et al. [10]	2025	Utilization of gold nanoparticles in plasmonic photothermal therapy for cancer treatment
Vishnusekhar [12]	2025	Review of the 'trio' of carbon nanomaterials (Fullerene, CNT, Graphene) and their discovery/ development
Das et al. [13]	2025	Comprehensive review on the use of carbon nanotubes for brain-targeted drug delivery
Kim and Cha [18]	2025	Development of extracellular-matrix-mimetic hydrogels using nanomaterials
Mimona et al. [23]	2025	Quantum dot nanomaterials are empowering advances in optoelectronic devices
Shukla et al. [24]	2025	Advancements in targeted quantum dot structures for enhanced cancer treatment
Ahmad et al. [29]	2025	Recent results in surfaces and interfaces regarding nanomedicine applications
Ioannou et al. [3]	2024	Application of nanotechnology in the diagnosis and treatment of antibiotic-resistant infections
Agboklu [9]	2024	The impact of AI on innovative nanotechnologies for advanced medical diagnosis
Patel et al. [25]	2024	Bioengineered metallic nanoparticles bridging the gap for medical applications (Nature to Nanomedicine)
Ramalingam et al. [6]	2023	Impact of nanotechnology on AI-based and conventional biosensing strategies for virus detection
Zenze et al.[16]	2023	Dendrimers acting as modifiers of inorganic nanoparticles for therapeutic cancer delivery
Fulton and Najahi-Missaoui [21]	2023	Current status and historical evolution of liposomes in cancer therapy
Nsairat et al. [20]	2022	Structure, composition, types, and clinical applications of liposomes

Opsonization, in which proteins wrap nanoparticles and cause macrophage clearance, is a component of immune responses. Despite reducing effects by PEGylation, repeated exposures cause anti-PEG antibodies. In vitro assays are done for the evaluation of hypersensitivity or complement activation [43].

Bioimaging techniques are utilized for tracking the nanoparticles' fate. Nanoparticles get accumulated in the liver and spleen, compelling designs for renal clearance to avoid persistence. Genotoxicity (comet assays), hemocompatibility tests, and cytotoxicity (MTT assays) are the most acceptable safety studies carried out, which adhere to ISO criteria. Chronic toxicity is evaluated in in vivo studies carried out in animals, whereas for human trials, FDA supervision is required. To date, Nanomaterial Registry databases facilitate data exchange and regulatory frameworks. These prioritize risk–benefit assessments. Moreover, advances in green synthesis strategies have been involved in lowering the pollutants. Finally, rigorous testing balances innovation with patient protection [44–46].

2 Role of Artificial Intelligence in Nanomedicine

Nanomedicine has become one of the flourishing fields working at the nexus of nanotechnology and healthcare. It has the potential to revolutionize diagnosis, treatment, and prevention of diseases. The integration of artificial intelligence (AI) into nanomedicine, which uses sophisticated computational methods, has enhanced the capabilities of nanotechnology. This section explores the multifaceted roles of AI in nanomedicine for drug formulation.

2.1 Machine Learning Algorithms for Drug Design

'Drug Discovery' consists of several steps, making the whole process lengthy, costly, and fraught with uncertainty. In this domain, machine learning (ML) algorithms have emerged as powerful tools, thus allowing researchers to accelerate the process of drug discovery. Huge datasets of biological activities, chemical substances, and pharmacokinetic features can be investigated using ML algorithms, which can identify patterns that may not be apparent through conventional methods [47].

One of the main utilities of ML in drug design is to predict the molecular bioactivity. Researchers can anticipate the efficacy of new chemical entities (NCEs) against specific biological targets by employing supervised learning techniques. For instance, algorithms based on random forests and support vector machines (SVMs) have been successfully employed for the prediction of the binding energies of the protein-NCE complexes, facilitating identification of promising potent lead compounds [48].

Furthermore, unsupervised learning methods like clustering algorithms can categorize the NCE based on their biological activity or chemical structures, assisting in the recognition of novel drug candidates. These algorithms can generate de novo designs of molecules in which AI generates NCEs that can exhibit desired potencies against a specific target [48]. Incorporation of ML into drug design enhances the process of drug design. Consequently, it also improves the precision in prediction which may result in drugs with less toxicity and enhanced efficacy [49–52].

2.2 *Deep Learning in Pathogen Detection*

Pathogen detection has also progressed by the application of deep learning, which is a subset of ML integrated with multi-layered neural networks [53]. Deep learning models are especially well-suited for recognizing infections in a variety of biological samples, such as blood, saliva, and tissue, because of their capacity to process and analyse complex data [54, 55]. Convolutional neural networks (CNNs), a popular method in deep learning, have been utilized for identifying infections by analysing the pictures from microscopy and other imaging methods. These models can identify intricate features from the images and predict the infected cells among the healthy cells precisely [56]. For example, CNNs were used to detect bacterial and viral infections based on the analysis report of the microscopic images of infected tissues, significantly cutting down on the time needed for diagnosis [57].

Subsequently, deep learning models were developed to analyse the genomic sequence of pathogenic species. Based on the large datasets of genetic information, these models enable rapid and precise identification of similar patterns indicative of specific pathogens [58]. A prompt diagnosis is essential for a successful public health response, which can be achieved by deep learning models. Additionally, an improved diagnostic precision is a result of the use of deep learning in pathogen detection makes it feasible to create point-of-care testing tools that can identify illnesses in environments with resource-limited settings.

2.3 *AI for Nanomaterial Optimization*

The generation of nanomaterials for their applications in medicine necessitates a deep, comprehensive understanding of their characteristics and interactions at the nanoscale. AI plays a pivotal role for researchers in engineering nanomaterials with specific characteristics required for their applications, such as drug delivery, imaging, and therapeutic interventions [59]. Based on the composition and structure of the nanomaterials, ML algorithms can be utilized for predicting their specific properties. For example, algorithms were used to design nanoparticles with ideal size, shape, and surface modifications, based on the previous reports, for improved drug delivery properties. By leveraging these predictive capabilities, researchers can reduce the trial-and-error approach traditionally associated with material development [60]. As demonstrated in (Fig. 1), ML can speed up the process of designing and synthesizing

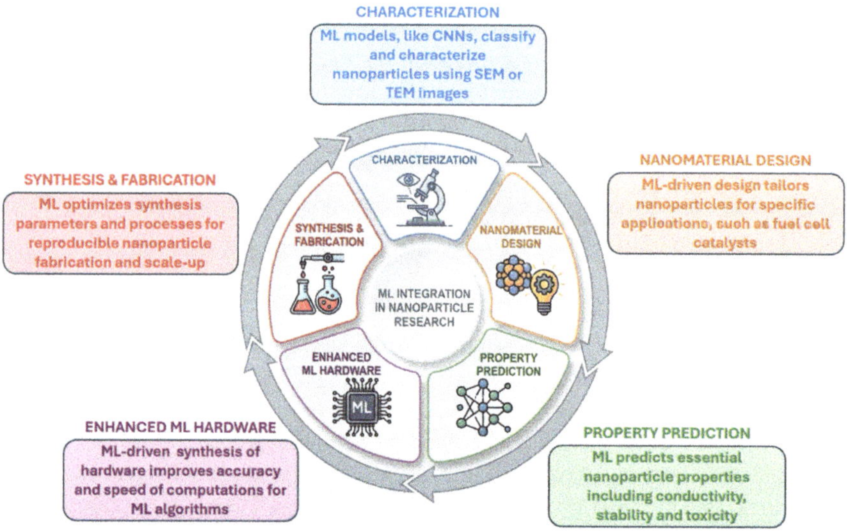

Fig. 1 Predictive model showcasing the convergence and collaboration of ML and nanotechnology

nanoparticles with desired properties using trained models, accelerating progress in the field.

Moreover, nanomaterials with specified functionalities can be designed and formulated using AI models. For example, reinforcement learning algorithms were used for the optimization of the nanoparticles synthetic procedure, which enabled the researchers to achieve nanoparticles with desired properties [61]. This dynamic approach not only accelerates the development of nanomaterials but also enhances their performance in clinical applications. The application of AI in nanomaterial optimization signifies advancement in the field, permitting rapid development of engineered nanotherapeutics with improved potencies [62].

2.4 Data Integration and Predictive Modelling

The advancement of nanomedicine research depends on the integration of various datasets. AI helps us in developing coherent predictive models established by the combination of data from multiple sources, like lab tests, empirical evidence, and clinical trials. These models can help guide decision-making in both research and clinical contexts by offering insights into the safety and effectiveness of nanomedicines [63]. AI-based predictive modelling assists in the identification of patient populations that are most likely to benefit from specific nanomedicines. AI systems can process the available patient data, such as genetic profiles, medical histories, and demographic information, and predict a probable treatment response. This has enabled the customization of treatments. This strategy is in line with precision medicine's tenets, which call for customized treatments based on each patient's unique traits [64].

Additionally, AI-driven data integration facilitates the identification of possible side effects associated with non-medical products. Safer design of nanotherapeutics can be achieved by drawing a correlation between the nanomaterial features and toxicological effects through the analysis of large datasets using AI [65]. A revolutionary development in predictive modelling provides us with the capacity to integrate and analyse thwarted datasets, empowering researchers to make learned choices that improve the effectiveness and safety of therapeutic interventions [66].

To summarize, AI has been identified as a multifaceted and transformative method in the field of nanomedicine. AI has high potential to modernize the landscape of healthcare by formulating deep learning models, improving disease identification, and using machine learning algorithms to expedite drug research. AI's performance in optimizing the properties of nanomaterials and integrating the various datasets for predictive modelling highlights today's technological potential in advancing medical research and medical research. Innovative solutions will be generated, proving the collaboration of AI with nanotechnology, a revolutionary development in the field of nanomedicine, to address some of the most pressing challenges in healthcare.

3 AI-Nanotech Synergy in Diagnostics, Nanotherapeutics, and Clinical Applications

The convergence of artificial intelligence with nanotechnology has fundamentally transformed clinical diagnostics and therapeutic strategies, enabling rapid, sensitive, and accurate disease detection through AI-enhanced biosensors, smart nanodevices, and personalized treatment platforms. This integration represents a paradigm shift from conventional laboratory-based diagnostics towards point-of-care testing (POCT), real-time monitoring, and precision medicine. AI-nanotech synergy achieves remarkable diagnostic sensitivity and specificity while simultaneously enabling targeted nanotherapeutics guided by machine learning optimization, particularly evident in managing infectious diseases, cancer, and antimicrobial resistance.

3.1 AI-Nanotech Synergy in Diagnostics

3.1.1 AI-Enhanced Biosensors and Wearables

Artificial Intelligence integration with nanosensors and wearable devices has created unprecedented capabilities for continuous, non-invasive health monitoring through physiological biomarker detection and real-time data analysis. Wearable biosensors empty electrochemical, optical, colorimetric and pressure based transduction mechanisms, utilizing nanomaterials such as graphene, gold nanoparticles (AuNPs), carbon nanotubes (CNTs), metal–organic frameworks (MOFs), and quantum dots to achieve sensitives in the femtomolar to picomolar range. AI algorithms enhance these systems by processing complex multivariate sensor data, reducing noise and interference, identifying subtle biomarker fluctuations and providing predictive analytics for early disease detection. Machine learning models, including convolutional neural networks (CNNs), recurrent neural networks (RNNs), support vector machines (SVMs), and random forests (RFs), analyse temporal patterns in physiological signals such as heart rate variability (HRV), oxygen saturation (SpO2), electrocardiographic (ECG) patterns, and temperature variations. For example, smartwatch-integrated HRV monitoring combined with machine learning algorithms successfully detected COVID-19 infection up to nine days before symptom onset with 81% accuracy and 88% pre-symptomatic detection rates. Wearable AI-driven glucose monitors demonstrate continuous glucose monitoring in diabetes management, delivering real-time alerts for hypoglycaemic/hyperglycaemic episodes while providing glucose trend forecasting through Long Short-Term Memory (LSTM) neural networks with root-mean-square errors of 19.1 mg/dL [67–70].

3.2 *Smart Nanodevices for Point-of Care-Testing*

Point-of-care testing (POCT) utilizing AI-enhanced nanosensors represents a revolutionary paradigm enabling rapid, decentralized diagnostics at clinical sites, homes, or remote locations. These devices integrate nanobiosensors with microfluidic platforms, lateral flow assays, and paper-based analytical devices (PADs) to achieve a detection limit below 10 femtomolar concentrations, enabling identification of pathogenic biomarkers within 5–60 min. Lateral flow assays incorporating gold nanoparticles demonstrate visible colorimetric detection without requiring specialized instrumentation, with COVID-19 antigen tests achieving 5–15 min turnaround times utilizing AuNP surface plasmon resonance shifts. AI-assisted optical nanosensors achieved 97.12% accuracy in detecting sexually transmitted viruses through machine learning analysis of photonic crystal sensors data. Electrochemical nanosensors functionalized with molecularly imprinted polymers (MIPs) and integrated with support vector machine classifiers achieved 95.2% accuracy for rapid COVID-19 diagnosis from respiratory samples, with the chemiresistive platform demonstrating 100% accuracy when analysing non-extracted viral RNA samples. Lab-on-chip platforms incorporating nanomaterial arrays enable simultaneous detection of multiple biomarkers, representing advances towards personalized medicine through multiplexed analysis. Smart nanodevices achieve WHO-defined ASSURED criteria (Affordable, Sensitive, Specific, User-friendly, Rapid, Equipment-free, Deliverable to end-user), democratizing diagnostic access across diverse healthcare settings [71–74].

3.3 Real-Time Monitoring of Infection Progression

AI-integrated nanosensors enable longitudinal tracking of infection dynamics through continuous biomarker monitoring, facilitating adaptive therapeutic interventions based on disease trajectory analysis. Nanosensor-based surveillance system detects pathogen-specific biomarkers, including viral nucleic acids, proteins, and host immunological responses, providing real-time infection status updates comparable to or exceeding conventional laboratory methods. For infectious disease, surface-enhanced Raman scattering (SERS) nanosensors integrated with ResNet-18 deep learning architecture achieved 94.52% diagnostic accuracy of COVID-19 detection while reducing assay time to 1–2 min through AI-based result verification. Machine learning algorithms analyse nanosensor output patterns to discriminate disease severity stages, enabling risk stratification for clinical prioritization, critical during pandemic surges when health care resource allocation determines patient outcomes. Wearable devices monitoring multiple physiological parameters (HRV, SPO_2, temperature, respiratory rate) integrated with fusion AI algorithms provide multimodal disease status assessment, detecting COVID-19 associated hypoxemia and autonomic dysfunction before symptomatic presentation. Real-time monitoring of antimicrobial resistance (AMR) markers via AI-enhanced nanosensors reduces susceptibility testing turnaround times from 72 h to under 4 h, enabling rapid antimicrobial stewardship decisions [75–80].

3.4 AI-Assisted Antiviral and Antimicrobial Nanomedicines

Artificial intelligence discovery of antimicrobial nanomaterials has accelerated the development of broad-spectrum antivirals and bactericidal agents capable of combating drug-resistant pathogens through machine learning screening of billions of chemical structures. Silver and copper nanoparticles exhibit inherent antimicrobial properties through reactive oxygen species (ROS) generation and membrane disruption. Mechanisms and AI optimization of nanoparticle concentration, size, and surface coatings maximize pathogen killing while minimizing host cell toxicity. Halicin, the first antibiotic discovered exclusively through AI screening, emerged from machine learning analysis of 2300 compounds tested against Escherichia coli, successfully inhibiting multidrug-resistant Acinetobacter baumannii both in *vitro* and in vivo—representing paradigm validation that AI accelerates antimicrobial discovery across the 'tree of life', including extinct organisms. Deep generative models design 'new-to-nature' antimicrobial molecules from genomic and proteomic data, with de la Funete's laboratory identifying functional antimicrobial peptides from Neanderthal and Denisovan protein sequences that demonstrated efficacy against clinical isolates. Nano-formulates of existing antivirals (dexamethasone, remdesivir) enhance bioavailability and tissue penetration, and AI predicts optimal lipid ratio for mRNA nanovaccines-the BNT162b2 COVID-19 vaccine employing lipid nanoparticles represents clinical validation of AI-designed immunological nanotherapeutics [68, 81, 82].

4 Case Studies and Clinical Applications

4.1 *COVID-19: AI-Driven Nanotech Solutions*

The COVID-19 pandemic catalysed the advancement of AI-nanotech platforms for simultaneous viral detection, therapeutics, and vaccination, demonstrating the clinical utility of integrated diagnostic-therapeutic ecosystems. Gold nanoparticles (AuNP)-based lateral flow immune-chromatographic assays provided rapid antibody detection within 5–15 min, fundamentally altering pandemic diagnostics by enabling decentralized screening without laboratory infrastructure. Sona Nanotech's platform detected COVID-19 biomarkers (IgG/IgM) through plasmonic colour shifts visible to unaided eyes [83]. AI-integrated nanosensors achieved 100% sensitivity and 90% accuracy in COVID-19 detection from saliva samples using bio-inspired peptide-based electrochemical biosensors, providing non-invasive, rapid screening amenable to population-scale testing [84]. Magnetic nanoparticles (MNP)-based RNA extraction platforms integrated with portable readout devices facilitated RT-PCR acceleration, reducing detection time from hours to minutes through AI-optimized magnetic field protocols [85]. Quantum dot (QD)-based fluorescence nanosensors achieved superior performance to conventional methods, with clinical evaluations across hundreds of samples demonstrating significantly reduced false-negative rates and enhanced early detection capabilities, establishing QD as a transformative pandemic response tool [86]. Lipid nanoparticles (LNP) formulations for mRNA COVID-19 vaccines were optimized through machine learning prediction of ionisable lipid combinations, enabling rapid vaccine development during pandemic emergency conditions. The Pfizer-BioNTech BNT 162b2 vaccine, achieving 95% efficacy, represented the successful translation of AI-designed nanotherapeutics. AI-assisted nanosensors combined ultralow detection limits (femtomolar concentrations), real-time data processing, and portable form factors, offering critical enhancements for variants differentiation and large-scale surveillance [83–88].

4.2 *Tuberculosis and Malaria Detection Using Smart Nanopores*

Nanobiosensor-based diagnostics revolutionized TB detection through rapid identification of *Mycobacterium tuberculosis genetic and proteomic* biomarkers, achieving sensitivity and specificity exceeding traditional microscopy and culture-based methods. Gold nanoparticle-based colorimetric assays for TB DNA detection delivered visual results within 2 min, enabling naked-eye evaluation without Spectrophotometric instrumentation, critical for resource-limited settings where TB burden remains highest. Graphene-enhanced nanosensors functionalized with DNA probes achieved femtomolar-level detection of TB-specific IS6110 sequences, with limits of detection (LOD) as low as 1.8 picomolar, enabling early infection diagnosis from minimal sputum samples [89]. Electrochemical nanobiosensors exploring self-assembled monolayers (SAMs) and MutS protein recognition elements achieved LOD of 1.8×10^{-11} μM, successfully detecting TB single nucleotide polymorphisms (SNPs) in clinical synovial fluid. Machine learning optimization of electrode materials (reduced graphene oxide ribbons with AuNP loading) enhanced conductivity and signal transduction, achieving 100% detection accuracy with XGBoost algorithms applied to terahertz metamaterial biosensor output [90]. Multiplexed nanoprobe platforms utilizing single gold nanoprobes achieved simultaneous detection of both *Mycobacterium tuberculosis* and *Plasmodium falciparum*, enabling combined screening of co-endemic pathogens in dual-infected populations. TB DNA detection platforms eliminated the need for polymerase chain reaction (PCR) amplification through direct nanosensor output, reducing turnaround time from hours to minutes while simplifying procedures for rural diagnostics facilities [89–91].

5 Challenges and Ethical Considerations in Nanotechnology for Healthcare

Although nanotechnology has numerous advances like improved diagnostics, tailored medication administration, and novel therapeutic approaches, it has a few limitations and challenges that still need to be addressed. The following are the three crucial areas that need further probe as we investigate the complexities surrounding the application of nanotechnology in healthcare: (i) algorithmic bias and data privacy; (ii) regulatory obstacles and clinical translation; and (iii) infrastructure, cost, and accessibility limitations. To guarantee the ethical development and application of nanotechnology in healthcare, each of these dimensions poses difficulties that must be overcome.

5.1 Data Privacy and Algorithmic Bias

Large volumes of datasets need to be collected frequently and analysed for the integration of nanotechnology into healthcare. Based on this data only, personalized treatment plans and improved patient diagnosis need to be developed using AI. However, this raises consciousness in the patients regarding their health data privacy owing to possible misuse of their health information [92]. The applications of ML algorithms in nanotechnology aspects to assess health data aggravate the ethical problems of data privacy. Moreover, algorithmic bias can creep into the results obtained due to inadvertently perpetuated biases present in the training data [93].

Inappropriate assumptions made during the ML process can lead to biased outcomes generated by the ML model. This can be manifested in various ways, like incorrect diagnosis, unfair treatment recommendations, or discrepancies in access to care [94]. For example, a predictive model may perform badly for people outside of a particular demographic group if the dataset used to train the model primarily contains data from that group. This can aggravate the health disparities and jeopardize the whole process of assured personalized medicine that nanotechnology seeks to deliver to the patients [95].

In order to tackle this bottleneck, healthcare practitioners, researchers, and policymakers must give top priority to data governance frameworks that guarantee strong data protection safeguards [96]. Moreover, efforts must be made to develop representative and diverse datasets for algorithm training to avoid algorithm bias. Consequently, continuous monitoring and auditing of algorithms for accuracy and fairness must be carried out for safeguarding against algorithmic bias, providing the best healthcare applications of nanotechnology [97].

5.2 Regulatory Hurdles and Clinical Translation

There are several regulatory obstacles faced during paving the pathway from laboratory research to clinical applications of nanotechnology in healthcare. The properties of nanomaterials vary from the bulk properties, which makes it complicated for the regulatory landscape. The current regulatory systems, like the U.S. Food and Drug Administration (FDA), do not cover the complexities associated with nanotechnology [98].

Nanoparticles' therapeutic applications and toxicity depend on their size, shape, surface chemistry, and aggregation state. Hence, developing consistent procedures for the evaluation and characterization of nanomaterials is essential for regulatory approval. Furthermore, extensive preclinical and clinical research is required before widespread use because the long-term consequences of nanomaterials on human health and the environment are still mostly unknown [99, 100].

Interdisciplinary collaboration between researchers, physicians, and regulatory bodies are the requirement of the process of clinical translational. Navigating the complexity of nanotechnology requires effective communication and teamwork, yet these exchanges are frequently constrained by institutional silos and conflicting goals. However, advances can be observed in this field with proper streamlining of the regulatory process and encouraging collaborative frameworks, which will expedite the usage of these novel therapies for patients more effectively [101].

5.3 *Cost, Accessibility, and Infrastructure Barriers*

Despite the enormous potential advantages of nanotechnology in healthcare, widespread adoption is severely hampered by the related expenses and infrastructure needs. Research, development, and manufacturing expenditures for nanotechnology-based solutions are frequently high. For small and medium-sized businesses (SMEs), which lack the resources of larger pharmaceutical corporations, these expenses may be unaffordable [102].

Furthermore, patients may not be able to get nanotechnology-based treatments due to their high costs, especially in environments with limited resources. Prioritizing cost-effective therapies becomes crucial as healthcare systems struggle with financial restrictions. If nanotechnology applications continue to be expensive, there is a chance that wealthy people will gain the most from them, worsening already-existing health disparities [103].

The adoption of nanotechnology in healthcare is also significantly impacted by infrastructure limitations. The creation and use of nanotechnology-based solutions require sophisticated production processes, specialized machinery, and qualified workers [104]. The deployment of these technologies can be hampered in many areas, especially in developing nations, by inadequate infrastructure. To guarantee that everyone can benefit from nanotechnology, it is imperative to close these infrastructure gaps. Stakeholders must work together to create sustainable business models that enable the conversion of nanotechnology advancements into reasonably priced healthcare solutions in order to overcome these obstacles. A comprehensive approach to increasing the accessibility of nanotechnology in healthcare must include public–private collaborations, government incentives, and infrastructural investments [105].

6 Conclusion

The integration of artificial intelligence with nanotechnology has fundamentally transformed clinical diagnostics and precision medicine, enabling rapid, sensitive, and accurate disease detection through AI-optimized biosensors while simultaneously guiding the development of targeted nanotherapeutics. AI-nanotech platforms have demonstrated transformative clinical impact in pandemic response, infectious disease detection, and antimicrobial resistance management, with particular achievements in COVID-19 diagnostics (5–15 min turnaround), TB detection (2-min visual assays), and AMR surveillance (4-h susceptibility profiling). Current challenges, including limited dataset availability, regulatory approval processes, data privacy concerns, and clinical validation requirements, necessitate continued multicentre research and standardization efforts, yet advances in federated learning, edge computing, and explainable AI provide pathways forward. Future directions emphasize multimodal biosensor fusion enabling simultaneous detection of multiple disease biomarkers, real-time treatment adaptation algorithms, and deployment of AI-nanoplatforms in resource-limited settings where diagnostics infrastructure remains inadequate, ultimately democratizing precision medicine globally.

References

1. Mardani M, Nasiri MJ (2025) The role of artificial intelligence in the diagnosis of infectious diseases: a transformative shift in global health. Arch Clinical Infectious Diseases 20(2)
2. The Hidden Crisis in Infectious Disease Diagnostics (2025) Micronbrane. https://micronbrane.com/hidden-crisis-in-infectious-disease-diagnostics/
3. Ioannou P, Baliou S, Simonis G (2024) Nanotechnology in the diagnosis and treatment of antibiotic-resistant infections. Antibiotics (Basel, Switzerland) 13(2):121. https://doi.org/10.3390/antibiotics13020121
4. Bloom DE, Black S, Rappuoli R (2017) Emerging infectious diseases: a proactive approach. Proc Natl Acad Sci 114(16):4055–4059
5. Cheah BCJ, Vicente CR, Chan KR (2025) Machine learning and artificial intelligence for infectious disease surveillance, diagnosis, and prognosis. Viruses 17(7):882. https://doi.org/10.3390/v17070882
6. Ramalingam M, Jaisankar A, Cheng L, Krishnan S, Lan L, Hassan A, Marrazza G et al (2023) Impact of nanotechnology on conventional and artificial intelligence-based biosensing strategies for the detection of viruses. Discover Nano 18(1):58
7. Shriwas PK, Sundaravadivazhagan B, Patil YM (2025) Role of AI in diagnosis of viral infection. In: Role of artificial intelligence, telehealth, and telemedicine in medical virology. Springer Nature Singapore, Singapore, pp 31–51
8. Yuan T, Yang Y, Zhan W, Rodriguez y Baena F, Dini D (2025) Infusion-based drug delivery to the brain: what is next? Expert Opinion Drug Delivery 1–6
9. Agboklu M (2024) The impact of artificial intelligence on innovative nanotechnologies for advanced medical diagnosis. J Nanotechnol Res 6:1–5
10. Badir A, Refki S, Sekkat Z (2025) Utilizing gold nanoparticles in plasmonic photothermal therapy for cancer treatment. Heliyon 11(4):e42738. https://doi.org/10.1016/j.heliyon.2025.e42738
11. Burduşel AC, Gherasim O, Grumezescu AM, Mogoantă L, Ficai A, Andronescu E (2018) Biomedical applications of silver nanoparticles: an up-to-date overview. Nanomaterials (Basel) 8(9):681. https://doi.org/10.3390/nano8090681
12. Vishnusekhar JA (2025) The trio of carbon nanomaterials: fullerene, CNT, and graphene—a review of their discovery and development in the last decade. Premier J Sci 9:100075
13. Das S, Roy S, Dinda SC, Bose A, Mahapatra C, Basu B, Prajapati B (2025) Carbon nanotubes in brain targeted drug delivery: a comprehensive review. Results Chem 15:102206. https://doi.org/10.1016/j.rechem.2025.102206
14. Bai RG, Muthoosamy K, Manickam S, Hilal-Alnaqbi A (2019) Graphene-based 3D scaffolds in tissue engineering: fabrication, applications, and future scope in liver tissue engineering. Int J Nanomed 14:5753–5783. https://doi.org/10.2147/IJN.S192779
15. Bakry R, Vallant RM, Najam-ul-Haq M, Rainer M, Szabo Z, Huck CW, Bonn GK (2007) Medicinal applications of fullerenes. Int J Nanomedicine 2(4):639–49. PMID: 18203430; PMCID: PMC2676811
16. Zenze M, Daniels A, Singh M (2023) Dendrimers as modifiers of inorganic nanoparticles for therapeutic delivery in cancer. Pharmaceutics 15(2):398. https://doi.org/10.3390/pharmaceutics15020398
17. Venditti I (2019) Morphologies and functionalities of polymeric nanocarriers as chemical tools for drug delivery: a review. J King Saud Univ Sci 31(3):398–411. https://doi.org/10.1016/j.jksus.2017.10.004

18. Kim DG, Cha GD (2025) Extracellular-matrix-mimetic hydrogels by using nanomaterials. Int J Mol Sci 26(11):4987. https://doi.org/10.3390/ijms26114987
19. García-Pinel B, Porras-Alcalá C, Ortega-Rodríguez A, Sarabia F, Prados J, Melguizo C, López-Romero JM (2019) Lipid-based nanoparticles: application and recent advances in cancer treatment. Nanomaterials (Basel) 9(4):638
20. Nsairat H, Khater D, Sayed U, Odeh F, Al Bawab A, Alshaer W (2022) Liposomes: structure, composition, types, and clinical applications. Heliyon 8(5):e09394. https://doi.org/10.1016/j.heliyon.2022.e09394
21. Fulton MD, Najahi-Missaoui W (2023) Liposomes in cancer therapy: How did we start and where are we now. Int J Mol Sci 24(7):6615. https://doi.org/10.3390/ijms24076615
22. Mukherjee S, Ray S, Thakur RS (2009) Solid lipid nanoparticles: a modern formulation approach in drug delivery system. Indian J Pharm Sci 71(4):349–358. https://doi.org/10.4103/0250-474X.57282
23. Mimona MA, Rimon MIH, Zohura FT, Sony JM, Rim SI, Arup MMR, Mobarak MH (2025) Quantum dot nanomaterials: empowering advances in optoelectronic devices. Chem Eng J Adv 21:100704. https://doi.org/10.1016/j.ceja.2025.100704
24. Shukla N, Cárdenas CY, Chanderiya A, Polozhentsev OE, Das R, Vyas S, Mukhanova E, Soldatov A, Belbekhouche S (2025) Advancements in targeted quantum dots structures for enhanced cancer treatment. Pharmaceutics 17(11):1396. https://doi.org/10.3390/pharmaceutics17111396
25. Patel J, Kumar GS, Roy H et al (2024) From nature to nanomedicine: bioengineered metallic nanoparticles bridge the gap for medical applications. Discover Nano 19:85. https://doi.org/10.1186/s11671-024-04021-9
26. Patra JK, Das G, Fraceto LF, Campos EVR, Rodriguez-Torres MDP, Acosta-Torres LS, Diaz-Torres LA, Grillo R, Swamy MK, Sharma S, Habtemariam S, Shin HS (2018) Nano based drug delivery systems: recent developments and future prospects. J Nanobiotechnology 16(1):71. https://doi.org/10.1186/s12951-018-0392-8
27. Miele E, Spinelli GP, Miele E, Di Fabrizio E, Ferretti E, Tomao S, Gulino A (2012) Nanoparticle-based delivery of small interfering RNA: challenges for cancer therapy. Int J Nanomedicine 7:3637–3657. https://doi.org/10.2147/IJN.S23696
28. Nosova AS, Koloskova OO, Nikonova AA, Simonova VA, Smirnov VV, Kudlay D, Khaitov MR (2019) Diversity of PEGylation methods of liposomes and their influence on RNA delivery. Medchemcomm 10(3):369–377. https://doi.org/10.1039/c8md00515j
29. Ahmad MZ, Mustafa G, Abdel-Wahab BA, Pathak K, Das A, Sahariah JJ, Kalita P, Alam A, Borthakur PP (2025) Results in Surfaces and Interfaces 19:10047
30. Lee Y, Thompson DH (2017) Stimuli-responsive liposomes for drug delivery. Wiley Interdiscip Rev Nanomed Nanobiotechnol 9(5). https://doi.org/10.1002/wnan.1450
31. Makadia HK, Siegel SJ (2011) Poly lactic-co-glycolic acid (PLGA) as biodegradable controlled drug delivery carrier. Polymers 3(3):1377–1397. https://doi.org/10.3390/polym3031377
32. Moise-Crintea A, Constantin AM, Jianu EM, Orlea IM, Manea M, Cojocariu RO, Carpa R, Borlea BA, Boznea CM, Coseriu RL, Sovrea A (2025) Advancing lung cancer treatment: a comprehensive review of photodynamic therapy and nanoparticle applications. Pharmaceutics 17(12):1579. https://doi.org/10.3390/pharmaceutics17121579
33. Kesharwani P, Iyer AK (2015) Recent advances in dendrimer-based nanovectors for tumor-targeted drug and gene delivery. In: Drug Discovery Today. https://doi.org/10.1016/j.drudis.2014.12.012
34. Ghezzi M, Pescina S, Padula C, Santi P, Del Favero E, Cantù L, Nicoli S (2021) Polymeric micelles in drug delivery: an insight of the techniques for their characterization and assessment in biorelevant conditions. J Control Release 10(332):312–336. https://doi.org/10.1016/j.jconrel.2021.02.031
35. Patel T, Clipstone C, Girija UV, Ahmad Z, Singh N (2025) Nanotoxicity of porous silica nanoparticles: physicochemical properties and mechanistic cellular endpoints. Nanomaterials (Basel) 15(23):1766. https://doi.org/10.3390/nano15231766

36. Liu X, Gao M, Bao J (2025) Precisely targeted nanoparticles for CRISPR-Cas9 delivery in clinical applications. Nanomaterials 15(7):540. https://doi.org/10.3390/nano15070540
37. Swierczewska M, Liu G, Lee S, Chen X (2012) High-sensitivity nanosensors for biomarker detection. Chem Soc Rev 41(7):2641–2655. https://doi.org/10.1039/c1cs15238f
38. Bhattacharjee G, Baruah N, Kumar A, Verma YK, Kundu LM, Jana S, Manna S (2026) Medical devices in modern healthcare. In: Jana S, Manna S (eds) Medical devices in modern healthcare: current research on new materials and engineering, device evaluation and regulatory implications, 1st edn. Academic Press, p 235
39. Safarkhani M, Aldhaher A, Heidari G, Zare EN, Warkiani ME, Akhavan O, Huh YS, Rabiee N (2024) Nanomaterial-assisted wearable glucose biosensors for noninvasive real-time monitoring: pioneering point-of-care and beyond. Nano Mater Sci 6(3):263–283. https://doi.org/10.1016/j.nanoms.2023.11.009
40. Li H, Li D, Chen H, Yue X, Fan K, Dong L, Wang G (2023) Application of silicon nanowire field effect transistor (SiNW-FET) biosensor with high sensitivity. Sensors 23(15):6808. https://doi.org/10.3390/s23156808
41. Chaudhary V, Bhadola P (2025) Artificial Intelligence-powered nanosensor platforms for non-invasive breathomic diagnostics. Nanotechnol Sci Appl 20(18):611–641. https://doi.org/10.2147/NSA.S546714
42. Zhang XQ, Xu X, Bertrand N, Pridgen E, Swami A, Farokhzad OC (2012) Interactions of nanomaterials and biological systems: implications to personalized nanomedicine. Adv Drug Deliv Rev 64(13):1363–1384. https://doi.org/10.1016/j.addr.2012.08.005
43. Shi D, Beasock D, Fessler A, Szebeni J, Ljubimova JY, Afonin KA, Dobrovolskaia MA (2022) To PEGylate or not to PEGylate: immunological properties of nanomedicine's most popular component, polyethylene glycol and its alternatives. Adv Drug Deliv Rev 180:114079
44. Silva DF, Melo ALP, Uchôa AFC, Pereira GMA, Alves AEF, Vasconcellos MC, Xavier-Júnior FH, Passos MF (2023) Biomedical approach of nanotechnology and biological risks: a mini-review. Int J Mol Sci 24(23):16719. https://doi.org/10.3390/ijms242316719
45. Pratiwi FW, Kuo CW, Chen BC, Chen P (2019) Recent advances in the use of fluorescent nanoparticles for bioimaging. Nanomedicine (Lond) 14(13):1759–1769. https://doi.org/10.2217/nnm-2019-0105
46. Jakic K, Selc M, Razga F, Nemethova V, Mazancova P, Havel F, Sramek M, Zarska M, Proska J, Masanova V, Uhnakova I, Makovicky P, Novotova M, Vykoukal V, Babelova A (2024) Long-term accumulation, biological effects and toxicity of BSA-coated gold nanoparticles in the mouse liver, spleen, and kidneys. Int J Nanomedicine 8(19):4103–4120. https://doi.org/10.2147/IJN.S443168
47. Roy Choudhury S, Singh K, Koner S, Ghosh J (2025) Technology, artificial intelligence, and the era of information in healthcare and allied sciences In: Sustainable development goals (SDG) and its intersection with health and well-being (Chapter 16). ISBN: 9798369397565 https://doi.org/10.4018/979-8-3693-9755-8.ch016
48. Sun L-Z, Kuang X-N, Lin S, Jin Q-H et al (2019) Nine heteroleptic copper(I)/silver(I) complexes prepared from phosphine and diimine ligands: syntheses, structures and terahertz spectra. Polyhedron 175:114177. https://doi.org/10.1016/j.poly.2019.114177
49. Visan AI, Negut I (2024) Integrating artificial intelligence for drug discovery in the context of revolutionizing drug delivery. Life 14(2):233. https://doi.org/10.3390/life14020233
50. Ghosh J, Roy A (2025) Reframing cancer as an immunological disorder through personalized nutrition and AI integration. In: AI-assisted computational approaches for immunological disorders (Chapter 12). ISBN: 9798369397268,https://doi.org/10.4018/979-8-3693-9725-1.ch012
51. Ghosh J, Andres F, Ali HH, Tanabe S (eds) (2025) Harnessing AI and machine learning for precision wellness IGI Global. https://doi.org/10.4018/978-8-3693-9521-9, ISBN: 9798369395219
52. Ishfaq M (ed) (2025) Computational methods in medicinal chemistry, pharmacology, and toxicology, 1st edn. Academic Press. https://www.elsevier.com/books

53. Al Meslamani AZ, Sobrino I, de la Fuente J (2024) Machine learning in infectious diseases: potential applications and limitations. Ann Med 56(1):2362869. https://doi.org/10.1080/07853890.2024.2362869
54. Ghosh J, Taneja J, Kant R (2025) Decoding host-pathogen interactions in Staphylococcus aureus: Insights into allelic variation and antimicrobial resistance prediction using artificial intelligence and machine learning-based approaches. bioRxiv. https://doi.org/10.1101/2025.02.18.638850
55. Taneja J et al (2025) Artificial Intelligence in predicting, diagnosing and preventing sexually transmitted infections (STIs). Venereology 4(2):5. https://doi.org/10.3390/venereology4020005
56. Liu M, Du X, Hu J, Liang X, Wang H (2024) Utilization of convolutional neural networks to analyze microscopic images for high-throughput screening of mesenchymal stem cells. Open Life Sci 19(1):20220859. https://doi.org/10.1515/biol-2022-0859
57. Abu-El-Ruz R, AbuHaweeleh MN, Hamdan A, Rajha HE, Sarah JM, Barakat K, Zughaier SM (2025) Artificial intelligence in bacterial infections control: a scoping review. Antibiotics 14(3):256. https://doi.org/10.3390/antibiotics14030256
58. Aarif KOM, Hasan VMY, Alam A, Pakruddin B et al (2025) Decoding DNA: deep learning's impact on genomic exploration. In: Deep learning in genetics and genomics. https://doi.org/10.1016/B978-0-443-27574-6.00005-9
59. Mishra D, Chaturvedi B, Soni V, Valecha D, Goel M, Ansari JR (2025) Impact of bridging the gap between artificial intelligence and nanomedicine in healthcare. Next Nanotechnol 8:100203. https://doi.org/10.1016/j.nxnano.2025.100203
60. Jinnouchi R, Minami S (2025) Machine learning force fields in electrochemistry: From fundamentals to applications. ACS Nano 19(22):22600–22644. https://doi.org/10.1021/acsnano.5c05553
61. Otyepka M, Pykal M, Otyepka M (2025) Advancing materials discovery through artificial intelligence. Appl Mater Today 47:102981. https://doi.org/10.1016/j.apmt.2025.102981
62. Olawade DB, Ige AO, Olaremu AG, Ijiwade JO, Adeola AO (2024) The synergy of artificial intelligence and nanotechnology towards advancing innovation and sustainability: a mini-review. Nano Trends 8:100052. https://doi.org/10.1016/j.nwnano.2024.100052
63. Chou WC, Canchola A, Zhang F, Lin Z (2025) Machine learning and artificial intelligence in nanomedicine. Wiley Interdiscip Rev Nanomed Nanobiotechnol 17(4):e70027. https://doi.org/10.1002/wnan.70027
64. Serrano DR, Luciano FC, Anaya BJ, Ongoren B, Kara A, Molina G, Ramirez BI, Sánchez-Guirales SA, Simon JA, Tomietto G, Rapti C, Ruiz HK, Rawat S, Kumar D, Lalatsa A (2024) Artificial Intelligence (AI) applications in drug discovery and drug delivery: revolutionizing personalized medicine. Pharmaceutics 16(10):1328. https://doi.org/10.3390/pharmaceutics16101328
65. Yapa P, Rajapaksha S, Munaweera I (2025) The integration of nanotechnology, nanomedicine, and artificial intelligence for advancements in healthcare: a conceptual review based on PRISMA method and future research directions. Next Res 2:100330. https://doi.org/10.1016/j.nexres.2025.100330
66. Sovacool BK, Baum CM, Low S, Fritz L (2024) The sociotechnical dynamics of blue carbon management: testing typologies of ideographs, innovation, and co-impacts for marine carbon removal. Environ Sci Policy 155:103730. https://doi.org/10.1016/j.envsci.2024.103730
67. Yin S (2025) Artificial intelligence-assisted nanosensors for clinical diagnostics: current advances and future prospects. Biosensors 15(10):656. https://doi.org/10.3390/bios15100656
68. Thryve Health (2024) AI-powered biosensors: transforming wearable health monitoring with real-time insights, predictive analytics, and personalized healthcare. Retrieved from https://www.thryve.health/blog/ai-powered-biosensors-wearable-health
69. Kim et al (2024) Advancements in nanobiosensor technologies for in-vitro diagnostics and point-of-care testing. Biosensors 15(11). PMC11609235
70. Kang et al (2021) Applications of nanotechnology in virus detection, tracking, imaging, and sensing. In: WIREs nanomedicine and nanobiotechnology. PMC7995016

71. Soumma SB (2025) AI-powered wearable sensors for health monitoring and diagnostics. In: Healthcare technology letters. https://doi.org/10.1016/j.htl.2025.10.14
72. Secchi V (2025) Advanced techniques and nanotechnologies for point-of-care testing. In: Frontiers in nanotechnology. https://doi.org/10.3389/fnano.2024.1465429
73. Kumar A (2025) Artificial intelligence-based wearable sensing technologies for the management of cancer, diabetes, and COVID-19. Biosensors 15(11):756. https://doi.org/10.3390/bios15110756
74. Chokkareddy R (2022) Smart nanodevices for point-of-care applications. Current Anal Chem 10(2)
75. Gulati S (2025) Nanosensors in healthcare: transforming real-time monitoring and diagnostics. In: Royal society of chemistry progress in materials. https://doi.org/10.1039/d5pm00125k
76. Smith AA (2023) Reshaping healthcare with wearable biosensors. Nature 615:475–488. https://doi.org/10.1038/s41598-022-26951-z
77. (2023) Nanoparticle-based drug delivery systems targeting cancer cell metastasis. J Nanobiotechnology. PMC10350659
78. Ahmad M (2023) Artificial intelligence in personalized medicine. In: Nature reviews molecular cell biology. PMC10617817
79. Rana M (2023) Nanoparticle-enhanced drug delivery systems for targeted cancer therapy. Emanresearch Publishers Biosensors 5(5406):4
80. Medhi R (2020) Nanoparticle-based strategies to combat COVID-19. ACS Appl Nano Mater 3(9):8557–8580. https://doi.org/10.1021/acsanm.0c01978
81. Li Q, Wang Q, Yang X, Wang K, Zhang H, Nie W (2017) High sensitivity surface plasmon resonance biosensor for detection of microRNA and small molecule based on graphene oxide-gold nanoparticles composites. Talanta 1(174):521–526. https://doi.org/10.1016/j.talanta.2017.06.048
82. de la Fuente-Nunez C (2022) Antibiotic discovery with machine learning. Nat Biotechnol 40(6):833–834. https://doi.org/10.1038/s41587-022-01327-w
83. Gholizadeh O, Yasamineh S, Amini P, Afkhami H, Delarampour A, Akbarzadeh S, Karimi Matloub R, Zahedi M, Hosseini P, Hajiesmaeili M, Poortahmasebi V (2022) Therapeutic and diagnostic applications of nanoparticles in the management of COVID-19: a comprehensive overview. Virol J 19(1):206
84. (2025) Nanobiosensors detect tuberculosis nucleic acid efficiently. ACS Appl Bio Mater 8(2), 652–672. https://doi.org/10.1021/jacsau.4c01206
85. Arnold A, McLellan S, Stokes JM (2025) How AI can help us beat AMR. NPJ Antimicrob Resist 3(1):18. https://doi.org/10.1038/s44259-025-00085-4
86. Veigas B, Pedrosa P, Carlos FF, Mancio-Silva L, Grosso AR, Fortunato E, Mota MM, Baptista PV (2015) One nanoprobe, two pathogens: gold nanoprobes multiplexing for point-of-care. J Nanobiotechnology 7(13):48. https://doi.org/10.1186/s12951-015-0109-1
87. Dikamu M, Syraji Y, Pr JKG, Raza A, Ezez D (2025) Nanotechnology in COVID-19 prevention, diagnosis, and treatment: a comprehensive review. Discov Nano 20(1):225. https://doi.org/10.1186/s11671-025-04412-6.
88. Lee S, Park JS, Woo H, Yoo YK, Lee D, Chung S, Yoon DS, Lee KB, Lee JH (2024) Rapid deep learning-assisted predictive diagnostics for point-of-care testing. Nat Commun 15(1):1695. https://doi.org/10.1038/s41467-024-46069-2
89. Rizvi SAA, Saleh AM (2018) Applications of nanoparticle systems in drug delivery technology. Saudi Pharm J 26(1):64–70. https://doi.org/10.1016/j.jsps.2017.10.012
90. Fatema SA (2024) Advances in nanomaterial-based sensors for early disease detection. Int J Adv Res Sci Commun Technol. Article 19895. https://doi.org/10.48175/IJARSCT-19895
91. Lan J (2022) Overview of application of nanomaterials in medical domain. Contrast Media Mol Imaging 4(2022):3507383. https://doi.org/10.1155/2022/3507383
92. Harishbhai Tilala M, Kumar Chenchala P, Choppadandi A, Kaur J, Naguri S, Saoji R, Devaguptapu B (2024) Ethical considerations in the use of artificial intelligence and machine learning in health care: a comprehensive review. Cureus 16(6):e62443. https://doi.org/10.7759/cureus.62443

93. Ahmed MM, Okesanya OJ, Oweidat M, Othman ZK, Musa SS, Lucero-Prisno Iii DE (2025) The ethics of data mining in healthcare: challenges, frameworks, and future directions. BioData Min 18(1):47. https://doi.org/10.1186/s13040-025-00461-w
94. Hanna MG, Pantanowitz L, Jackson B, Palmer O, Visweswaran S, Pantanowitz J, Deebajah M, Rashidi HH (2025) Ethical and bias considerations in artificial intelligence/machine learning. Mod Pathol 38(3):100686. https://doi.org/10.1016/j.modpat.2024
95. Arora A, Alderman JE, Palmer J, Ganapathi S, Laws E, McCradden MD, Oakden-Rayner L, Liu X et al (2023) The value of standards for health datasets in artificial intelligence-based applications. Advance online publication, Nature Medicine. https://doi.org/10.1038/s41591-023-02608-w
96. Celi LA, Cellini J, Charpignon ML, Dee EC, Dernoncourt F, Eber R, Mitchell WG, Moukheiber L, Schirmer J, Situ J, Paguio J, Park J, Wawira JG, Yao S et al (2022) Sources of bias in artificial intelligence that perpetuate healthcare disparities—a global review. PLOS Digit Health 1(3):e0000022. https://doi.org/10.1371/journal.pdig.0000022
97. Chen RJ, Wang JJ, Williamson DFK, Chen TY, Lipkova J, Lu MY, Sahai S, Mahmood F (2023) Algorithmic fairness in artificial intelligence for medicine and healthcare. Nat Biomed Eng 7(6):719–742. https://doi.org/10.1038/s41551-023-01056-8
98. Rodríguez-Gómez FD, Monferrer D, Penon O, Rivera-Gil P (2025) Regulatory pathways and guidelines for nanotechnology-enabled health products: a comparative review of EU and US frameworks. Frontiers Med 12, Article 1544393. https://doi.org/10.3389/fmed.2025.1544393
99. Chaudhuri T, Ghosh J (2025) Future trends in artificial intelligence for medical applications. In: Deep learning in medical signal and image processing (Chapter 20). ISBN: 9798369398173. https://doi.org/10.4018/979-8-3693-9816-6.ch020
100. Xuan L, Ju Z, Skonieczna M, Zhou PK, Huang R (2020) Nanoparticles-induced potential toxicity on human health: applications, toxicity mechanisms, and evaluation models. MedComm 4(4):e327. https://doi.org/10.1002/mco2.327
101. Rai A, Singh A, Menon I, Singh J, Rai V, Aswal GS (2018) Molar incisor hypomineralization: prevalence and risk factors among 7–9 years old school children in Muradnagar, Ghaziabad. Open Dent J 28(12):714–722. https://doi.org/10.2174/1745017901814010714
102. Ghauri S, Mazzarol T, Soutar GN (2023) Networking benefits for SME members of co-operatives. J Co-operative Organization Manage 11(2):100213. https://doi.org/10.1016/j.jcom.2023.100213
103. Ma X, Tian Y, Yang R et al (2024) Nanotechnology in healthcare, and its safety and environmental risks. J Nanobiotechnol 22:715. https://doi.org/10.1186/s12951-024-02901-x
104. Malik S, Muhammad K, Waheed Y (2023) Emerging applications of nanotechnology in healthcare and medicine. Molecules 28(18):6624. https://doi.org/10.3390/molecules28186624
105. Pokrajac L, Abbas A, Chrzanowski W, Dias GM, Eggleton BJ, Maguire S, Maine E, Malloy T, Nathwani J, Nazar L, Sips A, Sone J, van den Berg A, Weiss PS, Mitra S (2021) Nanotechnology for a sustainable future: addressing global challenges with the International Network4Sustainable Nanotechnology. ACS Nano 15(12):18635–18650. https://doi.org/10.1021/acsnano.1c08588

Precision Nanomedicines in Treating Autoimmune Disorders

Oumayma Abbassi, Chahrazad Belkhiri, Salah-eddine Azizi, Mohammed Dalli, and Bouchra Legssyer

Abstract The immune system's aberrant targeting of self-tissues causes autoimmune disorders, which result in chronic inflammation that can affect one or more organs. These disorders are caused by a breakdown of immune tolerance, which is typified by the generation of autoantibodies and the activation of autoreactive T cells. They are impacted by a complex interaction between environmental triggers and genetic susceptibility. The main goals of current treatment approaches are to suppress immune responses and reduce clinical symptoms; however, they do not offer a permanent cure and are frequently linked to significant side effects and limited efficacy, especially with longer-term use and more recent immunomodulatory therapies. In this regard, nanomedicine has become a novel and promising strategy for the treatment of autoimmune disorders. Polymeric nanoparticles and liposomal formulations are examples of nanoscale drug delivery systems that allow the targeted and regulated delivery of immunosuppressive and immunomodulatory drugs. These tactics lower systemic toxicity, increase therapeutic precision, improve drug bioavailability, and aid in the restoration of immunological tolerance. Thermo-responsive liposomal methotrexate formulations and PLGA-based nanoparticles for cyclosporine A delivery are two notable examples of how nanomedicine can improve safety and efficacy in the treatment of autoimmune diseases. This chapter aims to provide a comprehensive overview of precision nanomedicine approaches in the treatment of autoimmune disorders, with a particular focus on the design, mechanisms, and therapeutic advantages of nanocarrier-based drug delivery systems. The

O. Abbassi · C. Belkhiri · B. Legssyer
Laboratory for Agricultural Production Improvement, Biotechnology and Environment, Mohammed First University, Oujda, Morocco

S. Azizi (✉) · M. Dalli (✉)
Higher Institute of Nursing Professions and Health Techniques, Oujda, Morocco
e-mail: azizi.salah-eddine@ump.ac.ma

M. Dalli
e-mail: m.dalli@ump.ac.ma

M. Dalli
Laboratory of Bioresources, Biotechnology, Ethnopharmacology and Health, Faculty of Sciences, University Mohammed First, Oujda, Morocco

J. Chen (ed.), *AI-Driven Nanotechnology in Precision Medicine and Healthcare*, Smart Nanomaterials Technology, https://doi.org/10.1007/978-981-92-1690-1_10

chapter highlights how nanomedicine can overcome the limitations of conventional therapies by enabling targeted immunomodulation, minimizing adverse effects, and paving the way toward more personalized and effective management strategies for autoimmune diseases.

Keywords Autoimmune diseases · Nanomedicines for autoimmune diseases · Precision nanomedicine · Molecular mechanisms · Current immunotherapies · Loss of immune tolerance

1 Introduction

Two basic components make up the immune system: the adaptive and the innate systems. The body's initial line of defense is the innate immune system, which is non-specific and lacks memory. It comprises cellular components, including white blood cells, as well as physical and chemical barriers, including the epidermis, tears, and saliva. In contrast, the adaptive immune system is highly specific-to-specific antigens and has the capacity to form memory, enabling a more effective response to repeated exposure. In order to identify and combat pathogens, this system depends on B lymphocytes and T lymphocytes [28]. Autoimmune disorders (ADs) arise from the body's immune responses against its own components, leading our body's immune system to attack its own cells, tissues, and organs [122].

Autoimmune diseases (ADs) are commonly classified into systemic forms, such as rheumatoid arthritis, and organ-specific forms, such as type 1 diabetes mellitus [30, 62]. In this context, nanomedicine offers a promising avenue for autoimmune disease management, overcoming conventional therapies' drawbacks like systemic toxicity, poor bioavailability, and absent antigen specificity [49]. It leverages nanoscale delivery systems, such as polymeric nanoparticles, liposomes, lipid carriers, dendrimers, inorganic nanoparticles, and exosomes, for precise immune cell and tissue targeting, enhanced pharmacokinetics, and controlled drug release [23, 108]. By encapsulating immunosuppressants, biologics, antigens, or nucleic acids, these technologies reduce systemic exposure and off-target inflammation while increasing efficacy in disease models through site-specific delivery [43]. Advanced nanomedicine strategies—like tolerogenic nanoparticles presenting self-antigens for immune tolerance, gene delivery systems with siRNA/miRNA to target pathogenic pathways, and theranostic nanoparticles which are based on a strategy merging diagnostic imaging and targeted therapy to enable personalized and real-time ADs management. For example, in autoimmune models, tolerogenic nanoparticles suppress autoreactive reactions and stimulate regulatory T cells to restore immunological balance rather than just controlling inflammation [31, 69]. These strategies include nanocarrier delivery of siRNA/miRNA to target aberrant immune signaling, enhancing specificity and precision immunomodulation for systemic lupus erythematosus (SLE), rheumatoid arthritis (RA), and multiple sclerosis (MS). By selectively modulating pathogenic pathways and promoting regulatory responses for personalized autoimmune therapies that improve long-term outcomes and quality of life [103].

This chapter aims to present a comprehensive overview of precision nanomedicine strategies for the treatment of autoimmune disorders, emphasizing the design principles, underlying mechanisms and therapeutic benefits of nanocarrier-based drug delivery systems. It explores how nanomedicine addresses the shortcomings of conventional therapies by enabling targeted immunomodulation, reducing systemic toxicity and adverse effects, and supporting the development of more personalized and effective therapeutic approaches for autoimmune diseases.

2 Autoimmune Diseases

According to [124], autoimmunity is the immune system's response to self-structures, which includes the generation of autoantibodies and the activation of self-reactive T cells. Physiological and pathological autoimmunity are the two categories into which it can be divided (Table 1). Typically, physiological autoimmunity is benign, temporary, and helps maintain immunological homeostasis without leading to clinical illness. Pathological autoimmunity, on the other hand, results in tissue damage and autoimmune disease when immunological tolerance is compromised [122] (Table 1). Thus, contributing to the rise of multiple autoimmune diseases, which may affect only certain tissues of many organs throughout the body. An autoimmune disease can be identified as a condition in which the immune system functions abnormally [43, 124] (Table 1).

3 Molecular and Cellular Mechanisms Involved in Autoimmune Diseases

The pathophysiology of autoimmune diseases is complex. The generation of autoantibodies by B cells, self-reactive T cell activation, cytokine dysregulation, immunological tolerance deficiencies, and environmental or genetic variables that cause aberrant immune responses against self-antigens are important pathways [14, 122] (Fig. 1).

The generation of autoantibodies involves a range of mechanisms and factors. Genetic predisposition is crucial since some genes, especially some human leukocyte antigen (HLA) genes, enhance the probability of individuals developing autoantibodies [8, 16, 127]. Wang et al. [127] found that having relatives with autoimmune diseases raises the risk, with monozygotic twins showing a greater prevalence than dizygotic twins. One of the processes that results in the generation of autoantibodies is the loss of immunological tolerance. The immune system typically keeps the body from reacting to its own antigens by maintaining self-tolerance. However, this tolerance is compromised in autoimmune disorders, allowing self-reactive B cells to

Table 1 Classification of autoimmune disease

Autoimmune disease		Disease mechanism	References
Systematic autoimmune diseases			
Rheumatic diseases	Rheumatoid arthritis	Autoantibodies (RF, ACPA), T-cell activation (Th1/Th17), citrullination of self-proteins	[72]
Connective tissue disorders	Systemic lupus erythematosus	Autoantibodies against nuclear components, immune complex deposition, and IFN pathway activation	[8, 62]
	Primary Sjögren's syndrome	Autoantibodies, inflammation of lacrimal/salivary glands	[127]
	Systemic sclerosis	Autoantibodies against Scl-70 or centromere structures; fibrosis	[24]
Organ-specific autoimmune diseases			
Neurological disorder	Multiple sclerosis	Autoreactive T cells targeting myelin (MOG) inflammation/demyelination in the CNS	[127]
Endocrine disorder	Type 1 diabetes	Autoreactive T cells destroy pancreatic β-cells, and autoantibodies like GAD-65	[127]
Dermatological diseases	*Pemphigus vulgaris*	Autoantibodies against keratinocyte surface antigens cause blistering	[16, 63]
	Vitiligo	Autoantibodies and autoreactive T cells against melanocytes	[127]
	Psoriasis	Inflammation driven by autoreactive T cells (Th17 predominance)	

evade typical regulatory checkpoints and become activated [16, 55, 62]. A very small percentage of self-reactive B cells can "leak out," go to the periphery, and generate very little autoantibody, even in healthy people [122, 127]. These cells are essential because they make autoantibodies, present antigens, and release inflammatory cytokines [47, 105]. Deficits in enzymes like DNase 1 and serum amyloid protein (SAP) or complement proteins like C1q and C4, which normally remove immune complexes and apoptotic debris, can cause excessive B cell activation and autoantibody production, which frequently results in a systemic lupus erythematosus (SLE) phenotype [32].

Environmental factors, including infections, xenobiotics, and certain pharmaceuticals, can trigger the production of autoantibodies [143] (Fig. 1). One key mechanism is molecular mimicry, in which pathogens such as viruses and bacteria express molecules resembling host proteins, leading to cross-reactive immune responses [53, 62]. For instance, *Streptococcus pyogenes* can generate antibodies that recognize both

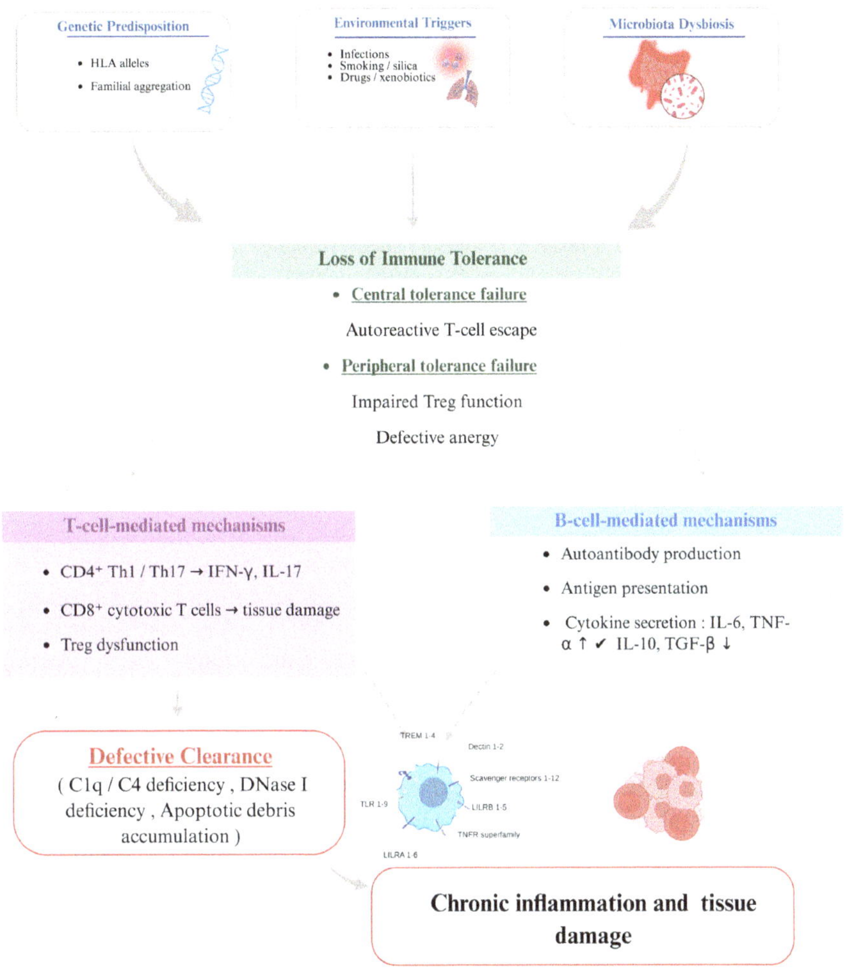

Fig. 1 Cellular and molecular mechanism included in AD

bacterial antigens and self-antigens such as myosin, contributing to cardiac tissue damage [27].

Similarly, gut microbial dysbiosis may promote autoantibody production through cross-recognition of microbial antigens [41]. Environmental xenobiotics, including smoking and silica, can cause tissue injury, cellular toxicity, and chronic inflammation, particularly in the lungs [97]. Such inflammatory environments can modify self-proteins, rendering them "foreign" to the immune system and triggering autoantibody generation [76, 97] (Fig. 1).

Furthermore, environmental chemicals, including pesticides and industrial solvents such as trichloroethylene, have been linked to the development of autoantibodies, potentially through modulation of immune regulation or epigenetic mechanisms [96]. The pathogenesis of autoimmune diseases involves multiple pathological features, largely driven by distinct cellular mechanisms, including both T and B lymphocytes. T cells play a central role in autoimmune pathogenesis due to a breakdown in the immune system's ability to distinguish self from non-self [12, 114]. Normally, central tolerance in the thymus eliminates self-reactive T cells and supports the generation of regulatory T cells (Tregs) [111, 136]. When central tolerance fails, self-reactive T cells with pathogenic potential can migrate to the periphery, necessitating peripheral tolerance mechanisms [44, 82]. Peripheral tolerance functions through processes that limit self-reactive T cells outside the thymus, including anergy, suppression by Tregs, clonal deletion, and T cell ignorance [106].

In autoimmune diseases, these peripheral checkpoints often fail, allowing autoreactive CD4 + T helper (Th) cells to become activated and differentiate into pathogenic subsets [114]. For example, Th1 and Th17 cells produce pro-inflammatory cytokines that drive tissue inflammation and destruction, a mechanism prominently observed in autoimmune diseases of the central nervous system (CNS) (Alberti and Handel 2021; Rosenblum et al. 2015). CD8 + cytotoxic T cells can also directly damage host tissues by targeting self-cells (Alberti and Handel 2021). A key contributor to this dysfunction is the decline or absence of Tregs, which are essential for suppressing autoreactive T cells and maintaining immune homeostasis (Rosenblum et al. 2015; [111]).

On the other hand, beyond their traditional duty of producing antibodies, B cells have a significant role in immunological control. Through a number of processes, including the production of autoantibodies, the presentation of self-antigens, and the release of inflammatory cytokines, they contribute to autoimmune pathogenesis [42, 46, 52, 105]. B cells employ their B cell receptors to identify autoantigens, digest them, and present antigen-derived peptides on MHC class II molecules to activate T cells. B cells are professional antigen-presenting cells (APCs) [35, 40, 118]. This effective antigen presentation is especially important when autoantigen levels are low in conditions like multiple sclerosis, rheumatoid arthritis, and type 1 diabetes because it stimulates autoimmune responses and self-reactive T cell activation [38, 56]. Together with B cells, other professional APCs like dendritic cells and macrophages present self-antigens and generate cytokines that control T cell polarization and function—processes crucial for triggering and sustaining autoimmunity [35, 58, 116]. Autoimmune diseases are made worse by cytokine imbalance. In addition to pro-inflammatory cytokines like IL-2, IL-4, TNF-α, IL-6, IFN-γ, and IL-12, B cells also produce immunosuppressive cytokines, including TGF-β1 and IL-10 [77, 78, 81]. Together with cytokines from other APCs, an imbalance in these B cell-derived cytokines maintains immune responses against self-tissues, resulting in persistent tissue damage and chronic inflammation [90, 123].

4 Current Therapies for Autoimmune Diseases

The current treatments for autoimmune disorders target certain immune components or suppress the immune system as a whole in an effort to control symptoms, reduce inflammation, and limit the progression of the disease [43, 62]. Traditionally, treatment began with conventional immunosuppressants, such as corticosteroids and nonsteroidal anti-inflammatory medications (NSAIDs), which primarily aim to reduce inflammation and alleviate symptoms [17, 36]. Corticosteroids are highly effective anti-inflammatory medications that decrease pro-inflammatory proteins and inhibit inflammatory mediators by diffusing into cells, attaching to glucocorticoid receptors, and controlling gene expression [36]. They are keystone for handling multiple autoimmune diseases among rheumatoid arthritis, autoimmune bullous diseases, inflammatory bowel disease, and rheumatoid arthritis [1, 7, 19, 83, 107]. Corticosteroids drastically lowered mortality rates from 70 to 30% in pemphigus vulgaris [117]. However, there is a significant danger of major side effects from long-term use, especially when using high doses, including severe infections, gastric ulcers, hyperglycemia, hypertension, cognitive impairment, osteoporosis, weight gain, hirsutism, and acne [1, 83]. The use of steroid-sparing medication is frequently required because corticosteroid monotherapy is rarely curative [19, 107]. Moreover, NSAIDs like ibuprofen, naproxen, and acetylsalicylate are used as first treatments to reduce inflammation and relieve pain, especially in diseases like rheumatoid arthritis [17]. Despite being effective for symptom control, they ordinarily do not alter the disease course or provide long-term solutions; employing them can have side effects [7]. In addition to immunosuppressants, biological agents represent a notable advancement in the treatment of these diseases, providing targeted therapeutic intervention unlike broad-acting conventional [11, 67, 75, 84]. TNF-α inhibitors, including infliximab, adalimumab, etanercept, certolizumab, pegol, and golimumab, function by neutralizing tumor necrosis factor—alpha, a pivotal pro-inflammatory cytokine implicated in the pathogenesis of various autoimmune disorders, such as rheumatoid arthritis (RA), psoriatic arthritis, ankylosing spondylitis, and inflammatory bowel disease [37, 61],P. [71, 74, 75]. These therapies reduce inflammation, promote a higher and improved way of living, and potentially affect how inflammatory arthritis and inflammatory bowel diseases [37, 99]. IL-6 inhibitors, such as tocilizumab and sarilumab, block the IL-6 receptor, thereby preventing it from initiating inflammation and autoimmune responses [64, 67, 75]. Tocilizumab demonstrates therapeutic efficacy in rheumatoid arthritis, systemic juvenile idiopathic arthritis, and scleritis by inhibiting inflammatory pathways and reducing the incidence of disease exacerbations [64, 99, 115]. B cell depleting agents, within which rituximab (anti-CD-20), operate by specifically targeting and removing B cells [20, 22, 139]. Though highly effective, these targeted treatments may cause undesirable consequences, such as increased vulnerability to infections, paradoxical autoimmune responses, and, though infrequently, hematologic abnormalities or increased malignancy risk [34].

Chimeric antigen receptor (CAR) T-cell therapy is a new treatment for autoimmune diseases that involves modifying a patient's own T cells. The altered cells can subsequently identify and eliminate the cells responsible for the autoimmune response [91, 126]. This therapy initiates with the extraction of T cells from the patient's blood; these cells are genetically engineered to express a CAR capable of recognizing antigens, such as CD19 on B cells. A significant number of therapeutic cells are subsequently produced by the expansion of the altered T cells. Once the cells are fully developed, they are reinfused into the patient. Upon administration, the CAR T cells recognize and target cells that express the target antigen, causing an immune response against those cells [91, 125]. Earlier clinical investigations and case series have shown promising results in refractory autoimmune disorders, including systemic lupus erythematosus (SLE), rheumatoid arthritis (RA), systemic sclerosis, idiopathic inflammatory myositis, myasthenia gravis, and Sjögren's disease [15, 54, 79, 88, 110, 133].

Some blood malignancies can be effectively treated with CAR T-cell therapy; nevertheless, there are certain, occasionally dangerous side effects that can occur. These include neurological issues that require careful surveillance and care, such as cytokine release syndrome (CRS) and immunological effector cell-associated neurotoxicity (ICANS) [73, 120, 132]. The most common undesirable effect of CAR T-cell therapy is CRS, which is brought on by a potent and extensive inflammatory response that follows T-cell activation and expansion [60, 73, 135]. An increase in pro-inflammatory cytokines causes a " cytokine storm," which is the result of this activation [60, 120, 135].

ICANS is the second most frequent side effect; it can occur concurrently with or after CRS [87, 120, 135]. Early signs such as tremor, difficulty in writing, and impaired verbal expression can rapidly escalate into more severe conditions, including seizures, coma, and cerebral edema [51, 87, 95, 120, 131, 135].

Despite these advances, the limitations and the side effects of these therapies underscore the need for new approaches, of which nanomedicine is gaining increasing attention [130].

5 Nanomedicines for Autoimmune Diseases

It is important to note that traditional therapies like immunosuppressants and biologicals can control symptoms; they usually have adverse effects such as systemic toxicity, incomplete efficacy, and severe effects [2, 3, 70, 89]. Nanomedicine offers a new path toward a potential therapeutic approach by utilizing nanoscale materials to boost drug delivery and precisely control immune responses [2, 89]. For example, nanocarriers, including liposomes and polymeric nanoparticles, are employed to deliver immunosuppressive drugs, cytokines, or biologics specifically to inflamed tissues, reducing systemic exposure and side effects [50, 93] (Fig. 2). Tolerogenic nanoparticles aim to restore immune tolerance by presenting antigens to antigen-presenting cells, promoting regulatory T cells, and suppressing autoreactive responses without

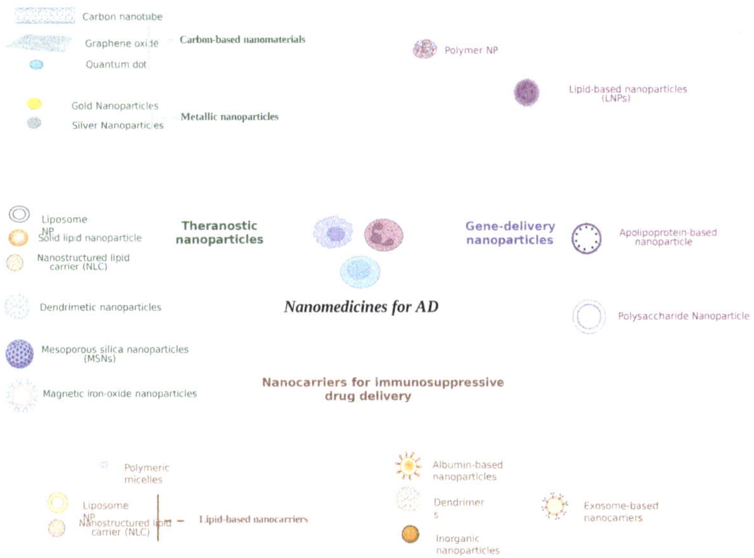

Fig. 2 Nanomedicine approaches in autoimmune diseases

global immunosuppression [13, 109, 122] (Fig. 1). Nanovaccines transport antigens and adjuvants to modulate immunity and reinstate tolerance, currently investigated in autoimmune diseases such as multiple sclerosis [24, 86] (Fig. 2). Theranostic nanoparticles combine therapeutic and diagnostic functions, offering potential for immune monitoring and targeted therapy [86], while gene delivery nanoparticles enable precise modulation of immune cell gene expression via siRNA or miRNA, supporting targeted immune regulation [39, 69] (Fig. 2).

5.1 Nanocarrier for Immunosuppressive Drug Delivery

Nanocarriers improve the transport of immunosuppressive drugs by improving their bioavailability and assuring targeted distribution to immune cells, especially in autoimmune inflammatory conditions such as rheumatoid arthritis (RA), systemic lupus erythematosus (SLE), and multiple sclerosis (MS) [69, 119]. Nanocarriers provide a sophisticated method to enhance the therapeutic efficacy of immunosuppressive agents by improving their stability and bioavailability in the target site. Agents including cyclosporine A, tacrolimus, methotrexate, and leflunomide are protected from enzymatic degradation and have their circulation duration extended when they are encapsulated in lipid-or polymer-based systems [134]. These nanocarriers are transforming immunosuppressive therapy for autoimmune diseases by protecting nucleic acid-based medications, enabling their targeted distribution to immune cells linked to the disease, and reducing systemic toxicity generally [66]. Liposomes, polymeric nanoparticles (NPs), micelles, albumin NPs, dendrimers, nanostructured carriers, inorganic particles, and exosomes are examples of prominent nanocarrier systems [21].

Synthetic polymeric blocks are used to create polymeric nanoparticles (NPs), which extend the half-life of circulation and reduce phagocytic absorption and therapeutic moiety inactivation. Therapeutic drugs can be delivered and targeted by them [85]. Because they can improve the solubility and absorption of medications that don't dissolve well in water, stop drugs from breaking down in the body, allow for a steady and controlled release of medication, and enable targeted delivery by attaching specific molecules to their surface, NPs are widely used in nanomedicine [80]. Furthermore, through the increased permeability and retention (EPR) effect, their nanoscale dimensions enable passive accumulation in sick tissues, enhancing therapeutic selectivity and reducing systemic toxicity [48, 145]. Many natural polymers, including chitosan, alginate, and guar gum, or synthetic polymers like poly(lactic-co-glycolic acid) (PLGA) and poly-caprolactone (PCL), can be used to manufacture NPs [80].

Polymeric nanoparticles come in two varieties. Nanocapsules are another name for the reservoir type. The medication molecules in this type are dispersed or dissolved inside a polymer-shelled core. This small "reservoir inside the particle holds the medication" [98]. The second kind are nanospheres or matrix-type nanoparticles. In this instance, the entire nanoparticle is a solid mixture of polymer and drug because the entire polymer matrix traps the drug molecules [21]. Synthetic polyester-based polymers are one of the most used materials for the nanoparticles design. It involves polylactic acid (PLA) and poly (lactic-co-glycolic acid) (PLGA) (H. [69]) (Fig. 2). This polymer is approved by the FDA, biodegradable, and characterized by erosion rate [100].

For example, cyclosporin (CyA) is a potent calcineurin inhibitor that is utilized for a variety of conditions, including autoimmune illnesses. Its limited therapeutic index and significant plasma protein binding cause systemic toxicity. Potential encapsulation in nanocarriers increases CyA's solubility and enables active targeting [59]. PLGA, especially to reduce systemic toxicity and increase oral bioavailability. This type of medication is used to treat a number of autoimmune conditions, including psoriasis, nephrotic syndrome, and RA [4]. PLGA nanoparticles have been extensively employed to improve the oral bioavailability of CyA while lowering systemic toxicity. Moreover, polylactide (PLA), a component of PLGA, can encapsulate CyA by covalently attaching the drug to the polymer backbone and forming nanoparticles via nanoprecipitation. In this approach, CyA serves as the initiator for ring-opening polymerization of lactide (LA) with a (BDI)ZnN(TMS)$_2$ catalyst, generating a hydrolyzable CyA-PLA conjugate. When combined with PLA-mPEG, the conjugate self-assembles into PEG-coated nanoparticles (~80 nm). Particle size can be tuned by adjusting the solvent or polymer concentration. PEGylation stabilizes the nanoparticles, and the ester linkage ensures gradual, controlled drug release without an initial burst. These CyA-PLA nanoparticles provide prolonged intracellular release and reduced systemic toxicity while maintaining the immunosuppressive activity of CyA [9].

This drug can also be encapsulated in PEG-PLA; this encapsulation preserves its strong immunosuppressive effect, facilitates regulated release, and enhances targeting to lymphatic tissues[121]. Actually, the study of L. Tang et al., in 2012, in a mixed lymphocyte reaction (MLR) assay, CyA/PEG-PLGA nanoparticles suppress T-cell proliferation with comparable efficacy to free CyA, showing identical IC_{50} values in both mixed lymphocyte reaction and CD3/CD8 stimulation assays. Additionally inhibit IFN- γ production at low concentration (approximately 10 ng/ml) [121]. Liposomes are another type of carrier that can be used to deliver immunosuppressive drugs that are characterized by phospholipid vesicles capable of encapsulating immunosuppressors in their aqueous core for hydrophilic drugs or within the bilayer for lipophilic drugs, safeguarding the contents from degradation and facilitating controlled release [5].

In inflamed autoimmune tissues, the vasculature exhibits increased permeability; liposomes < 200 nm employ this enhanced permeability and retention (EPR) effect to accumulate passively where they are internalized by local macrophages and dendritic cells, which subsequently release the medication intracellularly [5, 94].

Numerous liposomal strategies formulated for rheumatoid arthritis (RA) focus on achieving adapted management of immunosuppressive agents [45]. Methotrexate (MTX), a conventional disease-modifying anti-rheumatic drug (DMRD), is frequently encapsulated in liposomes to improve its therapeutic efficacy in rheumatoid arthritis (RA) [112, 144].

The development of specialized liposomal methotrexate (MTX) formulations leverages multiple mechanisms to improve targeted delivery. For instance, pH-sensitive liposomes composed of components such as dioleoylphosphatidylethanolamine (DOPE) and cholesteryl hemisuccinate (CHEMS) exploit the acidic microenvironment characteristic of inflamed rheumatoid arthritis (RA) joints to achieve controlled, site-specific drug release [45].

Several studies showed that Dextan sulfate modified pH-sensitive layered double hydroxide nanocomposites loaded with MTX (LDH-MTX-DS) and used in an animal model of arthritis. The collected results indicated an elevation of cellular uptake and controlled release in acidic environments, improving therapeutic effect [129].

Other strategies have utilized ultradeformable liposomes loaded with methotrexate (MTX) for transdermal delivery. These formulations can penetrate the skin efficiently and have been shown to reduce inflammatory cytokine levels in arthritic models, providing a promising non-invasive therapeutic option for rheumatoid arthritis and related conditions. Thermo-responsive flexible liposomes improve rheumatoid arthritis (RA) treatment outcomes by enhancing joint accumulation and accelerating methotrexate (MTX) release through the use of heat sensitivity, frequently in conjunction with microwave hyperthermia. Furthermore, certain liposomal systems are actively functionalized to target afflicted tissues or immune cells. For instance, scavenger receptors (SR-A) that are overexpressed on active macrophages can be bound by flexible nanoliposomes coated with dextran sulfate,

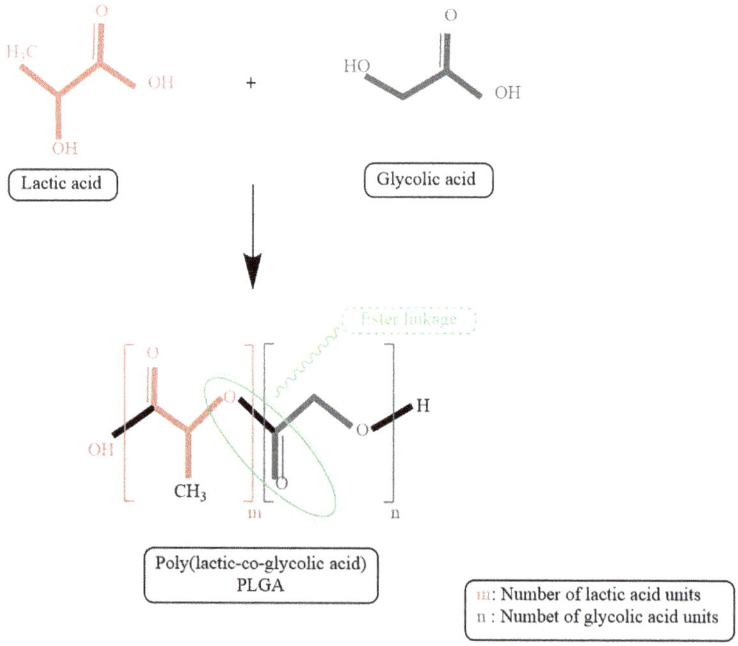

Fig. 3 Poly (lactic-co-glycolic acid) (PLGA) formation from lactic acid and glycolic acid via ester linkages

facilitating effective cellular uptake and boosting the effectiveness of RA treatment [146] (Fig. 3).

5.1.1 Theranostic Nanoparticles

As reported by Clemente-Casares et al. [26], theranostic NP is considered a promising strategy in the field of nanomedicine. The term "theranostic" combines **"therapy"** and **"diagnostics"** [142]. It is a nanoscale carrier that combines a therapeutic agent with a diagnostic imaging element. By combining high-resolution imaging with targeting immunomodulation, it has become an effective tool for addressing the heterogeneity of autoimmune disorders [10, 119].

Metallic nanoparticles, such as gold (AuNPs) and silver (AgNPs), combine strong X-ray absorption with localized surface plasmon resonance (LSPR), enabling photoacoustic (PA) and CT imaging of inflamed tissues [10, 52]. Fabricated with core–shell or star morphologies and coated with polymers, silica, or PEG, they achieve stability, stealth circulation, and functionalization for drug or antigen delivery.

5.2 Gold Nanoparticles (AuNPs)

In autoimmune models, AuNPs conjugated with antigens and tolerogenic ligands promote dendritic cell-mediated tolerance, increasing regulatory T cells ($FoxP3^+$ Tregs) and reducing disease progression in conditions like multiple sclerosis and type 1 diabetes [68]. In rheumatoid arthritis, they accumulate in inflamed synovium, scavenge reactive oxygen species, inhibit NF-κB signaling, lower pro-inflammatory cytokines, and limit angiogenesis and tissue damage. Intra-articular administration of small AuNPs reduces joint swelling, oxidative stress, and inflammatory infiltration, while larger nanoparticles enhance cellular uptake and mitigate cytokine spikes.

5.3 Silver Nanoparticles (AgNPs)

AgNPs are recognized for their anti-inflammatory and immunomodulatory properties, useful in nanotheranostic applications. They inhibit pro-inflammatory cytokine release from activated macrophages and promote an anti-inflammatory phenotype, making them effective in rheumatoid arthritis (RA) and systemic lupus erythematosus (SLE) [10, 101]. AgNPs influence both innate and adaptive immune cells, affecting T-cell differentiation into Th2 and Th17 subsets, and can stimulate mast cells via scavenger receptors, leading to cytokine release. Folic acid-modified AgNPs (FA-AgNPs) specifically target pro-inflammatory M1 macrophages, inducing apoptosis and repolarization into anti-inflammatory M2 macrophages, demonstrating therapeutic efficacy in RA models [141]. Their surface-plasmon resonance also enables optical imaging techniques, such as SERS, for monitoring particle localization in inflamed tissues [24].

5.4 Magnetic Nanoparticles (Fe_3O_4 / γ-Fe_2O_3)

Magnetic iron-oxide nanoparticles are versatile theranostic agents in autoimmune disorders due to their dual diagnostic and therapeutic roles [29, 102]. They serve as T2 MRI contrast agents, allowing visualization of inflammation in conditions such as multiple sclerosis (MS) and type 1 diabetes (T1D) [26]. Therapeutically, superparamagnetic iron oxide nanoparticles (SPIONs) can co-deliver autoantigens (e.g., MOG_{35-55}) with tolerogenic molecules, reprogramming dendritic cells to a tolerogenic phenotype and promoting regulatory T-cell proliferation, effectively inhibiting disease progression in EAE and T1D models [26, 65]. They can also modulate macrophage polarization, reducing recurrence in CNS autoimmune models [140].

5.5 Mesoporous Silica Nanoparticles (MSNs)

MSNs offer a substrate with a large surface area where antigens, immunomodulators, or siRNA can be incorporated. In RA and lupus models, they can successfully inhibit cytokine storms by releasing their cargo in response to the acidic environments of inflamed joints [137].

5.6 Carbon-Based Nanomaterials

Due to their high functionalization potential and photothermal characteristics, graphene oxide (GO), carbon nanotubes (CNTs), and graphene quantum dots (GQDs) are becoming popular for targeted immunotherapy [113]. Functionalized CNTs reduce the severity of EAE by modulating APCs to affect Th17 development [122]. GO nanosheets improve MS-like symptoms, control intracellular calcium signaling, and lessen astrocyte reactivity [33]. While decreased GO maintains MDSC immunosuppressive function and supports tolerance in MS animals, chitosan-functionalized GO (GO-CS) increases immunoadjuvant activity [18, 138]. In Graft-Versus-Host Disease (GVHD) and autoimmune models, GO-based neuropeptide administration and GQDs further encourage tolerogenic DCs and Treg proliferation, which lessens symptoms [122, 128]. Particle size and functionalization have a significant impact on immunological results; larger GO nanosheets promote the establishment of DC-T cell synapses [147].

5.7 Lipid-Based and Dendrimeric Systems

In models of experimental autoimmune encephalomyelitis (EAE), lipid nanoparticles—including solid lipid nanoparticles and nanostructured lipid carriers—improve the oral bioavailability of glucocorticoids and maintain remission. TNF-α in RA has been targeted by cationic liposomes [69]. Preclinical investigations on T1D and MS have demonstrated the effectiveness of functionalizing dendrimeric systems (e.g., PAMAM dendrimers) with self-antigens and tolerogenic adjuvants to deliver precise dosages to dendritic cells, producing anergy or regulatory T-cell differentiation [25].

5.8 Gene Delivery Nanoparticles

In autoimmune illnesses, specific immune response control is made possible by gene delivery nanoparticles containing siRNAs, miRNAs, or mRNA [57, 104]. Nucleic acids are protected, cellular absorption is improved, and cytosolic release for gene silencing is made possible by encapsulation in lipid, polymer, or apolipoprotein nanoparticles. In lupus and arthritis models, for example, mRNA-LNPs generating anti-CD19 antibodies specifically decrease pathogenic B cells, lowering inflammation and tissue damage without impacting T cells [49]. In RA and ulcerative colitis, PD-L1 mRNA-LNPs encourage tolerogenic APCs to inhibit T-cell activation [92]. Systems based on chitosan or ASO target colonic macrophages and are effective in IBD animals by suppressing pro-inflammatory cytokines, including TNF and IL-12 [6, 122].

6 Conclusion

This chapter examines autoimmune illnesses and connects new treatments to basic immunological dysregulation. Autoreactive B and T cells can cause tissue injury and chronic inflammation when central and peripheral tolerance are compromised. Although biologics and conventional medications enhance results, they frequently have limited efficacy and induce toxicity. With nanocarriers, gene vectors, and multifunctional nanoparticles increasing efficacy while lowering adverse effects, nanomedicine offers precision delivery, immune regulation, and theranostic capabilities. These methods have the potential to rebalance the immune system and offer safer, more individualized treatments, setting the stage for the development of autoimmune medicines in the future.

References

1. Abhyankar A, Tapper EB, Bonder A (2014) Immunosuppressive therapy in immune-mediated liver disease in the non-transplanted patient. Pharmaceuticals 7:18–28. https://doi.org/10.3390/ph7010018
2. Addissouky TA (2024) Nanocarrier-mediated drug delivery in rheumatoid arthritis : overcoming therapeutic limitations through targeted approaches. Clinical Orthopedics Trauma Care 6(7). https://doi.org/10.31579/2694-0248/108
3. Ahuja NK, Rajawat JS (2021) Novel nano therapeutic materials for the effective treatment of rheumatoid arthritis-recent insights. Int J Appl Pharmaceutics 13(6):31–40. https://doi.org/10.22159/ijap.2021v13i6.42912
4. Al-lawati H (2017) Nanomedicine for immunosuppressive therapy : achievements in pre-clinical research nanomedicine for immunosuppressive therapy : achievements in pre-clinical. Expert Opin Drug Deliv 15(4):397–418. https://doi.org/10.1080/17425247.2018.1420053
5. Van Alem CMA, Metselaar JM, Kooten CV, Rotmans JI (2021) Recent advances in liposomal-based anti-inflammatory therapy. Pharmaceutics 13:1–28
6. Andretto V, Dusi S, Zilio S, Repellin M, Kryza D, Ugel S, Lollo G (2023) Tackling TNF-α in autoinflammatory disorders and autoimmune diseases: From conventional to cutting edge in biologics and RNA- based nanomedicines. In: Advanced drug delivery reviews, 201 0(August). https://doi.org/10.1016/j.addr.2023.115080
7. Arnold M, Schrieber L, Brooks P (1988) Immunosuppressive drugs and corticosteroids in the treatment of rheumatoid arthritis. Drugs 36:340–363

8. Ayensu WK, Tchounwou PB, McMurray RW (2004) Molecular and cellular mechanisms associated with autoimmune diseases. Int J Environ Res Public Health 1(1):39–73. https://doi.org/10.3390/ijerph2004010039
9. Azzi J, Tang L, Moore R, Tong R, El Haddad N, Akiyoshi T, Mfarrej B, Yang S, Jurewicz M, Ichimura T, Lindeman N, Cheng J, Abdi R (2010) Polylactide-cyclosporin A nanoparticles for targeted immunosuppression. FASEB J 24(1). https://doi.org/10.1096/fj.10-154690
10. Barik P, Mondal S (2025) Immunomodulatory effects of metal nanoparticles: current trends and future prospects. Nanoscale 17. https://doi.org/10.1039/D5NR01030F
11. Behrangi E, Moodi F, Jafarzadeh A, Goodarzi A (2024) Paradoxical and bimodal immune-mediated dermatological side effects of TNF-α inhibitors: a comprehensive review. Skin Res Technol Official J Int Soc Bioengineering Skin (ISBS) Int Soc Digital Imaging Skin (ISDIS) Int Soc Skin Imaging (ISSI) 30(5):e13718. https://doi.org/10.1111/srt.13718
12. Benlaribi R, Gou Q, Takaba H (2022) Thymic self-antigen expression for immune tolerance and surveillance. Inflammation Regeneration 42(1):1–10. https://doi.org/10.1186/s41232-022-00211-z
13. Benne N, ter Braake D, Stoppelenburg AJ, Broere F (2022) Nanoparticles for inducing antigen-specific t cell tolerance in autoimmune diseases. Front Immunol 13:1–15. https://doi.org/10.3389/fimmu.2022.864403
14. Berry CT, Frazee CS, Herman PJ, Chen S, Chen A, Kuo Y, Ellebrecht CT (2025) Current advancements in cellular immunotherapy for autoimmune disease. Springer
15. Bhandari S, Bhandari S, Bhandari S (2023) Chimeric antigen receptor T cell therapy for the treatment of systemic rheumatic diseases: a comprehensive review of recent literature. Ann Med Surg 85(7):3512–3518. https://doi.org/10.1097/MS9.0000000000000891
16. Bolon B (2012) Cellular and molecular mechanisms of autoimmune disease. Toxicol Pathol 40(2):216–229. https://doi.org/10.1177/0192623311428481
17. Bullock J, Rizvi SAA, Saleh AM, Ahmed SS, Do DP, Ansari RA, Ahmed J (2019) Rheumatoid arthritis: a brief overview of the treatment. Med Princ Pract 27(6):501–507. https://doi.org/10.1159/000493390
18. Camacho-toledano C, Machín-díaz I, Lebrón-galán R, González-mayorga A, Palomares FJ, Serrano MC, Clemente D (2024) Graphene oxide films as a novel tool for the modulation of myeloid-derived suppressor cell activity in the context of multiple sclerosis† Celia. Nanoscale 16:7515–7531. https://doi.org/10.1039/d3nr05351b
19. Castiblanco C, Foster CS (2014) Review of systemic immunosuppression for autoimmune uveitis, pp 17–36. https://doi.org/10.1007/s40123-014-0023-x
20. Cendon S, Souza A, Santiago MB (2022) Vasculitis induced by biological agents used in rheumatology practice : a systematic review 37(2):300–310. https://doi.org/10.46497/ArchRheumatol.2022.9049
21. Chamundeeswari M, Jeslin J, Lal M (2019) Nanocarriers for drug delivery applications. Environ Chem Lett 17(2):849–865. https://doi.org/10.1007/s10311-018-00841-1
22. Chen T, Chang K, Su K (2022) Effects of biological/targeted therapies on bone mineral density in inflammatory arthritis
23. Cheng Z, Fobian SF, Gurrieri E, Amin M, Agostino VGD, Falahati M, Zalba S, Debets R, Garrido MJ, Saeed M, Seynhaeve ALB, Balcioglu HE, Hagen TLM (2024) Lipid-based nanosystems : the next generation of cancer immune therapy. J Hematol Oncol 1–39. https://doi.org/10.1186/s13045-024-01574-1
24. Chountoulesi M, Demetzos C (2020) Promising nanotechnology approaches in treatment of autoimmune diseases of central nervous system. Brain Sci 10(6). https://doi.org/10.3390/brainsci10060338
25. Cifuentes-Rius A, Desai A, Yuen D, Johnston APR, Voelcker NH (2021) Inducing immune tolerance with dendritic cell-targeting nanomedicines. Nat Nanotechnol 16(1):37–46. https://doi.org/10.1038/s41565-020-00810-2
26. Clemente-Casares X, Santamaria P (2014) Nanomedicine in autoimmunity. Immunol Lett 158(1–2):167–174. https://doi.org/10.1016/j.imlet.2013.12.018

27. Cusick MF, Libbey JE, Fujinami RS (2012) Molecular mimicry as a mechanism of autoimmune disease. Clin Rev Allergy Immunol 42(1):102–111. https://doi.org/10.1007/s12016-011-8294-7
28. Cuthrell KM, Tzenios N, Umber J (2022) Burden of Autoimmune Disorders; A review. Asian J Immunol 6(3):1–13
29. Dadfar SM, Roemhild K, Drude NI, von Stillfried S, Knüchel R, Kiessling F, Lammers T (2020) Iron Oxide nanoparticles: diagnostic, therapeutic and theranostic applications. Adv Drug Deliv Rev 138:302–325. https://doi.org/10.1016/j.addr.2019.01.005.Iron
30. Dalli M, Daoudi NE, Abrigach F, Azizi S, Bnouham M, Kim B, Gseyra N (2022) In vitro α -amylase and hemoglobin glycation inhibitory potential of Nigella sativa essential oil, and molecular docking studies of its principal components. October, 1–13. https://doi.org/10.3389/fphar.2022.1036129
31. Dalli M, El Guerraf A, Azizi S, Benataya K, Azghar A, Mi-kyung J, Maleb A, Bonglee K, Gseyra N (2022) Loaded n-Hydroxyapatite/SSG 3D scaffolds as a drug delivery system of Nigella sativa fractions for the management of local antibacterial infections, vol 10, pp 1–14
32. Davidson A, Diamond B (2001) Autoimmune diseases. New England J Med Rev 345(5):340–350
33. Di Mauro G, Amoriello R, Lozano N, Carnasciali A, Guasti D, Becucci M, Cellot G, Kostarelos K, Ballerini C, Ballerini L (2023) Graphene Oxide nanosheets reduce astrocyte reactivity to inflammation and ameliorate experimental autoimmune encephalomyelitis. ACS Nano 17(3):1965–1978. https://doi.org/10.1021/acsnano.2c06609
34. Drosos A, Pelechas E, Kaltsonoudis E, Markatseli TE, Voulgari PV (2021) Biologic therapies and autoimmune phenomena. Mediterranean J Rheumatol 32(2):96–103
35. Eiz-Vesper B, Schmetzer HM (2020) Antigen-presenting cells: Potential of proven und new players in immune therapies. Transfusion Med Hemotherapy 47(6):429–431. https://doi.org/10.1159/000512729
36. Elkoshi Z (2024) Autoimmune diseases refractory to corticosteroids and immunosuppressants. Front Immunol 15:1–9. https://doi.org/10.3389/fimmu.2024.1447337
37. Evangelatos G, Bamias G, Kitas GD, Kollias G, Sfikakis PP (2022) The second decade of anti-TNF-a therapy in clinical practice: new lessons and future directions in the COVID-19 era. Rheumatol Int 42(9):1493–1511. https://doi.org/10.1007/s00296-022-05136-x
38. Fahlquist-Hagert C, Wittenborn TR, Terczyńska-Dyla E, Kastberg KS, Yang E, Rallistan AN, Markett QR, Winther G, Fonager S, Voss LF, Pedersen MK, van Campen N, Ferapontov A, Jensen L, Huang J, Nieland JD, van der Poel CE, Palmfeldt J, Carroll MC, Degn SE et al (2023) Antigen presentation by B cells enables epitope spreading across an MHC barrier. Nature Commun 14(1). https://doi.org/10.1038/s41467-023-42541-7
39. Fan YN, Zhao G, Zhang Y, Ye QN, Sun YQ, Shen S, Liu Y, Xu CF, Wang J (2023) Progress in nanoparticle-based regulation of immune cells. Med Rev 3(2):152–179. https://doi.org/10.1515/mr-2022-0047
40. Fasano R, Malerba E, Prete M, Solimando AG, Buonavoglia A, Silvestris N, Leone P, Racanelli V (2022) Impact of antigen presentation mechanisms on immune response in autoimmune hepatitis. Front Immunol 12:1–11. https://doi.org/10.3389/fimmu.2021.814155
41. Garabatos N, Santamaria P (2022) Gut microbial antigenic mimicry in autoimmunity. Frontiers Immunol 13. https://doi.org/10.3389/fimmu.2022.873607
42. Getahun A, Cambier JC (2019) Non-antibody-secreting functions of B cells and their contribution to autoimmune disease. Annu Rev Cell Dev Biol 35:337–356. https://doi.org/10.1146/annurev-cellbio-100617-062518
43. Gharagozloo M, Majewski S, Foldvari M (2015) Therapeutic applications of nanomedicine in autoimmune diseases: From immunosuppression to tolerance induction. Nanomedicine Nanotechnol Biol Med 11(4):1003–1018. https://doi.org/10.1016/j.nano.2014.12.003
44. Gholijani N, Daryabor G, Kahmini FR (2024) T cell-intrinsic peripheral tolerance: a checkpoint target to treat autoimmunity. J Cellular Immunol 6(2):87–97. https://doi.org/10.33696/immunology.6.194

45. Girase R, Gujarathi NA, Sukhia A, Sai S, Kota N, Patil TS, Aher AA (2025) Targeted nanoliposomes for precision rheumatoid arthritis therapy: a review on mechanisms and in vivo potential. Drug Deliv 32(1). https://doi.org/10.1080/10717544.2025.2459772
46. Gómez-Manríquez J, Hernández-Bello J, Muñoz-Valle JF, Sifuentes-Franco S, Graciano-Machuca O, Morales-Núñez JJ (2025) B cell development: transcriptional regulation and immunological mechanisms in homeostasis. Front Immunol 16:1–16. https://doi.org/10.3389/fimmu.2025.1593338
47. Goodnow CC (2007) Multistep pathogenesis of autoimmune disease. Cell 130(1):25–35. https://doi.org/10.1016/j.cell.2007.06.033
48. Goyal R, Macri LK, Kaplan HM, Kohn J (2016) Nanoparticles and nanofibers for topical drug delivery, pp 77–92. https://doi.org/10.1016/j.jconrel.2015.10.049.Nanoparticles
49. Guo C, Tang Y, Zeng L, You X, Luo S, Du Y, Wang L, Wang L, Wang J, Chen J, Zhou Y (2025) Therapeutic potential of lipid nanoparticle-encapsulated CD19-targeting mRNAs in Lupus and Rheumatoid Arthritis. Adv Sci 2501628:1–15. https://doi.org/10.1002/advs.202501628
50. Gupta N, Shaik AR, Chettupalli AK, Bukke SPN, Goruntla N (2025) Nanomedicine for the management of rheumatoid arthritis: current status and prospective therapeutic approaches. In: Discover medicine, vol 2, issue 1. Springer Nature Singapore. https://doi.org/10.1007/s44337-025-00316-7
51. Gust J, Ponce R, Liles WC, Garden GA, Turtle CJ (2020) Cytokines in CAR T Cell-associated neurotoxicity. Front Immunol 11:1–23. https://doi.org/10.3389/fimmu.2020.577027
52. Gutiérrez-cruz SG, Muñoz-diosdado A, Gutiérrez-Calleja RA, Rodríguez-Cortés O, Ortiz-Reyez AE, Flores-Mejía R (2025) Influence of physicochemical factors on the interaction of metallic nanoparticles with immune system cells. Frontiers 1–25. https://doi.org/10.3389/fnano.2024.1496230
53. Habibi MA, Nezhad Shamohammadi F, Rajaei T, Namdari H, Pashaei MR, Farajifard H, Ahmadpour S (2023) Immunopathogenesis of viral infections in neurological autoimmune disease. BMC Neurol 23(1):1–15. https://doi.org/10.1186/s12883-023-03239-x
54. Haghikia A, Hegelmaier T, Wolleschak D, Böttcher M, Pappa V, Motte J, Borie D, Gold R, Feist E, Schett G, Mougiakakos D (2024) Clinical efficacy and autoantibody seroconversion with CD19-CAR T cell therapy in a patient with rheumatoid arthritis and coexisting myasthenia gravis. Ann Rheum Dis 83(11):1597–1598. https://doi.org/10.1136/ard-2024-226017
55. Harley ITW, Allison K, Scofield RH (2022) Polygenic autoimmune disease risk alleles impacting B cell tolerance act in concert across shared molecular networks in mouse and in humans. Front Immunol 13:1–30. https://doi.org/10.3389/fimmu.2022.953439
56. Hawker K (2009) B cells as a target of immune modulation. Ann Indian Acad Neurol 12(4):221–225. https://doi.org/10.4103/0972-2327.58275
57. Hofstraat SRJ, Anbergen T, Zwolsman R, Deckers J, van Elsas Y, Trines MM, Versteeg I, Hoorn D, Ros GWB, Bartelet BM, Hendrikx MMA, Darwish YB, Kleuskens T, Borges F, Maas RJF, Verhalle LM, Tielemans W, Vader P, de Jong OG, van der Meel R et al (2025) Nature-inspired platform nanotechnology for RNA delivery to myeloid cells and their bone marrow progenitors. Nat Nanotechnol 20(4):532–542. https://doi.org/10.1038/s41565-024-01847-3
58. Hopp AK, Rupp A, Lukacs-Kornek V (2014) Self-antigen presentation by dendritic cells in autoimmunity. Frontiers Immunol 5:1–14. https://doi.org/10.3389/fimmu.2014.00055
59. Huang Y, Tang H, Meng X, Zhao Z, Liu Y, Liu D, Chen B, Zou Z (2023) Development of large hollow particles for pulmonary delivery of Cyclosporine A. Pharmaceutics 15:2204
60. Hughes AD, Teachey DT, Diorio C (2024) Riding the storm: managing cytokine-related toxicities in CAR-T cell therapy. Seminars Immunopathol 46(3–4):1–19. https://doi.org/10.1007/s00281-024-01013-w
61. Jang DI, Lee AH, Shin HY, Song HR, Park JH, Kang TB, Lee SR, Yang SH (2021) The role of tumor necrosis factor alpha (Tnf-α) in autoimmune disease and current tnf-α inhibitors in therapeutics. Int J Mol Sci 22(5):1–16. https://doi.org/10.3390/ijms22052719

62. Jhu S (2024) Autoimmunity: mechanisms. Implications Adv Treatment 14:548–550
63. Juarranz Y (2021) Molecular and cellular basis of autoimmune diseases. Cells
64. Kastrati K, Aletaha D, Burmester GR, Chwala E, Dejaco C, Dougados M, Mcinnes IB, Ravelli A, Sattar N, Stamm TA, Takeuchi T (2022) A systematic literature review informing the consensus statement on efficacy and safety of pharmacological treatment with interleukin-6 pathway inhibition with biological DMARDs in mediated inflammatory diseases. RMD Open 8:1–23. https://doi.org/10.1136/rmdopen-2022-002359
65. Kenison JE, Jhaveri A, Li Z, Khadse N, Tjon E, Tezza S, Nowakowskac D, Plasencia A Jr, Sherr DH, Quintana FJ (2020) Tolerogenic nanoparticles suppress central nervous system inflammation. Immunol Inflammation 117(50):32017–32028. https://doi.org/10.1073/pnas.2016451117
66. Kim S, Choi B, Kim Y, Shim G (2023) Immune-modulating lipid nanomaterials for the delivery of biopharmaceuticals. Pharmaceutics 15(6). https://doi.org/10.3390/pharmaceutics15061760
67. Kourbeti IS, Boumpas DT (2005) Biological therapies of autoimmune diseases, pp 41–46
68. Koushki K, Shahbaz SK, Keshavarz M, Bezsonov EE, Sathyapalan T, Sahebkar A (2021) Gold nanoparticles: multifaceted roles in the management of autoimmune disorders. Biomolecules 11:1–39
69. Li H, Yang Y-G, Sun T (2022) Nanoparticle-based drug delivery systems for induction of tolerance and treatment of autoimmune diseases. Frontiers Bioengineering Biotechnology 10:1–14. https://doi.org/10.3389/fbioe.2022.889291
70. Li J, Li W, Zhuang L (2024) Natural biomimetic nano-system for drug delivery in the treatment of rheumatoid arthritis: a literature review of the last 5 years. Front Med 11:1–11. https://doi.org/10.3389/fmed.2024.1385123
71. Li P, Zheng Y, Chen X (2017) Drugs for autoimmune inflammatory diseases: From small molecule compounds to Anti-TNF Biologics. Front Pharmacol 8:1–12. https://doi.org/10.3389/fphar.2017.00460
72. Lin YJ, Anzaghe M, Schülke S (2020) Update on the pathomechanism, diagnosis, and treatment options for rheumatoid arthritis. Cells 9(4). https://doi.org/10.3390/cells9040880
73. Liu D, Zhao J (2018) Cytokine release syndrome: grading, modeling, and new therapy. J Hematol Oncol 11(1):1–7. https://doi.org/10.1186/s13045-018-0653-x
74. Lopetuso LR, Cuomo C, Mignini I, Gasbarrini A, Papa A (2023) Focus on Anti-Tumour Necrosis Factor (TNF)-α-related autoimmune diseases. Int J Mol Sci 24(9). https://doi.org/10.3390/ijms24098187
75. Luchetti MM, Benfaremo D, Gabrielli A (2017) Biologics in inflammatory and immunomediated arthritis. Curr Pharm Biotechnol 18(12):989–1007. https://doi.org/10.2174/1389201019666171226151852
76. Lugli EB, Correia RESM, Fischer R, Lundberg K, Bracke KR, Montgomery AB, Kessler BM, Brusselle GG, Venables PJ (2015) Expression of citrulline and homocitrulline residues in the lungs of non-smokers and smokers: implications for autoimmunity in rheumatoid arthritis. Arthritis Res Ther 17(1):1–8. https://doi.org/10.1186/s13075-015-0520-x
77. Lund FE (2008) Cytokine-producing B lymphocytes—key regulators of immunity. Curr Opin Immunol 20(3):332–338. https://doi.org/10.1016/j.coi.2008.03.003
78. Luu VP, Vazquez MI, Zlotnik A (2014) B cells participate in tolerance and autoimmunity through cytokine production. Autoimmunity 47(1):1–12. https://doi.org/10.3109/08916934.2013.856006
79. Mackensen A, Müller F, Mougiakakos D, Böltz S, Wilhelm A, Aigner M, Völkl S, Simon D, Kleyer A, Munoz L, Kretschmann S, Kharboutli S, Gary R, Reimann H, Rösler W, Uderhardt S, Bang H, Herrmann M, Ekici AB, Schett G et al (2022) Anti-CD19 CAR T cell therapy for refractory systemic lupus erythematosus. Nat Med 28(10):2124–2132. https://doi.org/10.1038/s41591-022-02017-5
80. Madawi EA, Raad A, Jayoush A, Rawas-qalaji M, Thu HE, Khan S, Sohail M, Mahmood A, Hussain Z (2023) Polymeric nanoparticles as tunable nanocarriers for targeted delivery of drugs to skin tissues for treatment of topical skin diseases. Pharmaceutics 15:657

81. Mazumder B (2018) B cells responses and cytokine production are regulated by their immune microenvironment. Physiol Behav 74(2):318–326. https://doi.org/10.1016/j.cyto.2015.02.007.B
82. Meng X, Layhadi JA, Keane ST, Cartwright NJK, Durham SR, Shamji MH (2023) Immunological mechanisms of tolerance: central, peripheral and the role of T and B cells. Asia Pac Allergy 13(4):175–186. https://doi.org/10.5415/apallergy.0000000000000128
83. Meurer M (2012) Immunosuppressive therapy for autoimmune bullous diseases. Clin Dermatol 30(1):78–83. https://doi.org/10.1016/j.clindermatol.2011.03.013
84. Min HK, Kim SH, Kim HR, Lee SH (2022) Therapeutic utility and adverse effects of biologic disease-modifying anti-rheumatic drugs in inflammatory arthritis. Int J Mol Sci 23(22). https://doi.org/10.3390/ijms232213913
85. Mishra B, Patel BB, Tiwari S (2010) Colloidal nanocarriers: a review on formulation technology, types and applications toward targeted drug delivery. Nanomedicine Nanotechnol Biol Med 6(1):9–24. https://doi.org/10.1016/j.nano.2009.04.008
86. Mobeen H, Safdar M, Fatima A, Afzal S, Zaman H, Mehdi Z (2022) Emerging applications of nanotechnology in context to immunology: a comprehensive review. Frontiers Bioengineering Biotechnology 10:1–20. https://doi.org/10.3389/fbioe.2022.1024871
87. Möhn N, Bonda V, Grote-Levi L, Panagiota V, Fröhlich T, Schultze-Florey C, Wattjes MP, Beutel G, Eder M, David S, Körner S, Höglinger G, Stangel M, Ganser A, Koenecke C, Skripuletz T (2022) Neurological management and work-up of neurotoxicity associated with CAR T cell therapy. Neurol Res Pract 4(1):1–10. https://doi.org/10.1186/s42466-021-00166-5
88. Müller F, Taubmann J, Bucci L, Wilhelm A, Bergmann C, Völkl S, Aigner M, Rothe T, Minopoulou I, Tur C, Knitza J, Kharboutli S, Kretschmann S, Vasova I, Spoerl S, Reimann H, Munoz L, Gerlach RG, Schäfer S, Schett G (2024) CD19 CAR T-cell therapy in autoimmune disease—a case series with follow-up. N Engl J Med 390(8):687–700. https://doi.org/10.1056/NEJMoa2308917
89. Nasra S, Bhatia D, Kumar A (2022) Recent advances in nanoparticle-based drug delivery systems for rheumatoid arthritis treatment. Nanoscale Adv 4(17):3479–3494. https://doi.org/10.1039/d2na00229a
90. Nikolajczyk BS (2010) B cells as under-appreciated mediators of non-autoimmune inflammatory disease. Curr Opin Neurol 50(3):234–242. https://doi.org/10.1097/00019052-199606000-00020
91. Oh S, Payne AS (2022) Engineering cell therapies for autoimmune diseases: From preclinical to clinical proof of concept. Immune Netw 22(5):1–16
92. Papatriantafyllou M (2025) Lipid nanoparticles with PDL1-encoding mRNA spread tolerance. Nat Rev Rheumatol 21(6):317. https://doi.org/10.1038/s41584-025-01251-8
93. Parvin N, Joo SW, Mandal TK (2025) Biodegradable and stimuli-responsive nanomaterials for targeted drug delivery in autoimmune diseases. J Functional Biomaterials 16(1). https://doi.org/10.3390/jfb16010024
94. Peer D, Karp JM, Hong S, Farokhzad OC, Margalit R, Langer R (2007) Nanocarriers as an emerging platform for cancer therapy. Nat Nanotechnol 2:751–760
95. Pensato U, Muccioli L, Zinzani P, D'Angelo R, Pierucci E, Casadei B, Dicataldo M, De Matteis S, Cortelli P, Bonifazi F, Guarino M (2022) Fulminant cerebral edema following CAR T-cell therapy: case report and pathophysiological insights from literature review. J Neurol 269(8):4560–4563. https://doi.org/10.1007/s00415-022-11117-8
96. Pestka JJ, Pollard KM, Rosenspire AJ (2021) Editorial: the role of the environment in autoimmunity. Front Immunol 12:10–12. https://doi.org/10.3389/fimmu.2021.641171
97. Pollard KM, Cauvi DM, Mayeux JM, Toomey CB, Peiss AK, Hultman P, Kono DH (2021) Mechanisms of environment-induced autoimmunity. Ann Rev Pharmacol Toxicol 61:135–157
98. Prabhu RH, Patravale VB, Joshi MD (2015) Polymeric nanoparticles for targeted treatment in oncology: current insights. Int J Nanomedicine 1001–1018
99. Pugliesi A, de Oliveira AB, Oliveira AB, Xavier R, da Mota LMH, Bertolo MB, Gonzalez-Gay MA, Citera G, de Carvalho LSF (2023) Compared efficacy of rituximab, abatacept, and

tocilizumab in patients with rheumatoid arthritis refractory to methotrexate or TNF inhibitors agents: a systematic review and network meta-analysis. Adv Rheumatol 63(1). https://doi.org/10.1186/s42358-023-00298-z
100. Puricelli C, Gigliotti CL, Stoppa I, Sacchetti S, Pantham D, Scomparin A, Rolla R, Pizzimenti S, Dianzani U, Boggio E, Sutti S (2023) Use of poly lactic-co-glycolic acid nano and micro particles in the delivery of drugs modulating different phases of inflammation
101. Rabiee N, Ahmadi S, Iravani S, Varma RS (2022) Functionalized silver and gold nanomaterials with diagnostic and therapeutic applications. Pharmaceutics
102. Rahman M (2023) Nanotheranostics magnetic resonance imaging and iron-oxide nanoparticles in the era of personalized medicine. Nanotheranostics 7(8). https://doi.org/10.7150/ntno.86467
103. Ren S, Xu Y, Dong X, Mu Q, Chen X, Yu Y, Su G (2024) Nanotechnology-empowered combination therapy for rheumatoid arthritis: principles, strategies, and challenges. J Nanobiotechnol 22(1):431. https://doi.org/10.1186/s12951-024-02670-7
104. Ren T, Ma L, Fu P, Zhang C (2025) Nanoparticles used for the delivery of RNAi-based therapeutics. Pharmaceutics 17(11). https://doi.org/10.3390/pharmaceutics17111502
105. Hampe C (2012) B cells in autoimmune diseases. Scientifica (Cairo) 38(5):18. https://doi.org/10.6064/2012/215308
106. Salaman MR, Gould KG (2020) Breakdown of T-cell ignorance: the tolerance failure responsible for mainstream autoimmune diseases? J Transl Autoimmunity 3:100070. https://doi.org/10.1016/j.jtauto.2020.100070
107. Schiavo AL, Puca RV, Ruocco V, Ruocco E (2010) Adjuvant drugs in autoimmune bullous diseases, efficacy versus safety: facts and controversies. Clin Dermatol 28(3):337–343. https://doi.org/10.1016/j.clindermatol.2009.06.018
108. Sen S, Xavier J, Kumar N, Ahmad MZ, Ranjan OP (2023) Exosomes as natural nanocarrier-based drug delivery system: recent insights and future perspectives. 3 Biotech 13(3):101. https://doi.org/10.1007/s13205-023-03521-2
109. Serra P, Santamaria P (2021) Peptide-MHC-based nanomedicines for the treatment of autoimmunity: engineering, mechanisms, and diseases. Front Immunol 11:1–9. https://doi.org/10.3389/fimmu.2020.621774
110. Sheng L, Zhang Y, Song Q, Jiang X, Cao W, Li L, Yi H, Weng X, Chen S, Wang Z, Wu W, Wang L, Zhao W, Yan Z (2023) Concurrent remission of lymphoma and Sjögren's disease following anti-CD19 chimeric antigen receptor-T cell therapy for diffuse large B-cell lymphoma: a case report. Front Immunol 14:1–7. https://doi.org/10.3389/fimmu.2023.1298815
111. Shirafkan F, Hensel L, Rattay K (2024) Immune tolerance and the prevention of autoimmune diseases essentially depend on thymic tissue homeostasis. Front Immunol 15:1–16. https://doi.org/10.3389/fimmu.2024.1339714
112. Smolen JS, Van Der Heijde D, Machold KP, Aletaha D, Landewé R (2014) Proposal for a new nomenclature of disease-modifying antirheumatic drugs. Ann Rheum Dis 73(3):3–5. https://doi.org/10.1136/annrheumdis-2013-204317
113. Soltani R, Guo S, Bianco A, Ménard-moyon C, Soltani R, Guo S, Bianco A, Carbon CM (2020) Carbon nanomaterials applied for the treatment of inflammatory diseases: preclinical evidence To cite this version : HAL Id : hal-03001720
114. Song Y, Li J, Wu Y (2024) Evolving understanding of autoimmune mechanisms and new therapeutic strategies of autoimmune disorders. Signal Transduction Targeted Therapy 9(1). https://doi.org/10.1038/s41392-024-01952-8
115. Sota J, Bruno MG (2021) Biologic therapies and small molecules for the management of non-infectious scleritis: a narrative review. Ophthalmol Therapy 10(4):777–813. https://doi.org/10.1007/s40123-021-00393-8
116. Stagg AJ, Knight SC (2001) Antigen-presenting Cells. In: Encyclopedia of life sciences. https://doi.org/10.1038/npg.els.0000903
117. Svecova D (2018) Current therapy in autoimmune bullous diseases current therapy in autoimmune bullous diseases. In: Autoimmune bullous diseases, pp. 171–203. https://doi.org/10.5772/intechopen.71541

118. Taher TE, Bystrom J, Ong VH, Isenberg DA, Renaudineau Y, Abraham DJ, Mageed RA (2017) Intracellular B lymphocyte signalling and the regulation of humoral immunity and autoimmunity. Clin Rev Allergy Immunol 53(2):237–264. https://doi.org/10.1007/s12016-017-8609-4
119. Talamini L, Matsuura E, De Cola L, Muller S (2021) Immunologically inert nanostructures as selective therapeutic tools in inflammatory diseases. Cells 10(3):1–18. https://doi.org/10.3390/cells10030707
120. Tallantyre EC, Evans NA, Parry-Jones J, Morgan MPG, Jones CH, Ingram W (2021) Neurological updates: neurological complications of CAR-T therapy. J Neurol 268(4):1544–1554. https://doi.org/10.1007/s00415-020-10237-3
121. Tang L, Azzi J, Kwon M, Mounayar M, Tong R, Yin Q, Moore R, Skartsis N, Fan TM, Abdi R, Cheng J (2012) Immunosuppressive activity of size-controlled PEG-PLGA nanoparticles containing encapsulated Cyclosporine A. J Transplantation, 9. https://doi.org/10.1155/2012/896141
122. Tang Y, Li L (2024) The application of nanovaccines in autoimmune diseases. Int J Nanomed 19(January):367–388. https://doi.org/10.2147/IJN.S440612
123. Vadasz Z, Haj T, Kessel A, Toubi E (2013) B-regulatory cells in autoimmunity and immune mediated inflammation. FEBS Lett 587(13):2074–2078. https://doi.org/10.1016/j.febslet.2013.05.023
124. Viswanath D (2013) Understanding autoimmune diseases—a review. J Dental Med Sci 6(6):8–15
125. Vormittag P, Gunn R, Ghorashian S, Veraitch FS (2018) A guide to manufacturing CAR T cell therapies. Curr Opin Biotechnol 53:164–181. https://doi.org/10.1016/j.copbio.2018.01.025
126. Vukovic J, Abazovic D, Vucetic D, Medenica S (2024) CAR-engineered T cell therapy as an emerging strategy for treating autoimmune diseases. October, 1–8. https://doi.org/10.3389/fmed.2024.1447147
127. Wang L, Wang FS, Gershwin ME (2015) Human autoimmune diseases: a comprehensive update. J Intern Med 278(4):369–395. https://doi.org/10.1111/joim.12395
128. Wang Q, Qin X, Fang J, Sun X (2021) Nanomedicines for the treatment of rheumatoid arthritis: state of art and potential therapeutic strategies. Acta Pharmaceutica Sinica B 11(5):1158–1174. https://doi.org/10.1016/j.apsb.2021.03.013
129. Wang X, Yang B, Xu X, Su M, Xi M, Yin Z (2020) Dextran sulfate—modified pH-sensitive layered double hydroxide nanocomposites for treatment of rheumatoid arthritis
130. Wang X, Sun S, He C, Li C, Bai W, Zhang Y, He M, Lv L, Dong Y, Ma P, Zhou Q, Zhan L (2021) A novel plasma membrane-phillic graphene oxide nanocarrier for neuropeptide delivery to generate tolerogenic dendritic cells in GVHD Immunotherapy. Res Square 30
131. Wang Z, Han W (2018) Biomarkers of cytokine release syndrome and neurotoxicity related to CAR-T cell therapy. Biomarker Res 6(1):1–10. https://doi.org/10.1186/s40364-018-0116-0
132. Willis MD, Robertson NP (2019) Neurotoxicity of novel cancer immunotherapies. J Neurol 266(8):2087–2089. https://doi.org/10.1007/s00415-019-09444-4
133. Wu D, Xu-Monette ZY, Zhou J, Yang K, Wang X, Fan Y, Young KH (2025) CAR T-cell therapy in autoimmune diseases: a promising frontier on the horizon. Front Immunol 16:1–23. https://doi.org/10.3389/fimmu.2025.1613878
134. Xiang G, Cui Y, Wang P, Feng Y, Zhang C, Lou J, Zhou X (2025) Nanomedicine targeting the PD-1/PD-L1 axis in autoimmune diseases: breaking conventional barriers to restore immune tolerance. J Nanobiotechnol 23(1):664. https://doi.org/10.1186/s12951-025-03766-4
135. Xiao X, Huang S, Chen S, Wang Y, Sun Q, Xu X, Li Y (2021) Mechanisms of cytokine release syndrome and neurotoxicity of CAR T-cell therapy and associated prevention and management strategies. J Experimental Clinical Cancer Res 40(1). https://doi.org/10.1186/s13046-021-02148-6
136. Xing Y, Hogquist KA (2012) T-Cell tolerance: central and peripheral. Cold Spring Harb Perpect Biol 4:1–15
137. Xu B (2023) Multifunctional mesoporous silica nanoparticles for biomedical applications. Signal Transduction Targeted Therapy 8. https://doi.org/10.1038/s41392-023-01654-7

138. Yan T, Zhang H, Huang D, Feng S, Fujita M, Gao X-D (2017) Chitosan-functionalized graphene oxide as a potential immunoadjuvant. Nanomaterials (Basel, Switzerland) 7(3). https://doi.org/10.3390/nano7030059
139. Yang B, Zhao M, Wu H, Lu Q (2020) A comprehensive review of biological agents for lupus: beyond single target 11:1–11. https://doi.org/10.3389/fimmu.2020.539797
140. Yang C, Li S, Wang L (2025) Engineered iron oxide nanoplatforms: reprogramming immunosuppressive niches for precision cancer theranostics. Mol Cancer 24. https://doi.org/10.1186/s12943-025-02443-2
141. Yang Y, Guo L, Wang Z, Liu P, Liu X, Ding J, Zhou W (2021) Targeted silver nanoparticles for rheumatoid arthritis therapy via macrophage apoptosis and re-polarization. Biomaterials 264:120390. https://doi.org/10.1016/j.biomaterials.2020.120390
142. Yasir M, Mishra R, Tripathi AS, Maurya RK, Shahi A, Zaki MEA, Hussain SA, Masand VH (2024) Discover nano theranostics : a multifaceted approach utilizing nano-biomaterials. Discover Nano 19:35. https://doi.org/10.1186/s11671-024-03979-w
143. Yasmeen F, Pirzada RH, Ahmad B, Choi B, Choi S (2024) Understanding autoimmunity: mechanisms, predisposing factors, and cytokine therapies. Int J Mol Sci 25(14). https://doi.org/10.3390/ijms25147666
144. Zeb A, Salman O, Yu C, Akram M, Kim H, Kim M, Kang J, Majid A, Chang S, Bae O, Kim J (2017) Enhanced anti-rheumatic activity of methotrexate-entrapped ultradeformable liposomal gel in adjuvant-induced arthritis rat model. Int J Pharm 525(1):92–100. https://doi.org/10.1016/j.ijpharm.2017.04.032
145. Zhang Z, Tsai P, Ramezanli T (2013) Polymeric nanoparticles-based topical delivery systems for the treatment of dermatological diseases. In: WIREs nanomedicine and nanobiotechnology. https://doi.org/10.1002/wnan.1211
146. Zhao Y, Han J, Zhang F, Liao T, Na R, He G, Ye W (2022) Flexible nano-liposomes-based transdermal hydrogel for targeted delivery of dexamethasone for rheumatoid arthritis therapy. Drug Deliv 29(1):2266–2277. https://doi.org/10.1080/10717544.2022.2096718
147. Zhou Q, Gu H, Sun S, Zhang Y, Hou Y, Li C, Zhao Y, Ma P, Lv L, Aji S, Sun S, Wang X, Zhan L (2021) Large-sized graphene oxide nanosheets increase DC-T-cell synaptic contact and the efficacy of DC vaccines against SARS-CoV-2. Adv Mater (Deerfield Beach, Fla.) 33(40):e2102528. https://doi.org/10.1002/adma.202102528

Smart Nanotechnology for Next-Generation Vaccine Development

Ritu Raj, Shweta Dang, and Reema Gabrani

Abstract The rising incidence of emerging and reemerging infectious diseases, together with the need for adaptable, rapid, and personalized vaccination platforms, has driven significant progress at the intersection of nanotechnology, synthetic biology, and artificial intelligence. Smart nanotechnology involves the use of a carefully designed nanoscale delivery system that allows for precise control of antigen delivery, immune system targeting, and helps overcome limitations associated with traditional vaccine methods. Coupled with AI-based tools for predicting epitopes, analyzing immune responses, and optimizing nanoparticle designs, these platforms promise to accelerate vaccine development, enabling personalized immunization (e.g., for cancer, rare infections, or rapidly evolving pathogens) and improving access to vaccines worldwide. This chapter provides an overview of how nanovaccines are designed, considering the type of nanomaterial, the mechanism of immune activation, and how AI is being integrated with nanomedicine. It introduces the main nanoparticle platforms (such as lipid, protein-based, polymeric, and virus-like particles), discusses their advantages and limitations, along with preclinical and clinical examples. Finally, it concludes by addressing future directions, including mucosal and needle-free delivery, multivalent/multiplex vaccines, self-amplifying RNA vaccines, theranostic "nanovaccines," and AI-guided personalized immunization strategies.

Keywords Artificial intelligence · Lipid nanoparticle · mRNA vaccine · Nanovaccine · Nanomedicine · Next-generation vaccines · Personalized vaccine · Virus-like particle

---

R. Raj · S. Dang · R. Gabrani (✉)
Department of Biotechnology, Jaypee Institute of Information Technology, Noida, Uttar Pradesh, India
e-mail: reema.gabrani@mail.jiit.ac.in

J. Chen (ed.), *AI-Driven Nanotechnology in Precision Medicine and Healthcare*, Smart Nanomaterials Technology, https://doi.org/10.1007/978-981-92-1690-1_11

1 Introduction

The rapid increase in emergence and re-emergence of infectious diseases, like antimicrobial-resistant pathogens, COVID-19, Ebola, and Zika, has highlighted the critical gaps in conventional immunization strategies and emphasizes the need for next-generation vaccine platforms [1, 2]. Traditional vaccines, including live attenuated, inactivated, or subunit formulations, have played an important role in global public health; however, despite their success, they face major limitations like prolonged development timelines, poor thermal stability, deficient immune response, and reduced effectiveness against rapidly mutating pathogens. In this context, smart nanotechnology has emerged as a promising strategy, enabling vaccine delivery systems to be precisely engineered with control over antigen presentation, immune modulation, and pharmacokinetics [3].

Nanotechnology-enabled vaccines use nanoscale carriers such as lipid nanoparticles, polymeric nanoparticles, virus-like particles (VLPs), self-assembling protein scaffolds, inorganic nanoparticles, and biomimetic exosomes that can protect antigens from being degraded, target immune cells, and manage more precise and effective immune responses [4]. The worldwide success of the mRNA-LNP vaccine for COVID-19 clearly demonstrated the real-world potential of nano-enabled platforms, setting standards for rapid development, large-scale manufacturing, and adaptability in pandemic readiness [5]. Smart nanomaterials can be engineered with targeting ligands, adjuvants, or stimuli-responsive components that respond to pH, enzymes, hypoxia, temperature, and redox potential, allowing the antigen to be released in a controlled manner and improve their cross-presentation via MHC I and II pathways [6, 7].

Beyond infectious diseases, nanotechnology is also enabling major advances in therapeutic cancer vaccines, personalized immunotherapies, and multi-epitope precision vaccines, integrating artificial intelligence, computational modeling, and systems biology to enhance vaccine design and their translation into clinical applications [8]. Furthermore, mucosal nanovaccines delivered intranasally, orally, or transdermally provide needle-free, patient-friendly options capable of triggering strong local and systemic immune response, thereby offering a promising way to transform vaccination strategies for global and low-resource settings [9].

This chapter discusses the science behind smart nanotechnology and how it is shaping the future of vaccines. It covers the types of materials used, design principles, and immunological mechanisms that support the use of smart nanotechnology for next-generation vaccines. It also addresses the key challenges related to translation, safety considerations, regulations, and future directions, highlighting personalized vaccinology, self-amplifying RNA systems, pan-pathogen vaccine strategies, and universal vaccines. Collectively, these advancements demonstrate the potential to create an adaptable and responsive global immunization ecosystem for the twenty-first century and decades ahead.

2 Key Nanomaterials and Platforms for Next-Generation Vaccines

The development of smart nanotechnology-based vaccines is driven by precisely engineered nanomaterials that function as carriers, adjuvants, or antigen-presenting scaffolds. These platforms offer flexibility to carry a diverse range of antigens, including proteins, peptides, nucleic acids, and viral vectors, which help the immune system to respond more effectively. They also enable controlled and efficient delivery.

2.1 Lipid Nanoparticles (LNPs)

Lipid nanoparticles are among the most significant nanoplatforms because of their successful application in mRNA vaccines against COVID-19 [10]. LNPs composed of ionizable lipids, phospholipids, cholesterol, and polyethylene glycol (PEG)-modified lipids, forming stable structures, are capable of encapsulating and protecting nucleic acids [11].

LNPs provide several key benefits as a carrier for nucleic acids. They are highly efficient at delivering messenger RNA (mRNA) and self-amplifying RNA (saRNA) into cells and shielding them from being degraded by enzymes during circulation in the body and cellular uptake [11]. LNPs are biodegradable and can be manufactured at a large scale, making them suitable for clinical translation and vaccine production [12]. Their design can be easily adjusted, which will help address new variants of the pathogen. By facilitating efficient antigen-expression in host cells, LNP-based systems trigger a strong immune response, making them suitable for pandemic readiness and personalized medicine (Fig. 1) [13].

2.2 Polymeric Nanoparticles

Biodegradable polymeric nanoparticles such as PLGA, PLA, PCL, chitosan, and PEGylated systems provide controlled and sustained antigen release [14]. Their composition and surface functionality enable targeted delivery, mucosal penetration, and combination therapy [15]. Polymeric nanoparticles have several advantages for vaccines and immunotherapy. One of their advantages is the ability to achieve adjustable and controlled release profiles [16]. Additionally, it enables the co-delivery of antigens and immunological adjuvants within the same carrier [17, 18]. These systems are generally biocompatible and safe, as many of the polymers used, like PLGA, chitosan, and PEG, are biodegradable and approved for clinical use. This makes them suitable for translational and clinical applications.

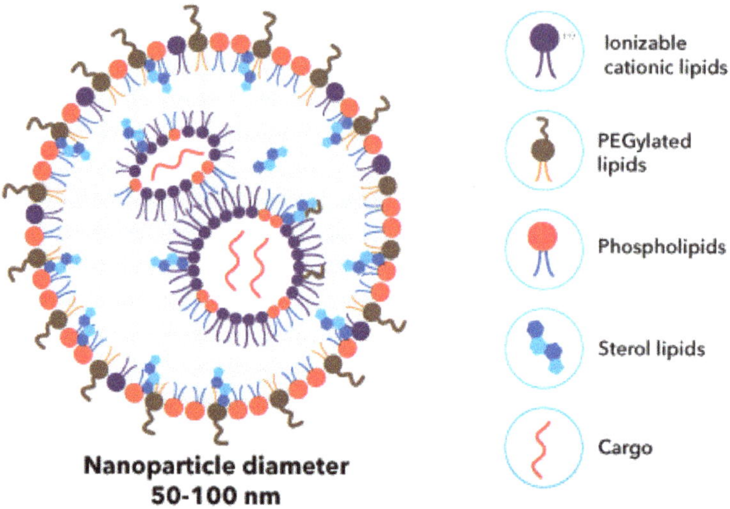

Fig. 1 Structure of a typical lipid nanoparticle (created with BioRender.com)

2.3 *Virus-Like Particles (VLPs)*

VLPs are self-assembled, non-infectious nanostructures mimicking natural viral morphology without containing replicating genetic material [19]. VLP-based vaccines are highly effective because they display antigens in a repetitive and ordered manner. This enables strong immunogenicity even in the absence of strong adjuvants [20]. VLPs can be efficiently produced using recombinant expression systems and allow the development of multivalent vaccine formulations [21]. VLP-based vaccines are already approved for hepatitis B and human papillomavirus, and new generations include modular and chimeric VLPs for emerging targets [22, 23].

2.4 *Inorganic Nanoparticles*

Inorganic materials such as gold nanoparticles, silica nanoparticles, calcium phosphate, quantum dots, and iron oxide nanoparticles offer unique optical, magnetic, and structural properties for customized delivery and immunomodulation [24]. These materials are highly stable and can be synthesized in a controlled and reproducible manner, allowing precise control over size, shape, and surface properties [25]. Their surface can be modified with antigens, targeting ligands, or immune-stimulating molecules, allowing targeted delivery and strong immune responses [26]. In addition, many inorganic nanoparticles support theranostic applications by combining therapeutic functions with imaging capabilities. Among these, gold nanoparticles, for example, are used as nanocarriers and immune stimulators in cancer vaccine research [27].

2.5 *Biomimetic and Cell-Derived Nanocarriers*

Biomimetic nanovaccines utilize biological membranes from red blood cells, cancer cells, platelets, or exosomes to achieve natural targeting properties and immune evasion [28]. They are membrane-derived nature, exhibiting good biocompatibility, and can achieve intrinsic targeting ability while reducing immune clearance. They exhibit cell-specific homing characteristics that enhance targeted delivery to the tissues [29]. Additionally, membrane-bound proteins and epitopes support efficient immune activation and enhance antigen presentation [30]. Extracellular vesicle-based nanovaccines have shown potential for cancer vaccine development and tumor immunotherapy [31].

2.6 *Peptide and Protein Self-assembling Nanostructures*

Peptides are considered highly promising biomaterials for regenerative scaffolds because communication within the extracellular matrix (ECM) is primarily governed by peptide-based signaling motifs [32]. Peptides and proteins can autonomously aggregate into nanoscale fibers, cages [33]. Strong antibody response and efficient B-cell receptor interactions depend on precise control over antigen geometry and spacing, which is made possible by this organized arrangement [34]. These self-assembling systems can also be made to carry more than one antigenic epitope, which makes them suitable for multivalent vaccine development [35]. Self-assembled peptide nanostructures offer several clear advantages over conventional delivery systems. They are chemically versatile, highly biocompatible, and can efficiently carry both hydrophobic and hydrophilic drugs. In addition, their ability to recognize and interact with specific molecular targets makes them especially effective and precise as delivery platforms [36]. These advantages have led to increasing usage of such constructs in universal influenza vaccine research [37].

2.7 *Emulsion-Based Nanocarriers*

Oil-in-water nanoemulsions (e.g., MF59 and AS03) represent clinically approved platforms that improve uptake of antigen and APC recruitment [38]. The nanoemulsion-based platform can induce a strong local immune response, therefore, increasing both humoral and adaptive immune responses to the antigens linked with it [39]. Importantly, both MF59 and AS03 have excellent safety profiles and are appropriate for large-scale vaccination. Their long history of regulatory approval and successful use in licensed influenza vaccines emphasizes their importance in current vaccine development [40].

3 Integration of Artificial Intelligence and Smart Nanovaccines

Artificial intelligence (AI) and machine learning (ML) are transforming vaccination research and development. When combined with smart nanovaccine platforms, nanoparticles can be created for targeted distribution, controlled release, and immune modulation [41]. AI accelerates rational design, optimizes formulation and production, anticipates immune responses, and enables personalized vaccination strategies [42]. This integration reduces the timeline for development, improves efficacy, and safety prediction [43]. The performance of nanovaccines depends on many interdependent variables like antigen display, adjuvant composition, charge, core material, size, shape, surface chemistry, and release kinetics. AI is good at modeling high-dimensional parameter spaces and finding effective solutions much more quickly than trial-and-error approaches [44]. For emerging pathogens or rapidly mutating viruses, AI can analyze genomic/proteomic data to prioritize epitopes for encapsulation or display on nanoparticles [45]. AI allows individual or cohort-level prediction of immune responses based on genetics, age, and co-morbidities, enabling tailored nanovaccine formulations or dosing regimens [46]. AI/ML models optimize process parameters like mixing, microfluidic flows, encapsulation efficiency, and detect anomalies in real time, improving batch consistency [47].

3.1 AI-Enabled Workflow for Smart Nanovaccine Development

3.1.1 Antigen Discovery and Epitope Prioritization

The integration of sequence databases, structural modeling, and machine learning-based classifiers enables the accurate prediction of antigenic regions, HLA-restricted T-cell epitopes, and conformational B-cell epitopes [48–50]. In 2021, Ras-Carmona et al. developed a user-friendly web tool (BCEPS) that uses machine learning models trained on 555 linearized conformational B-cell epitopes obtained from antibody–antigen protein structures to predict linear B-cell epitopes that are likely to be immunogenic and able to generate cross-reactive antibodies against native antigens [49].

Several studies on SARS-CoV-2 have demonstrated how machine learning can be used to better understand the specificity of immune responses against the virus. In a study, Hotop et al. used machine learning techniques to analyze data from peptide microarrays that were tested with serum samples from patients showing different levels of disease severity [51].

AI-driven modeling helps tackle the challenge of HLA polymorphism, which often limits vaccine effectiveness. By focusing on common HLA supertypes such as HLA-A2, HLA-A3, and HLA-B7, AI makes it possible to design multi-epitope Th-CTL peptide herpes vaccines that can trigger immune responses in diverse populations [42].

3.1.2 Nanoparticle Design Optimization

Surrogate models (Gaussian processes, neural networks) map nanoparticle physicochemical inputs to outputs like cellular uptake, endosomal escape, antigen presentation, and cytokine profiles [52]. Generative models like variational autoencoders, generative adversarial networks, propose novel nanoparticle compositions or ligand patterns predicted to maximize desired immune outcomes [53].

3.1.3 In Silico Immunogenicity and Safety Prediction

Multi-scale computational models combine molecular, cellular, and systemic features to predict immunogenic potency and adverse immune reactions such as excessive inflammation and complement activation [54]. Transfer learning from existing vaccine datasets allows faster prediction for new constructs (Fig. 2 and Table 1).

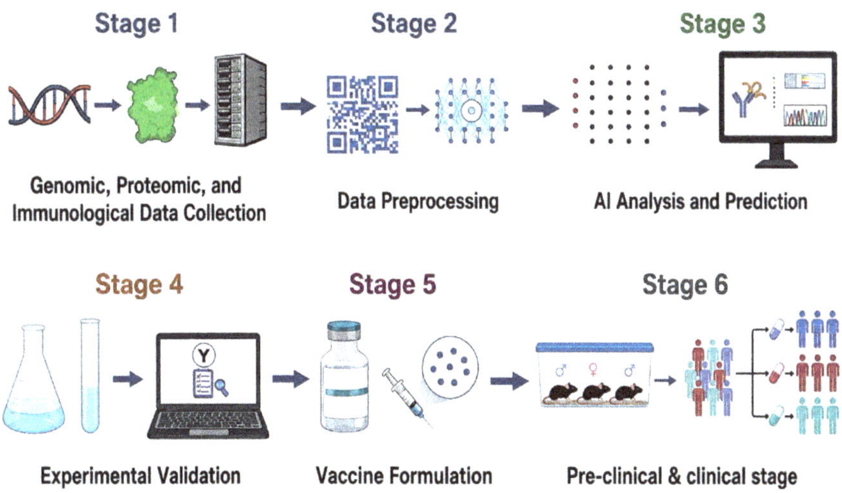

Fig. 2 AI-driven vaccine development workflow (created with BioRender.com)

Table 1 AI-models used in vaccine development study

AI-model	Application	References
VaxiJen	Antigen identification without sequence alignment	[55]
NetMHCpan-4.0 and NetMHCIIpan-4.1	NetMHCpan-4.0—Predicts peptide binding to MHC class II NetMHCIIpan-4.1—Predicts peptide binding to MHC class I	[56]
Vaxign-ML	Prediction of vaccine candidates for emerging pathogens	[57]
IntegralVac	Automated prediction of single and multivalent cancer and COVID-19 epitopes	[58]

4 Key Applications and Examples of Nanovaccines

4.1 Cancer Immunotherapy: Nanovaccines for Therapeutic or Preventive Cancer Vaccines

Cancer vaccines remain an important area of research because of their potential of activating the immune system that can specifically target tumor cells while reducing harm to healthy cells, and at the same time induce immunological memory against tumor-associated antigens (TAAs) [59]. Nanovaccine platforms like liposomes, polymeric nanoparticles, virus-like particles (VLPs), and protein nanocages have been extensively explored to deliver TAAs, neoantigens, or tumor-specific antigens (TSAs) to antigen-presenting cells (APCs), enhancing antigen uptake, presentation, and immune activation.

In a study done by Egelman et al. [60], it was reported that the peptide Coil29 self-assembled into alpha-helical nanofibers in which the C-termini were oriented toward the fiber core and the N-termini were exposed on the surface. Subsequent work by the Collier laboratory showed that attaching different peptide epitope $(KEYA)_{20}Q11$ to the N-terminus did not interfere with the nanofiber assembly and that immunization with these epitope-presenting nanofibers induced humoral and cellular responses against the linked antigens even in the absence of additional adjuvants [61].

As each tumor has its own unique set of mutations, personalized cancer vaccines that target tumor-specific neoantigen peptides are considered a promising therapeutic approach [62]. In a recent study, Liu et al. [63] reported a biomimetic nanovaccine of dendritic cell-derived artificial cytomembrane nanovesicles that could directly present neoantigens to $CD8^+$ T-cells without relying on APC delivery. It could co-deliver antigen epitopes, anti-PD-1 antibodies, and B7 costimulatory molecules, and targeted the lymphatic system.

In summary, nanoparticle-based vaccines have gained increasing attention in cancer immunotherapy over recent years, as they enable the efficient delivery of complex TAAs while promoting controlled immune response [64].

4.2 Nanovaccines Against Infectious Diseases

Nanovaccines have shown strong potential to activate the host immune system to fight disease and limit pathogen transmission. Several major infectious diseases like HIV, hepatitis C, malaria, and tuberculosis lack effective vaccination [41].

The mRNA vaccines came forward as an effective alternative to traditional vaccines during the SARS-CoV-2 pandemic. Moderna, in collaboration with the National Institute of Allergy and Infectious Diseases (NIAID), developed an mRNA-based vaccine in which lipid nanoparticle-encapsulated mRNA encoded the viral spike protein. This NP vaccine led to the activation of T- and B-cell immune responses [65, 66]. Similarly, Pfizer-BioNTech developed BNT162 vaccines, including BNT162 (b1, b2), which contain mRNA encoding the receptor-binding domain (RBD) trimer or the full-length membrane-anchored S protein [67]. Novavax employed a baculovirus–Sf9 system to produce a recombinant full-length SARS-CoV-2 S glycoprotein [41].

Nanomaterials are also being explored as promising tools for the development of an HIV vaccine. Their ability to strengthen the antibody responses, enhance antibody effector functions, and activate B cell precursors required to generate broadly neutralizing antibodies (bnAbs) [68]. Recently, Leggat et al. reported a phase I, first-in-human clinical trial of a germline-targeting nanoparticle vaccine based on an engineered outer domain of the HIV-1 envelope protein gp120, known as eOD-GT8 [69]. In a related study, Xin et al. developed an amphiphilic polymer-based HIV vaccine by conjugating poly(maleic anhydride-alt-1-octadecene) with polyethylene glycol, enabling the polymer to associate with the HIV-1 envelope trimer and form stable Env–nanoparticle complexes [70].

4.3 Nanovaccines for Allergic Diseases

Nanoparticle-based allergen vaccines are being actively explored as new options for allergen immunotherapy. They offer several advantages, like controlled particle size, surface modification, targeted delivery, flexibility to carry different antigens, and ease of production. In a study, Luo et al. [71] showed that a PLGA-based nanovaccine containing ovalbumen (OVA) and a ubiquitin E3 ligase (A20) effectively reduced airway allergic response in sensitized mice by promoting the development of regulatory T cells and increasing the production of IL-10. This suggests that the PLGA-OVA + A20 nanovaccine has potential for translation into treatments for allergic asthma. In another study, a novel approach came forward in which butyrate-releasing micelles were used to treat peanut allergy, which helped in re-establishing microbial balance and mucosal homeostasis within the gut [72].

4.4 Nanoscale Adjuvants

Adjuvants are commonly used in vaccines to strengthen the immune response to an antigen [73]. The addition of adjuvants to vaccines can help guide, amplify, and accelerate an immune response that is more effective and better suited for protection against specific infections or malignancies [74]. Adjuvants currently approved for use in vaccine development for humans in the United States and Europe comprise aluminum salts (alum), squalane-based oil-in-water emulsions like AF03, AS03, along with virosomes and AS04 (formulation that combines monophosphoryl lipid A with alum) [73]. In a study, Jangra et al. [75] developed an intranasal adjuvant combined with a nanoemulsion that was capable of activating TLRs and NLRP3 with an RNA-based RIG-I agonist. When combined with the SARS-CoV-2 spike protein, this formulation induced strong immune responses in mice, including pronounced Th1-biased cellular immunity and high levels of neutralizing antibodies against the viruses (Table 2).

5 Challenges and Future Directions

Although many nanocarriers, such as LNPs, polymeric nanoparticles, inorganic nanoparticles, and VLPs, have shown promising results, long-term safety profiles remain insufficiently understood. Potential concerns include accumulation in organs such as liver, spleen, or kidneys [76], induction of chronic inflammation or immunotoxicity [77], risk of unintended immune activation or autoimmunity, nanomaterial degradation, and metabolic clearance issues [78].

Producing nanovaccines with consistent particle size, surface charge, and antigen loading efficiency is a highly precise and complex process. During the production, several challenges, such as batch-to-batch variability, can impact vaccine efficacy and safety [79], high manufacturing costs and requirement for specialized equipment [80], complex cold-chain logistics for lipid and mRNA-based systems are observed [81–83].

Smart nanovaccines do not easily fit into traditional regulatory frameworks, which were designed for small molecules or biological vaccines. Major challenges include the lack of standardized evaluation protocols, insufficient regulatory harmonization across countries [84], and uncertainty in long-term monitoring and surveillance requirements.

Personalized immune responses vary widely depending on the factors, such as age, genetics, and comorbidities, which may affect the efficacy of nanovaccines.

Table 2 Nanoparticles in vaccine development and examples in clinical trials*

Nanovaccine name	Disease/ Application	NCT ID	Clinical phase
Nab-paclitaxel, durvalumab, and tremelimumab with or without personalized synthetic long peptide vaccine (neoantigen vaccine)	Triple negative breast cancer	NCT03606967	Phase 2
Tumor mRNA-loaded DOTAP liposome vaccine	Eye melanoma	NCT05264974	Phase 1
KRAS neoantigen mRNA vaccine (ABO2102)	Pancreatic neoplasm and solid tumors	NCT06577532	Early phase 1
426c.Mod.Core-C4b vaccine adjuvanted with 3 M-052-AF	HIV	NCT06613789	Phase 1
Moderna mRNA-1273.222	HIV	NCT05168813	Phase 2/3
UVAX-1107 and UVAX-1197 mixed with aluminum hydroxide (AH) and CpG 1018 adjuvants	HIV	NCT06541093	Early phase 1
Ad26.COV2.S (VAC31518, JNJ-78436735) vaccine, SARS-CoV-2 rS (CovovaxTM), BNT162b2 (Pfizer)	HIV	NCT05515042	Phase 2
NVX-CoV2373	HIV	NCT05112848	Phase 2
SpFN_1B-06-PL	SARS-CoV-2	NCT04784767	Phase 1
SARS-CoV-2 Rs with or without MATRIX-M™ Adjuvant	SARS-CoV-2	NCT04368988	Phase 1/2
NVX-CoV2373	SARS-CoV-2	NCT05875701	Phase 3
Matrix-M with RH5.2 VLP and/or R21	Malaria	NCT05357560	Phase 1
Recombinant Trivalent Nanoparticle Influenza Vaccine (Tri-NIV) with Matrix M1™ adjuvant	Influenza	NCT03293498	Phase 1/2
RSV + hMPV mRNA vaccine	RSV	NCT06686654	Phase 1/2
RSV LNP vaccine	RSV	NCT05639894	Phase 1/2

NCT National clinical trial, *RSV* respiratory syncytial virus, *VLP* virus-like particle, *HIV* human immunodeficiency virus, *SARS-CoV-2* severe acute respiratory syndrome coronavirus 2, *mRNA* messenger RNA, *LNP* lipid nanoparticle

* Data taken from National Library of Medicine (https://www.ClinicalTrials.gov)

Predicting biodistribution and interaction with biological systems remains a challenge due to complex nano–bio interface dynamics [85], and individual variability in immunity, age, genetics, etc. [86].

Logistic and socioeconomic constraints, such as high R&D investment and expensive clinical validation [87], limited nanotechnology expertise in developing nations, public perception and vaccine hesitancy related to emerging technologies, are often seen in the process.

Future next-generation delivery platforms may include self-adjuvanting nanoparticles, stimuli-responsive nanocarriers activated by pH, enzymes, temperature, or ROS, and single-dose vaccines with controlled release enabling booster-free immunization. The development of needle-free nanovaccine platforms, including injectable, inhalable, oral, mucosal, transdermal, and microneedle, enables effective immunization against respiratory diseases while reducing dependence on cold-chain storage. Integration of artificial intelligence and computational modeling accelerates predictive nanoparticle and antigen-adjuvant design, optimizes immune kinetics, and supports rapid vaccine prototyping during outbreaks [88].

6 Conclusion

Smart nanotechnology is reshaping vaccine development by enabling precise control over antigen delivery, immune activation, and safety profiles that are difficult to achieve with conventional vaccine platforms. By integrating advanced nanomaterials such as lipid nanoparticles, polymeric systems, virus-like particles, and self-assembling peptide structures with artificial intelligence-driven design and prediction tools, next-generation vaccines can be developed more rapidly and efficiently. These approaches enhance immunogenicity, support multivalent and personalized vaccine strategies, and improve population coverage by accounting for immune diversity. With continued progress in smart nanotechnology and artificial intelligence, vaccine development is becoming faster, safer, and more scalable, strengthening our ability to respond to infectious diseases, cancer, and future global health threats.

Abbreviations

AI	Artificial intelligence
APC	Antigen-presenting cell
bnAbs	Broadly neutralizing antibodies
DNA	Deoxyribonucleic acid
HLA	Human leukocyte antigen
LNP	Lipid nanoparticle
MHC	Major histocompatibility complex
ML	Machine learning
mRNA	Messenger RNA
OVA	Ovalbumin
PEG	Polyethylene glycol
PLGA	Poly (lactic acid-co-glycolic acid)
PLA	Poly (lactic acid)
PCL	Polycaprolactone
RIG-1	Retinoic acid-inducible gene I

RNA	Ribonucleic acid
saRNA	Self-amplifying RNA
TAA	Tumour-associated antigen
TSA	Tumour-specific antigen
TLR	Toll-like receptor
Th	T-helper
VLP	Virus-like particle

References

1. Pardi N, Hogan MJ, Porter FW, Weissman D (2018) MRNA vaccines—a new era in vaccinology. Nat Rev Drug Discov 17(4):261–279
2. Verbeke R, Lentacker I, De Smedt SC, Dewitte H (2021) The dawn of mRNA vaccines: the COVID-19 case. J Control Release 333:511–520
3. Hou X, Zaks T, Langer R, Dong Y (2021) Lipid nanoparticles for mRNA delivery. Nat Rev Mater 6(12):1078–1094
4. Sun Z, Zhao H, Ma L, Shi Y, Ji M, Sun X et al (2024) The quest for nanoparticle-powered vaccines in cancer immunotherapy. J Nanobiotechnol 22(1):61
5. Sadarangani M, Marchant A, Kollmann TR (2021) Immunological mechanisms of vaccine-induced protection against COVID-19 in humans. Nat Rev Immunol 21(8):475–484
6. Mi P (2020) Stimuli-responsive nanocarriers for drug delivery, tumor imaging, therapy and theranostics. Theranostics. 10(10):4557
7. Aljabali AA, Obeid MA, Bashatwah RM, Serrano-Aroca Á, Mishra V, Mishra Y et al (2023) Nanomaterials and their impact on the immune system. Int J Mol Sci 24(3):2008
8. Kumar A, Dixit S, Srinivasan K, Dinakaran M, Vincent PDR (2024) Personalized cancer vaccine design using AI-powered technologies. Front Immunol 15:1357217
9. Kim J, Eygeris Y, Gupta M, Sahay G (2021) Self-assembled mRNA vaccines. Adv Drug Deliv Rev 170:83–112
10. Wilson B, Geetha KM (2022) Lipid nanoparticles in the development of mRNA vaccines for COVID-19. J Drug Deliv Sci Technol 74:103553
11. Wei P-S, Thota N, John G, Chang E, Lee S, Wang Y et al (2024) Enhancing RNA-lipid nanoparticle delivery: organ-and cell-specificity and barcoding strategies. J Control Release 375:366–388
12. Yang L, Gong L, Wang P, Zhao X, Zhao F, Zhang Z et al (2022) Recent advances in lipid nanoparticles for delivery of mRNA. Pharmaceutics 14(12):2682
13. Swetha K, Kotla NG, Tunki L, Jayaraj A, Bhargava SK, Hu H et al (2023) Recent advances in the lipid nanoparticle-mediated delivery of mRNA vaccines. Vaccines 11(3):658
14. Athira T, Selvaraju K, Gowrishankar N (2023) Biodegradable polymeric nanoparticles: the novel carrier for controlled release drug delivery system. Int J Sci Res Arch. 8(1):630–637
15. Mangu K, Gudeli R, Rizwanullah M (2025) Recent advancement in polymeric nanoparticles for oral chemotherapy: transforming cancer treatment. BioImpacts 15:31117
16. Kamaly N, Yameen B, Wu J, Farokhzad OC (2016) Degradable controlled-release polymers and polymeric nanoparticles: mechanisms of controlling drug release. Chem Rev 116(4):2602–2663
17. Masood F (2016) Polymeric nanoparticles for targeted drug delivery system for cancer therapy. Mater Sci Eng C 60(Suppl C):569–578
18. Ben-Akiva E, Witte SE, Meyer RA, Rhodes KR, Green JJ (2019) Polymeric micro-and nanoparticles for immune modulation. Biomater Sci 7(1):14–30

19. He J, Yu L, Lin X, Liu X, Zhang Y, Yang F et al (2022) Virus-like particles as nanocarriers for intracellular delivery of biomolecules and compounds. Viruses 14(9):1905
20. Mohsen MO, Augusto G, Bachmann MF (2020) The 3Ds in virus-like particle based-vaccines:"design, delivery and dynamics. Immunol Rev 296(1):155–168
21. Sherry L, Bahar MW, Porta C, Fox H, Grehan K, Nasta V et al (2025) Recombinant expression systems for production of stabilised virus-like particles as next-generation polio vaccines. Nat Commun 16(1):831
22. Zhang X, Li S, Modis Y, Li Z, Zhang J, Xia N et al (2016) Functional assessment and structural basis of antibody binding to human papillomavirus capsid. Rev Med Virol 26(2):115–128
23. Zhang Y, Wang C, Zheng Y, Chen F, Feng Y, Fang L et al (2025) A chimeric Ad5-Envp-VLP vaccine platform confers broad-spectrum immunity against emerging and re-emerging pathogens. bioRxiv 2025.07.03.663092
24. Saikia N (2024) Inorganic-based nanoparticles and biomaterials as biocompatible scaffolds for regenerative medicine and tissue engineering: current advances and trends of development. Inorganics 12(11):292
25. Percoco RM, Schirinzi S, Mandriota G, Cortese B, Gigli G, Palamà IE et al (2025) Smart inorganic nanoparticles in nanomedicine: strategies for synthesis and functionalization. Int J Pharm 2025:126192
26. Hess KL, Medintz IL, Jewell CM (2019) Designing inorganic nanomaterials for vaccines and immunotherapies. Nano Today 27:73–98
27. Zdrehus R-S, Mocan T, Sabau LI, Matea CT, Tăbăran F, Pop T et al (2025) CEA-functionalized gold nanoparticles as a nanovaccine platform: in vitro evaluation of cytocompatibility, cellular uptake, and antigen processing. Vaccines 13(7):668
28. Chheda D, Shete S, Tanisha T, Bahadure SD, Sampathi S, Junnuthula V et al (2024) Multifaceted therapeutic applications of biomimetic nanovaccines. Drug Discov Today 29(6):103991
29. Hu T, Huang Y, Liu J, Shen C, Wu F, He Z (2023) Biomimetic cell-derived nanoparticles: emerging platforms for cancer immunotherapy. Pharmaceutics 15(7):1821
30. Wusiman D, Wang Y, Wang M, Wang J, Wu R, Tuo Z et al (2025) Biomimetic nanovaccines in cancer therapy: mechanisms, efficacy, and clinical translation. Mater Today Bio 2025:102116
31. Meng Y, Yao Z, Ke X, Hu M, Ren H, Gao S et al (2025) Extracellular vesicles-based vaccines: emerging immunotherapies against cancer. J Control Release 378:438–459
32. Kokkoli E, Mardilovich A, Wedekind A, Rexeisen EL, Garg A, Craig JA (2006) Self-assembly and applications of biomimetic and bioactive peptide-amphiphiles. Soft Matter 2(12):1015–1024
33. Colombo G, Soto P, Gazit E (2007) Peptide self-assembly at the nanoscale: a challenging target for computational and experimental biotechnology. Trends Biotechnol 25(5):211–218
34. Zottig X, Côté-Cyr M, Arpin D, Archambault D, Bourgault S (2020) Protein supramolecular structures: from self-assembly to nanovaccine design. Nanomaterials 10(5):1008
35. Zhang M-M, Zhu J, Wang Z-Y, Bai Y-L, Li H-B, Nashun B et al (2025) Self-assembled peptides: a new generation of vaccine adjuvant platform. Vaccines 13(12):1183
36. Habibi N, Kamaly N, Memic A, Shafiee H (2016) Self-assembled peptide-based nanostructures: smart nanomaterials toward targeted drug delivery. Nano Today 11(1):41–60
37. Deng L, Chang TZ, Wang Y, Li S, Wang S, Matsuyama S et al (2018) Heterosubtypic influenza protection elicited by double-layered polypeptide nanoparticles in mice. Proc Natl Acad Sci USA 115(33):E7758–E7767
38. Huang Z, Gong H, Sun Q, Yang J, Yan X, Xu F (2024) Research progress on emulsion vaccine adjuvants. Heliyon 10(3)
39. Cao W, Davis WG, Kim JH, De La Cruz JA, Taylor A, Hendrickson GR et al (2016) An oil-in-water nanoemulsion enhances immunogenicity of H5N1 vaccine in mice. Nanomed Nanotechnol Biol Med 12(7):1909–1917
40. Deng J, Cai W, Jin F (2014) A novel oil-in-water emulsion as a potential adjuvant for influenza vaccine: development, characterization, stability and in vivo evaluation. Int J Pharm 468(1–2):187–195

41. Abusalah MAH, Chopra H, Sharma A, Mustafa SA, Choudhary OP, Sharma M et al (2023) Nanovaccines: a game changing approach in the fight against infectious diseases. Biomed Pharmacother 167:115597
42. Benmohamed L, El Fatimi EH, Lekbach Y, Prakash S (2025) Artificial intelligence and machine learning in the development of vaccines and immunotherapeutics—yesterday, today, and tomorrow. Front Artif Intell 8:1620572
43. Olawade DB, Teke J, Fapohunda O, Weerasinghe K, Usman SO, Ige AO et al (2024) Leveraging artificial intelligence in vaccine development: a narrative review. J Microbiol Methods 224:106998
44. Jena GK, Patra CN, Jammula S, Rana R, Chand S (2024) Artificial intelligence and machine learning implemented drug delivery systems: a paradigm shift in the pharmaceutical industry. J Bio-X Res 7:0016
45. Bahrami Y, Bolideei M, Mohammadzadeh S, Gahrouei RB, Mohebbi E, Haider KH, et al (2025) Applications of artificial intelligence and nanotechnology in vaccine development. Int J Pharm 126096
46. Farahani AF, Kasraei N (2024) Evaluating the impact of artificial intelligence on vaccine development: lessons learned from the COVID-19 pandemic. MedRxiv 2024:2024.10.23.24315991
47. Shah P, Kendall F, Khozin S, Goosen R, Hu J, Laramie J et al (2019) Artificial intelligence and machine learning in clinical development: a translational perspective. NPJ Digital Med 2(1):69
48. Grewal S, Hegde N, Yanow SK (2024) Integrating machine learning to advance epitope mapping. Front Immunol 15:1463931
49. Ras-Carmona A, Pelaez-Prestel HF, Lafuente EM, Reche PA (2021) BCEPS: a web server to predict linear B cell epitopes with enhanced immunogenicity and cross-reactivity. Cells 10(10):2744
50. Backert L, Kohlbacher O (2015) Immunoinformatics and epitope prediction in the age of genomic medicine. Genome Med 7(1):119
51. Hotop S-K, Reimering S, Shekhar A, Asgari E, Beutling U, Dahlke C et al (2022) Peptide microarrays coupled to machine learning reveal individual epitopes from human antibody responses with neutralizing capabilities against SARS-CoV-2. Emerg Microbes Infect 11(1):1037–1048
52. Iaquinta S, Khazaie S, Albanna S, Fréour S, Jacquemin F (2024) Preprint machine learning for the sensitivity analysis of a model of the cellular uptake of nanoparticles for the treatment of cancer. Int J Numer Methods Biomed Eng 40(12):e3878
53. Liu YT, Zhang LL, Jiang ZY, Tian XS, Li PL, Wu PH et al (2025) Applications of artificial intelligence in biotech drug discovery and product development. MedComm 6(8):e70317
54. Wang Q, Liu Y, Li C, Xu B, Xu S, Liu B (2025) Machine learning-enhanced nanoparticle design for precision cancer drug delivery. Adv Sci 12(30):e0313
55. Doytchinova IA, Flower DR (2007) VaxiJen: a server for prediction of protective antigens, tumour antigens and subunit vaccines. BMC Bioinform 8(1):4
56. Reynisson B, Alvarez B, Paul S, Peters B, Nielsen M (2020) NetMHCpan-4.1 and NetMHCIIpan-4.0: improved predictions of MHC antigen presentation by concurrent motif deconvolution and integration of MS MHC eluted ligand data. Nucleic Acids Res 48(W1):W449–W54
57. Ong E, Wang H, Wong MU, Seetharaman M, Valdez N, He Y (2020) Vaxign-ML: supervised machine learning reverse vaccinology model for improved prediction of bacterial protective antigens. Bioinformatics 36(10):3185–3191
58. Suri S, Dakshanamurthy S (2022) IntegralVac: a machine learning-based comprehensive multivalent epitope vaccine design method. Vaccines 10(10):1678
59. Shetty S, Alvarado PC, Pettie D, Collier JH (2024) Next-generation vaccine development with nanomaterials: recent advances, possibilities, and challenges. Annu Rev Biomed Eng 26(1):273–306
60. Egelman E, Xu C, DiMaio F, Magnotti E, Modlin C, Yu X et al (2015) Structural plasticity of helical nanotubes based on coiled-coil assemblies. Structure 23(2):280–289

61. Votaw NL, Collier L, Curvino EJ, Wu Y, Fries CN, Ojeda MT et al (2021) Randomized peptide assemblies for enhancing immune responses to nanomaterials. Biomaterials 273:120825
62. Aldous AR, Dong JZ (2018) Personalized neoantigen vaccines: a new approach to cancer immunotherapy. Bioorg Med Chem 26(10):2842–2849
63. Liu C, Liu X, Xiang X, Pang X, Chen S, Zhang Y et al (2022) A nanovaccine for antigen self-presentation and immunosuppression reversal as a personalized cancer immunotherapy strategy. Nat Nanotechnol 17(5):531–540
64. Chen F, Wang Y, Gao J, Saeed M, Li T, Wang W et al (2021) Nanobiomaterial-based vaccination immunotherapy of cancer. Biomaterials 270:120709
65. WH Organization (2022) The COVID-19 vaccine tracker and landscape compiles detailed information of each COVID-19 vaccine candidate in development by closely monitoring their progress through the pipeline
66. Shin MD, Shukla S, Chung YH, Beiss V, Chan SK, Ortega-Rivera OA et al (2020) COVID-19 vaccine development and a potential nanomaterial path forward. Nat Nanotechnol 15(8):646–655
67. Jackson LA, Anderson EJ, Rouphael NG, Roberts PC, Makhene M, Coler RN et al (2020) An mRNA vaccine against SARS-CoV-2—preliminary report. N Engl J Med 383(20):1920–1931
68. Haynes BF, Wiehe K, Borrow P, Saunders KO, Korber B, Wagh K et al (2023) Strategies for HIV-1 vaccines that induce broadly neutralizing antibodies. Nat Rev Immunol 23(3):142–158
69. Leggat DJ, Cohen KW, Willis JR, Fulp WJ, Decamp AC, Kalyuzhniy O et al (2022) Vaccination induces HIV broadly neutralizing antibody precursors in humans. Science 378(6623):eadd6502
70. Xin X, Liu Y, Guo L, Wang H, Lu D, Chang Y et al (2023) Improvement of B cell responses by an HIV-1 amphiphilic polymer nanovaccine. Nano Lett 23(9):4090–4094
71. Luo X-Q, Zhong J-W, Qiu S-Y, Zhi M, Yang L-Q, Zhou Y-L et al (2020) A20-OVA nanoparticles inhibit allergic asthma in a murine model. Inflammation 43(3):953–961
72. Wang R, Cao S, Bashir MEH, Hesser LA, Su Y, Hong SMC et al (2023) Treatment of peanut allergy and colitis in mice via the intestinal release of butyrate from polymeric micelles. Nat Biomed Eng 7(1):38–55
73. Zhu M, Wang R, Nie G (2014) Applications of nanomaterials as vaccine adjuvants. Hum Vaccin Immunother 10(9):2761–2774
74. Coffman RL, Sher A, Seder RA (2010) Vaccine adjuvants: putting innate immunity to work. Immunity 33(4):492–503
75. Jangra S, Landers JJ, Rathnasinghe R, O'Konek JJ, Janczak KW, Cascalho M et al (2021) A combination adjuvant for the induction of potent antiviral immune responses for a recombinant SARS-CoV-2 protein vaccine. Front Immunol 12:729189
76. Jakic K, Selc M, Razga F, Nemethova V, Mazancova P, Havel F et al (2024) Long-term accumulation, biological effects and toxicity of BSA-coated gold nanoparticles in the mouse liver, spleen, and kidneys. Int J Nanomed 4103–4120
77. Hofer S, Hofstätter N, Punz B, Hasenkopf I, Johnson L, Himly M (2022) Immunotoxicity of nanomaterials in health and disease: current challenges and emerging approaches for identifying immune modifiers in susceptible populations. Wiley Interdiscip Rev Nanomed Nanobiotechnol 14(6):e1804
78. Calé A, Elblová P, Andělová H, Lunova M, Lunov O (2025) Analyzing molecular determinants of nanodrugs' cytotoxic effects. Int J Mol Sci 26(14):6687
79. Stevanović MM, Qian K, Huang L, Vukomanović M (2025) PLGA-based co-delivery nanoformulations: overview, strategies, and recent advances. Pharmaceutics 17(12):1613
80. Waugh S, Cameron CE (2024) Syphilis vaccine development: aligning vaccine design with manufacturing requirements. Hum Vaccin Immunother 20(1):2399915
81. Mangla B, Kumar P, Javed S, Pathan T, Ahsan W, Aggarwal G (2025) Regulating nanomedicines: challenges, opportunities, and the path forward. Nanomedicine 20(15):1911–1927
82. Uddin MN, Roni MA (2021) Challenges of storage and stability of mRNA-based COVID-19 vaccines. Vaccines 9(9):1033

83. Seo SH, Song MK (2025) Advancements and challenges in next-generation mRNA vaccine manufacturing systems. Clin Exp Vaccine Res 14(4):299
84. Kardani SL (2024) Nanocarrier-based formulations: regulatory challenges, ethical and safety considerations in pharmaceuticals. Asian J Pharm (AJP) 18(02)
85. Zhang X-Q, Xu X, Bertrand N, Pridgen E, Swami A, Farokhzad OC (2012) Interactions of nanomaterials and biological systems: Implications to personalized nanomedicine. Adv Drug Deliv Rev 64(13):1363–1384
86. Falahi S, Kenarkoohi A (2022) Host factors and vaccine efficacy: implications for COVID-19 vaccines. J Med Virol 94(4):1330–1335
87. Younis MA, Tawfeek HM, Abdellatif AA, Abdel-Aleem JA, Harashima H (2022) Clinical translation of nanomedicines: challenges, opportunities, and keys. Adv Drug Deliv Rev 181:114083
88. He J, Xiao W, Hua D, Liu M, Guo H, Xu L et al (2025) Nanovaccines: innovative advances from design strategies to clinical translation. Vaccines 13(9):900

Smart Nanotechnology for Medical Diagnostics

Shobana Sampath, Keerthana Sivakumar, and Shivani Sivaskandhamohan

Abstract Medical diagnostics is transforming with smart nanotechnology, which makes detection methods quicker, more precise, and less invasive. While the development of new nanomaterials has advanced significantly, the conversion of these advancements into useful, clinically usable diagnostic instruments has received less attention. This chapter attempts to highlight new advances in smart nanotechnology and investigate how these cutting-edge technologies improve the effectiveness, accuracy, and dependability of contemporary diagnostic processes. A qualitative research approach separates various smart nanoparticle technologies and prompts nanodevices and nanodetection systems to monitor cancer and infectious diseases alongside metabolic issues. The combination of speedy biological detection reactions with specific and sensitive properties results in lower testing errors and shortened test durations. Nanomaterials enable on-site diagnostic tests at either urban or remote sites due to their small programmable design that allows real-time analysis and mobility. The implementation of smart nanotechnology platforms in healthcare organizations with limited resources shows the potential to enhance therapeutic results while transforming regular testing procedures. Engineers must work together with medical staff and business partners to further strengthen safe and affordable nanotechnology applications in healthcare facilities.

Keywords Cancer · Nanoparticles · Nanomaterials · Infectious diseases

1 Introduction

Nanotechnology applies and designs materials that measure within nanometer dimensions from 1 to 100 nm to bring out unique physical and chemical properties resulting from increased surface area-to-volume ratios. Due to their exclusive physical and chemical properties, these systems enable strong biological molecule interactions

S. Sampath (✉) · K. Sivakumar · S. Sivaskandhamohan
Department of Biotechnology, Vel Tech Rangarajan Dr. Sagunthala R&D Institute of Science and Technology, Avadi, Chennai, Tamil Nadu, India
e-mail: swtshoba@gmail.com

J. Chen (ed.), *AI-Driven Nanotechnology in Precision Medicine and Healthcare*, Smart Nanomaterials Technology, https://doi.org/10.1007/978-981-92-1690-1_12

that enhance early disease biomarker identification, genetic mutation, and infectious pathogen detection. Smart nanotechnology delivers highly sensitive and selective diagnosis due to its distinctive operational characteristics. The simultaneous connection between nanoparticles and nanodevices to biomolecules enhances the detection signals needed for evaluating minimal biomarker amounts in blood or urine samples [1]. Portable point-of-care (PoC) diagnostic tools developed using nanotechnology make testing faster and easier to operate outside laboratory facilities while producing results at the point of care. Quantum dots can function as multiplexing probes through wavelength control because their size allows the release of distinct light frequencies for detecting multiple disease-related markers in one inspection. Numerous biomarkers are added to nanoparticle systems to boost their diagnostic features, which enhances the accuracy of MRI, CT, and PET imaging for medical detection [2]. Smart nanotechnology medical diagnosis has various applications, as nanotechnology demonstrates multiple uses for disease diagnostics across various medical disease groups. Medical diagnosis becomes more effective through tumor biomarker detection using nanobiosensors combined with nanowires, which have specific DNA probes or antibodies for advanced-stage cancer detection and improved clinical results. Nanoparticle-based assays enable fast pathogen identification of HIV as well as tuberculosis, malaria, and influenza while producing elevated diagnostic sensitivity compared to conventional methods. Genetic disorder screening employs nanopore sequencing together with nanoscale biosensors, which provide efficient DNA and RNA sequence analysis. Nanoparticles provide precise identification of autoantibodies and cytokines, together with inflammatory markers, when used in autoimmune and cardiovascular disease diagnostic processes. These precise results exceed conventional diagnostic assay capabilities.

Technologies and nanomaterials used in nanodiagnostics depend critically on multiple modern nanomaterials for achieving their success. The widespread employment of gold nanoparticles (AuNPs) in research occurs because they demonstrate optical characteristics while also maintaining biocompatibility. Nanotechnology enables the detection of cancer markers together with viral proteins through both colorimetric and fluorescence-based methods. Semiconductor nanocrystals known as quantum dots excel for multiplexed detection and cellular imaging because of their fluorescence, which depends on their size. The application of magnetic nanoparticles serves two purposes in biomedical fields: they improve magnetic resonance imaging quality, and they aid in making magnetic biosensors that perform non-invasive diagnostic tests [3]. Nanowires and nanobiosensors function through electrical signal change detection caused by binding target molecules, which proves beneficial for cancer biomarker detection, together with DNA mutation analysis.

Statistical nanotechnology demonstrates the extensive potential for medical diagnostics through precise early detection, along with person-centered treatment and improved healthcare service effectiveness. Various hurdles, including medium-term dependency maintenance and economic scalability of manufacturing processes, need solutions alongside achieving approval authorizations and integration into medical facilities. Interdisciplinary collaboration, along with innovation, will enable the solution of existing obstacles to fully harness the potential of nanodiagnostics [4].

2 New Emerging Smart Nanomaterials for Medical Diagnostic Applications

Smart nanomaterials enable outstanding advancements in disease diagnosis through their introduction into medical practices that enhance their speed and accuracy while reducing errors. Tiny nanoscale particles display particular features that allow them to interact precisely with biological molecules found in the human body. The unique characteristics permit nanomaterials to transform various diagnostic elements, such as biosensors and molecular imaging tools, which leads to improved detection efficiency.

The smallest among nanomaterials are quantum dots (QDs), which function as tiny, colorful semiconductor crystals that show light spectrum variations through size variation. Multiple biomarkers flying inside the body become evident through the QD flashlight system due to their continuous and intense fluorescence. The medical field extensively employs quantum dots for live cell imaging, along with using them to identify diseases in their early stages through highly precise cancer and genetic disorder detection protocols [5].

Moving on to the delightful category is the gold nanoparticle (GNP), which provides both visual appeal and key functionality to the table. The combination of surface plasmon resonance with biocompatibility creates highly observable optical properties that designate nanoparticles as effective agents for imaging, along with biomarker detection. The specific nanocages and nanoclusters of gold nanoparticles deliver the best advantages for performing drug delivery procedures and photothermal treatments, which prove valuable for detecting gastrointestinal cancer early on [6]. The diagnostic world gains its magic through the use of magnetic nanoparticles (MNs). The small particles made from iron enable precise medical interventions through magnetic control, making them effective for body-related molecule targeting and separation. These tiny particles serve as excellent instruments for MRI enhancement and treatment purposes, simultaneously allowing a "theranostics" combination of diagnosis and therapy. Magnetic nanoparticles shine as excellent diagnostic tools because they are both highly magnetically responsive and medically safe [2].

The domain of structure and strength belongs exclusively to carbon-based nanomaterials, including graphene and carbon nanotubes (CNTs). Safer biosensors require highly conductive materials with large surface areas, which these carbon-based nanomaterials deliver perfectly. These materials enable the detection of DNA and proteins as well as pathogens through a process that does not require additional annotation or procedural steps. These nanomaterials demonstrate excellent accuracy and speed, making them suitable for point-of-care tests as well as advanced bioimaging [7]. The last category comprises adaptable and elegant polymeric and hybrid nanostructures. These nanostructures take multiple assignments in the nanoworld. The release of diagnostic agents from polymeric nanoparticles occurs at specific locations during precise release times. The materials possess attributes that include stability, combined with biocompatibility and easy modifications. Such hybrid nanostructures consisting of materials paired with metals or magnets serve as

Table 1 Emerging smart nanomaterials, their features, and medical diagnostic applications

S. no.	Nanomaterial	Feature	Medical diagnostic application	References
1	Quantum dots	Size-tunable fluorescence, multiplexing	Detection of biomarkers, genetic diagnostics	[9]
2	Gold nanoparticles	Biocompatibility	Cancer imaging, biosensing	[10]
3	Magnetic nanoparticles	MRI contrast	Targeted drug delivery, biosensors	[11]
4	Carbon-based nanomaterials	High conductivity	Biosensor, rapid detection	[12]
5	Polymeric and hybrid nanoparticles	Biocompatibility	Theranostics, imaging	[13]

This table highlights novel smart nanomaterials used for medical diagnostics appears within this table. Advanced applications such as biosensing, imaging, and targeted detection result from the individual features that quantum dots, gold, magnetic, carbon-based, and polymeric nanoparticles each provide

comprehensive disease monitoring solutions that help treat cancer, along with heart and neurological cases.

Smart nanomaterials represent technical breakthroughs that conditionally impact medical detection and treatment methods and create opportunities for increased healthcare performance and patient optimism about future medical care (Table 1) [8].

3 Platform and Detection Mechanism Enabled via Nanotechnology

Medical diagnostics experienced a revolution through anonymous technology platforms, which developed modern, highly sensitive, and fast detection methods to diagnose various diseases. Biomolecule analysis occurs on a single-molecule basis through the use of nanoscale materials consisting of nanoparticles alongside nanowires and nanocantilevers that display distinctive physicochemical characteristics. The detection of disease biomarkers reaches superior levels as compared to conventional methods, resulting in the possibility of analyzing minimal sample sizes for early diagnosis and in vivo monitoring. The three fundamental categories of nanotechnology biomarker analytics consist of nanoparticle sensors that benefit from increased surface area for multiple bindings and nanowire detectors detecting electrical signals after biomarker detection, as well as lab-on-a-chip devices that integrate microfluidics and nanodetection systems to enhance molecular research. The detection platforms developed through nanotechnology advancements streamline off-site laboratory testing and provide multiple testing capabilities that advance medical diagnostics, particularly for cancer detection, infectious disease management, and genetic disorder screening [1].

3.1 Nano-Based Diagnostic Platform

Nanotechnology-based diagnostic systems use the specific qualities of nanomaterials to develop speedy yet highly sensitive and precise measurement tools for disease biomarker identification [1]. Advanced biosensing technologies reach higher detection levels through integration with nanoparticles consisting of gold nanostructures, along with quantum dots and magnetic nanostructures. Nanoparticles allow dense biomolecule binding across their large surface area, which results in more efficient multivalent interactions that improve test sensitivity, together with selectivity levels. The detection capabilities of nanobiosensors become enhanced when using nanowires, carbon nanotubes, and quantum dots together to deliver both label-free and labeled detection through optical, magnetic, and electrical signals [14]. The compact nature of lab-on-a-chip platforms makes the diagnostic process usable for on-site screening because nanomaterials allow the integration of all diagnostic functions onto a single small platform (Fig. 1) [15]. Quantum dot-based multiple sensing systems allow simultaneous monitoring of several disease indicators through single evaluations, which significantly enhances diagnostic rapidness [3].

The LOC device gathers disease-related molecules from samples, boosts the signal with the help of nanoparticles, and turns those signals into fluorescence or electrochemical readings that show up on a portable reader.

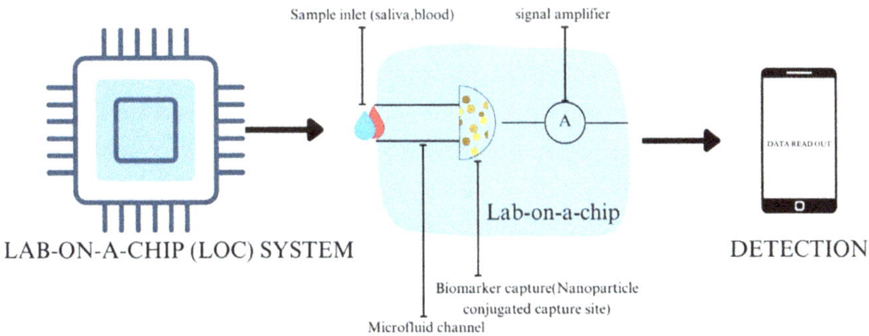

Fig. 1 Integrated lab-on-a-chip system for rapid detection of biomarkers

3.2 *Nano-Enabled Detection Mechanism*

Diagnostic performance gets major improvements from detection mechanisms that nanotechnology enables. Detectable optical signals occur from biomarker interactions with nanoparticles through three optical detection approaches consisting of surface plasmon resonance (SPR), fluorescence, and surface-enhanced Raman scattering (SERS). Gold nanoparticles exhibit color changes after they encounter target molecules, and quantum dots demonstrate stable size-based fluorescence detection, which serves for identifying and analyzing multiple targets (Fig. 2) [3]. The incorporation of carbon nanotubes and graphene materials in electrochemical detection enables sensitive electrical signal-based biosensing through unlabeled measurements of current and impedance data [14]. Magnetic detection through magnetic nanoparticles allows for biomarker identification without disturbing the human body by detecting changes in magnetic properties for MRI and magnetic biosensing purposes [16]. The combination of nanoscale mechanical and thermal sensors detects biochemical reaction-induced changes in mass as well as stiffness and heat that provide additional sensitivity to diagnostic systems of today.

Placing biomolecules (DNA, enzymes, antibodies) on gold nanoparticles makes them highly sensitive, as interactions with targets show up with fluorescence, changes in electrical signals, or color changes. We use detection modules to view and process the signal transduction tasks.

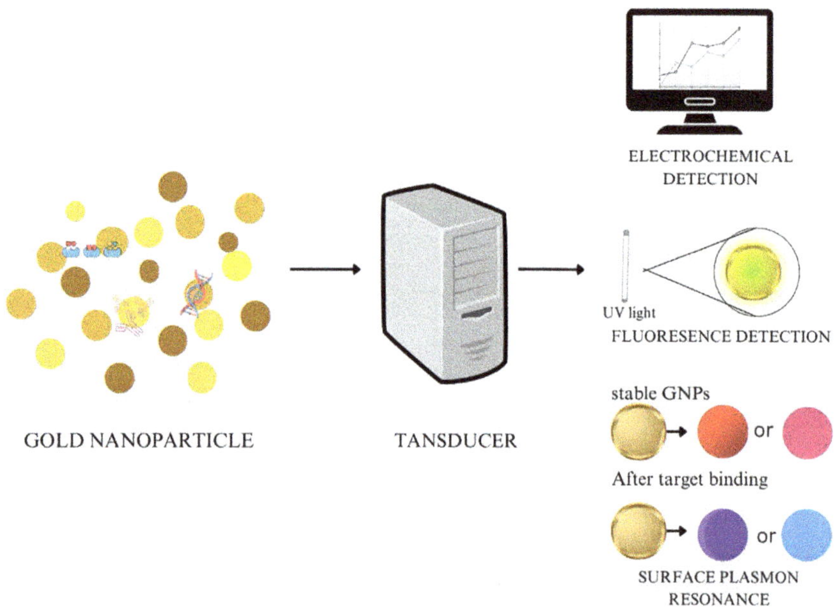

Fig. 2 Gold nanoparticle-connected biosensors used for multiplex detection

4 Smart Nanotechnology for Cancer and Infectious Disease Diagnosis

Active nanotechnology utilizes carefully designed natural response-triggered nanomaterials to deliver sensitive, precise, and fast disease detection capabilities that monitor cancer and various infectious pathogens. Smart nanomaterials that are subject to molecular-level control enable the improvement of early detection and real-time patient tracking as well as precise treatment delivery. The methodology improves over traditional diagnosis techniques because they demonstrate poor sensitivity levels and need invasive procedures [17].

Smart nanomaterials demonstrate great potential in diagnostic research through their four fundamental categories: nanoparticles, quantum dots, magnetic nanoparticles, and carbon-based nanomaterials. Such materials incorporate functional groups to detect specific disease markers while intensifying detection signals and delivering therapeutic functions for therapeutic-diagnostic combinations known as theranostics. Smart nanomaterials possess distinct optical and electrical properties, along with magnetic properties, which allow for secure biosensory and molecular detecting operations needed to achieve early disease recognition on a personalized level [18].

4.1 Cancer Diagnosis Using Smart Nanotechnology

Modern cancer diagnosis occurs through the use of smart nanotechnology, which combines biomarker sensitivity with enhanced imaging technology, detection sensors, and multitalented devices for continuous feedback verification. Nanobiosensors depend on nanomaterials such as magnetic nanoparticles, nanowires, and dendrimers to provide highly sensitive blood-based cancer protein testing through DNA probes and antibody-based detection systems. DNA-programmed gold nanoparticles, which incorporate magnetic nanoparticles, function as bio-barcode detectors of prostate-specific antigen (PSA) at exceptionally low attomolar levels exceeding standard detection approaches. DNA-directed antibody immobilization systems allow researchers to detect HER2 and CA-125 markers simultaneously in blood samples for the improvement of both early diagnosis and disease monitoring capabilities [19]. Nanoparticles represent the core imaging components because they function as contrast elements within MRI testing, PET analysis, and CT procedures. Magnetic nanoparticles offer improved contrast for magnetic resonance imaging to detect tumors and examine lymph nodes simultaneously, while silica-hybrid "C-dots" permit multifunctional imaging that helps surgeons accurately identify metastatic melanoma before surgery. Drug delivery and imaging functions join forces within a single therapeutic platform due to this combination of biological applications. The diagnostic possibilities of liquid biopsy advance because nanomaterials can extract ctDNA along with exosomes and circulating tumor cells (CTCs) from small blood volumes through nanowire and nanotrap hydrogel nanoparticle techniques, which monitor tumors continuously without surgery. Narrow nanodevices using both microfluidics and nanosensors enable quick point-of-care measurement of biochemical data through signal detection for multiple biomarker screening. Smart nanotechnology applications provide ongoing tumor intelligence as well as therapeutic data to develop individual medical solutions that detect cancer at early stages, help choose treatments, and improve overall patient health results [20]

4.2 Infectious Disease Diagnosis Using Smart Nanotechnology

Nanotechnology advances with sensitivity and speed enable rapid direct diagnosis of pathogens at low concentrations, including HIV, influenza, and COVID-19, along with tuberculosis (TB) and E. coli by using smart nanotechnology methods. The specific fingerprint features of nanomaterials enable detection at ultralow biomarker amounts because this makes early diagnosis and efficient disease control possible.

Nanoparticle-based lateral flow assays (LFAs) are present as popular point-of-care diagnostic tools through their application in diagnostics. The LFAs undergo upgrades with nanoparticles, specifically gold nanoparticles, to create better test response characteristics. Antigen test kits for COVID-19 detection use colloidal gold nanoparticles with antibody conjugates to detect viral proteins while giving instant visual results that do not require complicated equipment [21]. Gold nanorods with plasmonic activity have upgraded these systems through improved fluorescent signal detection, resulting in better sensitivity, which reaches lower detection thresholds than conventional LFAs by 30 times [22].

The use of smart nanodevices supports superior viral RNA and protein detection abilities despite their low concentration levels in clinical specimens. Copper in situ growth amplified by polyethyleneimine techniques produces better detection signals, which enable precise SARS-CoV-2 and E. coli pathogen identification with high repeatability and sensitivity levels [23]. Nanomaterial-supported LOC devices shorten the process of biomarker analysis and sample handling for real-time diagnostic testing that operates from small sample amounts at point-of-care sites [1]. The use of nanotechnology enables multiplexed detection through quantum dots because these materials have adjustable optical characteristics for the simultaneous identification of various pathogens and viral variants in unified testing systems. Medical professionals benefit from this technology for diagnosing multiple infections and emerging variant tracking because it strengthens their ability to make healthcare decisions and monitor public health trends [24].

4.3 Various Detection Mechanisms

Detection mechanism	Basic principle	Features	Application	References
Optical (fluroescence, SPR, SERS)	Light emission/ reflection/ scattering changes	Highly sensitive (SPR), molecular fingerprinting (SERS)	Cancer diagnosis	[25]
Electrochemical (impedance, current-based)	Biomolecular interactions trigger changes in electrical signals	MNP-enhanced signal amplification that is steady and sensitive	DNA methylation detection, cancer biomarkers like CEA	[26]
Magnetic (magnetic relaxation, magneto-resistive)	Variations in resistance or magnetic properties	Extremely sensitive	Planar-Hall magneto-resistive sensors combined with microfluidics and biosensors based on MNPs	[26, 27]
Magnetic/Thermal (nanocantilevers)	Physical modifications to the cantilever's resonance or deflection	Label-free, quantitative, real-time detection	Cantilevers coated with antibodies for cancer indicators and smartphone integration for individualized diagnosis	[28]

5 Applications in Monitoring Metabolic and Chronic Disease

Nanobiosensors make the best use of the outstanding properties of nanomaterials, namely a high surface-to-volume ratio and improved catalytic activity, to successfully detect glucose with a high sensitivity [29]. Typically, GOx, isolated, is immobilized onto materials such as graphene or carbon nanotubes for catalysis of glucose oxidation to generate measurable electrical or optical signals. For instance, the near-infrared fluorescent nanosensors based on GOx-modified carbon nanotubes permit non-invasive and fast glucose monitoring in biofluids such as saliva, tears, and interstitial fluid. These nanobiosensors have also been included in wearables of microneedle patches that non-invasively collect interstitial fluid and release insulin using PEG-functionalized micropumps, thus creating a closed-loop glucose management system in real time [30]. Data can be passed wirelessly to smartphones, the use of finger-prick tests is minimized, and individuals control diabetes better [31].

The use of nanotechnology has eased the development of closed-loop systems that combine continuous glucose sensing with automatic insulin delivery. Such systems typically employ microneedle arrays or implantable nanosensors with intelligent control circuits that make for a carefully calculated modulation of insulin release according to actual instant glucose levels. This approach increases accuracy, reduces the probability of hypoglycemia, and facilitates dynamic management of diabetes [32]. Nano-structured materials enhance the biocompatibility, durability, and functionality of these systems. Surface modifications and bio-inspired nanodesigns are beneficial owing to sensor stability, thus making them useful for long-term implantation or wearable applications. These are significant advances toward an entirely automated form of personalized diabetes care [31].

Mature nanotechnologies, such as the combination of surface-enhanced Raman scattering (SERS) with bimetallic nanostructures, make it possible to make ultra-sensitive and specific measurements of key metabolic markers such as cholesterol and triglycerides. These nanosensors can identify imbalances that are otherwise subtle in lipoproteins (HDL, LDL) and triglyceride levels that are vastly important in early diagnosis and intervention for metabolic syndrome, atherosclerosis, coronary artery disease, and type 2 diabetes. Because of their size, their nanoscale size—thousands of times narrower than a human hair, these structures can provide precise biochemical knowledge at a molecular level, which can lead to better assessment of risks and better clinical decision-making on time [33].

Nanotechnology has also turned non-invasive glucose and metabolic monitoring a reality through detection in other biofluids, other than blood, such as sweat, saliva, tears, and urine. This is an innovation that minimizes the number of painful blood draws, thus improving the level of comfort and complying with treatment [34].

Graphene-based wearable nanosensors that are developed by GraphWear are designed to be worn on the wrist or the abdomen. We have these devices that monitor glucose through sweat analysis constantly, and they also transmit real-time data to smartphones, which allows for quick alerts and better glycemic control. Given enhanced stability, miniaturization, and user-friendly interfaces, this health equipment is revolutionizing, smooth real-time health monitoring, and improving chronic disease management [30].

6 Benefits of Smart Nanodiagnostic Systems

Research results from authoritative organizations point out the many good features of nanodiagnostic systems in medical analysis. Thanks to being very sensitive and specific, these systems are able to detect just a few trillionths (attomolar) or millionths (picomolar) of biomarkers using nanoparticles such as gold nanoparticles and magnetic beads. Consequently, such assays are able to tell apart pathogens or cancers that are similar, making the outcomes more accurate than with ELISA [35]. Because of cantilevers and field-effect transistors, point-of-care diagnosis is possible as these devices rapidly and accurately detect substances in real time [36].

Contemporary smart nanodiagnostic systems tend to be smaller and better fit for use outside a lab. At sites where labs are few, such as in the field, lab-on-a-chip and handheld biosensors are very helpful. Antibodies or aptamers on nanoparticles ensure both detection accuracy and lower the chance of errors. A number of nanodiagnostic devices can spot various biomarkers at once in a single assay. As a result, we are able to tell if a person has more than one infection or disease, such as COVID-19 and the flu, at the same time. As there are fluorescence, surface plasmon resonance, electrochemical, magnetic, and colorimetric signals that these systems can generate, they can be applied to many types of diagnostic tasks. Many advanced systems improve in accuracy when pH or temperature varies since they can adjust automatically to stay within the required parameters. Since you only use a little bit of saliva, blood, or urine for these, they are less uncomfortable for the patient, so they are more likely to stick with the scheduled tests [37].

Being very sensitive, these tools help us find diseases early and watch for further changes or responses to treatments. The big outlays for the first research and production may be high, but once the devices are mass-produced, healthcare could become more affordable and require less testing in the future. Altogether, these advantages make smart nanodiagnostic systems a key part of future healthcare that addresses prevention, personalization, and availability [35].

7 Challenges and Limitations in Clinical Translation

Despite its great potential for medical diagnostics, many challenges still keep it from being used in hospitals. Even though nanodiagnostic agents are powerful, it is still difficult to explain how they will work and interact with the body in animals and humans because their impacts on people are not always expected from animal studies. It is not simple to construct and complete these systems since they need to be the same every time and closely regulated to work well in large numbers [38]. Concern exists among experts that nanoparticles may not interfere with our immune and bodily systems, so their long-term effects have to be thoroughly studied before they can be used clinically. The lack of clear guidelines for nanomedicines, since they do not truly fit into the usual categories for medicine and devices, makes it harder to determine their regulation, their assessment, and the steps for reviewing them. Scaling up a new drug from clinical trials to production for the market can be difficult because batch changes might affect both how effective and safe the drugs are. Intellectual property problems, the expensive development process, and a legal requirement to prove these approaches are safer than existing ones mean that many cannot adopt them. All these factors together usually mean that introducing smart nanodiagnosis systems into regular medical care takes a lot of time and effort [38].

7.1 Manufacturing and Technical Difficulties

Smart nanotechnology being used in medicine is slowed down by the difficulty of resolving technical and manufacturing problems. Recreating the same process for making nanomaterials and scaling up is very hard because even the tiniest changes in these materials can affect their safety and usefulness. The difficulty in making nanoparticles results in minor changes from one batch to another and this can lead to both unreliable experiments and unintended actions by the cells, so strong control is necessary [1].

It is also hard to add biomolecules like antibodies or aptamers to the outside of nanoparticles with the help of the process known as surface functionalization. Functionalized nanoparticles can stop working or lose their stability when the temperature or pH is changed, whether during making, keeping, or shipment. Keeping nanoparticles for a lengthy period can make them deteriorate or lessen their effects, which reduces how long the diagnostic devices can be used [39]. Then, using nanomaterials in advanced diagnostic systems, such as lab-on-a-chip devices, brings new engineering issues. When using nanomaterials with devices such as microfluidics or electronics, since they often have different properties and need to be carefully matched so they operate correctly and signals are sent without problems. Because of all these moving parts, there is a risk that devices may not work properly or as intended in clinical care [40]. Furthermore, it is still a big problem that there is no global agreement on how to develop nanodiagnostic systems. At present, no common guidelines exist for characterizing, testing, and validating nanodiagnostic devices, which makes it hard to compare findings between different laboratories, holds up regulatory approval, and complicates certification. Countries need to come together to ensure that standards for their manufacturing, quality, and risks are set correctly so nanodiagnostic systems can be transferred to clinical use safely [41].

7.2 *Clinical, Safety, and Regulatory Obstacles*

Smart nanotechnology for medical diagnostics cannot be widely used because there are many rules, experiments, and safety concerns to deal with. We have to ensure biosafety and toxicity since we are still in the dark about what nanoparticles could do to our bodies in the future, like piling up in tissues or causing the immune system to act. Full research on animals and humans is required by the FDA and EMA before they can understand how the drug moves in the body and interacts with various chemicals, so the process of approving a drug takes time and effort. Nanotechnology needs to be produced and checked in the same way as regular chemicals for stability and consistency [42].

It is very difficult for these smart diagnostic platforms to get accepted by regulatory bodies. Although regulatory agencies require companies to share a lot of data about safety, impacts, and reliable production, new systems based on nanotechnology usually fail to reach the required standards because clear guidelines are lacking and they are too complex. So, it tends to take more time for new nanomedicines to start clinical trials, or such trials may not take place at all, because laws are not the same everywhere [43]. It's also true that problems with infrastructure and finances greatly hinder the adoption of nanodiagnostic systems. The first step to creating these platforms is very costly because they are made with sophisticated devices and need experts to conduct various diagnostic tests. If manufacturing is not enough locally, more products need to be imported, which raises expenses and damages the environment. That is why such treatments must be given in hospitals with the necessary equipment, which leads to them being seldom found in developing countries [44]. Also, inadequate healthcare systems, a lack of knowledge among those who make decisions and those who treat patients, and difficulties in fitting new treatments into existing plans make it hard to use these medicines on a large scale in areas with few resources.

Also, when smart nanodiagnostics use AI and IoT, it adds additional ethical and privacy problems. Since patient information is now shared quickly online, hospitals and medical offices must focus on improved cybersecurity and clear ethical rules when more of these technologies are added to healthcare. Equality issues arise since advanced nanotechnology might widen the distance between rich and poor areas if it is not available to all. Problems with nanomaterial waste and its effects on nature should be solved so that development stays environmentally friendly. These systems will only be safe and fair for patients in clinics after their regulatory, clinical, systemic, and ethical challenges have been addressed [45].

8 Case Studies and Commercially Available Nanodiagnostics

An example of clinical diagnostics is the Luminex Viegrine System, which uses gold nanoparticles and detects many different types of nucleic acids at the same time. This system employs proprietary NanoGrid™ technology, which utilizes gold nanoparticles (13–20 nm in diameter) functionalized with specific oligonucleotides or antibodies to target DNA, RNA, or proteins of interest. The goal is to make the target nucleic acids and the capturing probes on the microarray match which allows gold nanoparticles to cling to them. These probes enable signal amplification through silver staining, producing gold-silver aggregates that generate a measurable colorimetric change or surface plasmon resonance (SPR) effect. BPA is able to spot and strengthen RNA, so it reduces the processing time to less than 2 h after samples are handled [45].

Clinically, the Verigene System is FDA-approved and widely used in hospitals for diagnosing infectious diseases such as methicillin-resistant Staphylococcus aureus (MRSA) and bloodstream infections. All the steps of isolation, purification and hybridization are performed by the machines automatically, keeping the operation steady and easier for the operator. Because of multiplexing, the system allows doctors to identify several pathogens simultaneously, aiding their diagnostic procedure and helping them care for patients on the spot. Even though it requires advanced tools and usually appears in the richest hospitals, it is still widely used for diagnosing infections at health centers because it is dependable and offers fast outcomes [46]. Thus, relying on gold nanoparticle probes through the Verigene System allows molecular diagnostics to perform more precisely and cover more tests since the optical characteristics of gold and automation are used to give FDA-approved results.

8.1 Detection Platforms Based on Quantum Dots

Due to the brightness, stability, and narrow-spectrum emission when the dots are sized, quantum dots have made cancer imaging and detection easier. The use of QD Vision allows researchers to detect several important cancer markers in a single picture very clearly and accurately. Because QDs can carry many fluorescent labels at one time, they make detection both more sensitive and specific than traditional fluorophores can. Attaching ligands to them, such as antibodies and peptides, makes them target both the tumor and the areas nearby more accurately and securely. In this way, QDs can detect tumors both in the body and in experiments, track what is happening in the tumor environment, and support early cancer identification and therapy choice for patients [47].

As long as cancer biomarker imaging was not advanced, researchers used quantum dots to clearly show both tumors and the blood vessels supplying them by using fluorescence and imaging methods. Since they give off near-infrared (NIR) light, these dyes are able to penetrate tissue more deeply and help reduce the background autofluorescence, making it simpler to operate and remove tumors accurately. Tests show that QDs can observe tumor implants and metastasis signs in lymph nodes that other techniques usually overlook. It is also possible to design multifunctional nanoplatforms so that they can be used both for cancer screening and for targeting therapy [48].

Due to their high stability, brightness, and ability to do multiple things at once, QDs can be applied in tests that doctors can quickly use in wearable health monitors. They could keep an eye on several cancer biomarkers without requiring invasive procedures, which could improve the chances of early detection and monitor patients remotely from outside hospitals. More efforts have been put into reducing the toxicity and improving the compatibility of QDs since these issues are important for clinical use [49]. On the whole, quantum dot-based systems are helpful for early cancer screening, clear imaging, and can be seamlessly applied in portable devices, which provides many advantages over traditional approaches.

9 Integration with AI, IOT, and Point-of-Care Systems

Where nanotechnology, AI, IoT, and PoC systems are united, they help medical diagnostics progress swiftly, so that diseases can be identified early, adapted treatments can be given, and healthcare becomes more effective. In general, integration in nanotechnology makes it possible for sensors to check for diseases as early as their molecular and cellular stages [50]. AI's role helps these devices quickly and accurately process lots of difficult biomedical information. AI powered by machine and deep learning can make predictions, detect patterns, and decide based on information collected by sensors. AI helps to set up sensors more effectively, which reduces the number of tests required and increases the accuracy of diagnoses [51]. From IoT and IoMT, wireless sensors and point-of-care (PoC) devices can be used to collect data in real-time. IoT allows doctors to check their patients' health instantly, look at their vital signs, and provide care quickly through telemedicine services. Because of working together with nanotechnology, sensors, AI, and the internet, illnesses can be found earlier and doctors respond promptly [52].

AI uses inputs from nanodevices combined with the IoT to generate treatment options for each person and help them heal with less risk of side effects. As a result of IoT-enabled wearable nanobiosensors and portable devices, it is possible to check vital signs and biomarkers anywhere which improves care in places with limited resources and makes fewer hospital visits needed. Thanks to AI and IoT, nanotechnology makes it easier for drugs to be delivered only where they are needed which improves people's outcomes and leads to fewer issues [2].

10 Conclusion

Due to advances in smart nanotechnology, medical diagnostics now have access to new tools and systems beyond what traditional methods could provide. With their integration, functional nanomaterials such as gold nanoparticles, quantum dots, magnetic nanoparticles, carbon nanostructures, and polymeric hybrids have led to better sensitivity, specificity, and speed in detecting biomarkers. Since these materials are so small, they make it possible to detect small amounts of substances, measure several substances at once and monitor ongoing processes, so they work well in early cancer screening, infectious disease monitoring, following metabolic issues and watching over chronic conditions.

Thanks to nanotechnology, lab-on-a-chip devices, nanobiosensors, and smart systems, diagnostics can now take place closer to the patient through point-of-care (PoC) solutions. Merging these platforms with the help of fluorescence, SPR, SERS, electrochemical and magnetic approaches leads to rapid, accurate, and simple testing using small amounts of sample. Besides, integrating platforms like EMR or EHR with technology like artificial intelligence (AI), Internet of Things (IoT) and wearable sensors allows 24/7 monitoring, treatment checks, and individualized medical care. At the same time, making nanodiagnostic systems available in healthcare settings is still far from straightforward. Issues linked to making, adjusting and assembling nanoparticles must be fixed so that the process becomes consistent, scalable, and stable. Additional issues like unclear regulations, worries about biosafety and missing standardized ways to validate the technology stop its widespread use in hospitals. Also, the expensive initial setup and equipment needed may make it difficult for places with little resources to use biology technologies.

In spite of this, the potential for smart nanodiagnostics is still very strong. As scientists, engineers, clinicians, and policymakers keep collaborating and more money is put into translational research, many of these challenges will be addressed. Nanotechnology combined with digital health technology is creating a new wave of diagnostics that works more quickly, is smarter, very gentle on patients, and focuses on their needs. This way, smart nanotechnology acts as a main pillar for the advancement of personalized and precise medicine.

References

1. Thwala LN, Ndlovu SC, Mpofu KT, Lugongolo MY, Mthunzi-Kufa P (2023) Nanotechnology-based diagnostics for diseases prevalent in developing countries: current advances in point-of-care tests. Nanomaterials 13(7):1247. https://doi.org/10.3390/nano13071247
2. Malik S, Muhammad K, Waheed Y (2023) Emerging applications of nanotechnology in healthcare and medicine. Molecules 28(18):6624. https://doi.org/10.3390/molecules28186624
3. Santhoshakumari T, Bhuvaneswari C, Suchitra MR (2024) Nanotechnology and medical diagnostics: a systematic review. Nanotechnol Percept. https://doi.org/10.62441/nano-ntp.vi.1160
4. Nanotechnology in diagnosis: a review. https://www.scirp.org/journal/paperinformation?paperid=78183. Accessed 1 May 2025
5. Abdellatif AAH, Younis MA, Alsharidah M, Al Rugaie O, Tawfeek HM (2022) Biomedical applications of quantum dots: overview, challenges, and clinical potential. Int J Nanomed 17:1951–1970. https://doi.org/10.2147/IJN.S357980
6. Yang Z et al (2022) The applications of gold nanoparticles in the diagnosis and treatment of gastrointestinal cancer. Front Oncol 11:819329. https://doi.org/10.3389/fonc.2021.819329
7. ScienceDirect. https://www.sciencedirect.com/science/article/pii/S2589965124001673. Accessed 01 May 2025
8. Harun-Ur-Rashid M, Jahan I (2024) Innovative hybrid nanostructures: pioneering advances in modern therapy. Front Nanotechnol 6:1458894. https://doi.org/10.3389/fnano.2024.1458894
9. Pisanic TR, Zhang Y, Wang TH (2014) Quantum dots in diagnostics and detection: principles and paradigms. Analyst 139(12):2968–2981. https://doi.org/10.1039/c4an00294f
10. Dykman LA, Khlebtsov NG (2011) Gold nanoparticles in biology and medicine: recent advances and prospects. Acta Nat 3(2):34–55
11. Magnetic nanoparticles in human health and medicine. Wiley Online Books. https://onlinelibrary.wiley.com/doi/book/https://doi.org/10.1002/9781119754725. Accessed 1 May 2025
12. Maiti D, Tong X, Mou X, Yang K (2019) Carbon-based nanomaterials for biomedical applications: a recent study. Front Pharmacol 9:1401. https://doi.org/10.3389/fphar.2018.01401
13. Innovative hybrid nanostructures: pioneering advances in modern therapy. Frontiers. https://doi.org/10.3389/fnano.2024.1458894/full
14. Nanomaterials in biosensors: fundamentals and applications. PMC. https://pmc.ncbi.nlm.nih.gov/articles/PMC7152100/. Accessed 1 May 2025
15. Pede M, Deokar S, Kawade DR (2024) Nanotechnology in medicine: a comprehensive review of current applications and future perspectives. IJPS J. https://doi.org/10.5281/zenodo.14259472
16. Nanotechnology in medicine: drug delivery. Open Medscience. https://openmedscience.com/nanotechnology-in-medicine-revolutionising-drug-delivery-and-diagnostics/. Accessed 1 May 2025
17. Application of nanotechnology in cancer diagnosis and therapy—a mini-review. PMC. https://pmc.ncbi.nlm.nih.gov/articles/PMC7646098/. Accessed 1 May 2025
18. Tantray J, Patel A, Parveen H, Prajapati B, Prajapati J (2025) Nanotechnology-based biomedical devices in the cancer diagnostics and therapy. Med Oncol Northwood Lond Engl 42(2):50. https://doi.org/10.1007/s12032-025-02602-x

19. Nanotechnology and early cancer detection and diagnosis. NCI. https://www.cancer.gov/nano/cancer-nanotechnology/detection-diagnosis. Accessed 2 May 2025
20. Kashyap BK, Singh VV, Solanki MK, Kumar A, Ruokolainen J, Kesari KK (2023) Smart nanomaterials in cancer theranostics: challenges and opportunities. ACS Omega 8(16):14290–14320. https://doi.org/10.1021/acsomega.2c07840
21. Zhang Y, Chai Y, Hu Z, Xu Z, Li M, Chen X, Liu J (2022) Recent progress on rapid lateral flow assay-based early diagnosis of COVID-19. Front Bioeng Biotechnol 10:866368. https://doi.org/10.3389/fbioe.2022.866368/full
22. Ultrasensitive lateral-flow assays via plasmonically active antibody-conjugated fluorescent nanoparticles. Nat Biomed Eng. https://www.nature.com/articles/s41551-022-01001-1. Accessed 2 May 2025
23. Nanotechnology's frontier in combatting infectious and inflammatory diseases: prevention and treatment. Signal Transduct Target Therapy. https://www.nature.com/articles/s41392-024-01745-z. Accessed 2 May 2025
24. Nanotechnology fundamentals applied to clinical infectious diseases and public health. Open Forum Infectious Diseases, Oxford Academic. https://academic.oup.com/ofid/article/8/12/ofab583/6432367?login=false. Accessed 2 May 2025
25. Electrochemical protein biosensors for disease marker detection: progress and opportunities. Microsyst Nanoeng. https://www.nature.com/articles/s41378-024-00700-w. Accessed 9 May 2025
26. Laroui H, Rakhya P, Xiao B, Viennois E, Merlin D (2013) Nanotechnology in diagnostics and therapeutics for gastrointestinal disorders. Dig Liver Dis Off J Ital Soc Gastroenterol Ital Assoc Study Liver 45(12):995–1002. https://doi.org/10.1016/j.dld.2013.03.019
27. Hermsen SAB, Roszek B. Lab-on-a-chip devices for clinical diagnostics
28. Onyemaobi IM et al (2025) Nanomaterials and clinical SERS technology: broad applications in disease diagnosis. J Mater Chem B 13(9):2890–2911. https://doi.org/10.1039/D4TB02525C
29. Recent progress in nanobiosensors for precise detection of blood glucose level. PMC. https://pmc.ncbi.nlm.nih.gov/articles/PMC8786499/. Accessed 9 May 2025
30. A wearable, rapidly manufacturable, stability-enhancing microneedle patch for closed-loop diabetes management. Microsyst Nanoeng. https://www.nature.com/articles/s41378-024-00663-y. Accessed 9 May 2025
31. Revolutionizing diabetes management: nanostructured biosensors for real-time monitoring. JoNSNEA. https://journals.stmjournals.com/jonsnea/article=2024/view=152081/. Accessed 9 May 2025
32. Liu Y, Yang L, Cui Y (2024) A wearable, rapidly manufacturable, stability-enhancing microneedle patch for closed-loop diabetes management. Microsyst Nanoeng 10(1):1–15. https://doi.org/10.1038/s41378-024-00663-y
33. IIT-Guwahati develops advanced nanotechnology for cholesterol, triglyceride detection—ThePrint—PTIFeed. https://theprint.in/india/iit-guwahati-develops-advanced-nanotechnology-for-cholesterol-triglyceride-detection/2433493/. Accessed 9 May 2025
34. Makaram P, Owens D, Aceros J (2014) Trends in nanomaterial-based non-invasive diabetes sensing technologies. Diagnostics 4(2):27–46. https://doi.org/10.3390/diagnostics4020027
35. Alharbi KK, Al-sheikh YA (2014) Role and implications of nanodiagnostics in the changing trends of clinical diagnosis. Saudi J Biol Sci 21(2):109–117. https://doi.org/10.1016/j.sjbs.2013.11.001
36. Preethi M, Roy L, Lahkar S, Borse V (2022) Outlook of various diagnostics and nanodiagnostic techniques for COVID-19. Biosens. Bioelectron. X 12:100276. https://doi.org/10.1016/j.biosx.2022.100276
37. Jackson TC, Patani BO, Ekpa DE (2017) Nanotechnology in Diagnosis: A Review. Adv. Nanoparticles 6(3):3. https://doi.org/10.4236/anp.2017.63008
38. Hua S, de Matos MBC, Metselaar JM, Storm G (2018) Current trends and challenges in the clinical translation of nanoparticulate nanomedicines: pathways for translational development and commercialization. Front Pharmacol 9:790. https://doi.org/10.3389/fphar.2018.00790

39. Desai N (2012) Challenges in development of nanoparticle-based therapeutics. AAPS J 14(2):282–295. https://doi.org/10.1208/s12248-012-9339-4
40. Lastra LS, Sharma V, Farajpour N, Nguyen M, Freedman KJ (2021) Nanodiagnostics: a review of the medical capabilities of nanopores. Nanomedicine Nanotechnol Biol Med 37:102425. https://doi.org/10.1016/j.nano.2021.102425
41. Pede M, Deokar S, Kawade DR (2024) Nanotechnology in medicine: a comprehensive review of current applications and future perspectives. Int J Pharm Sci. https://doi.org/10.5281/zenodo.14259472
42. Foulkes R, Man E, Thind J, Yeung S, Joy A, Hoskins C (2020) The regulation of nanomaterials and nanomedicines for clinical application: current and future perspectives. Biomater Sci 8(17):4653–4664. https://doi.org/10.1039/D0BM00558D
43. Allan J et al (2021) Regulatory landscape of nanotechnology and nanoplastics from a global perspective. Regul Toxicol Pharmacol 122:104885. https://doi.org/10.1016/j.yrtph.2021.104885
44. Regulation of nanomedicines|Possible challenges. https://www.pharmafocusasia.com/strategy/regulation-nanomedicines. Accessed 2 Jun 2025
45. Kaushik A (2019) Biomedical nanotechnology related grand challenges and perspectives. Front Nanotechnol. https://doi.org/10.3389/fnano.2019.00001
46. https://www.artisantg.com/info/Luminex_Nanosphere_VERIGENE_Datasheet_202493014112.pdf
47. Quantum dots for cancer research: current status, remaining issues, and future perspectives. PMC. https://pmc.ncbi.nlm.nih.gov/articles/PMC3643664/. Accessed 2 Jun 2025
48. Bio-conjugated quantum dots for cancer research: detection and imaging. Frontiers. https://doi.org/10.3389/fonc.2021.749970/full. Accessed 2 Jun 2025
49. Wang L-W, Peng C-W, Chen C, Li Y (2015) Quantum dots-based tissue and in vivo imaging in breast cancer researches: current status and future perspectives. Breast Cancer Res Treat 151(1):7–17. https://doi.org/10.1007/s10549-015-3363-x
50. Emerging applications of nanotechnology in healthcare and medicine. PMC. https://pmc.ncbi.nlm.nih.gov/articles/PMC10536529/. Accessed 2 Jun 2025
51. Yammouri G, Ait Lahcen A (2024) AI-reinforced wearable sensors and intelligent point-of-care tests. J Pers Med 14(11):1088. https://doi.org/10.3390/jpm14111088
52. Artificial Intelligence (AI) and Internet of Medical Things (IoMT) assisted biomedical systems for intelligent healthcare. PMC. https://pmc.ncbi.nlm.nih.gov/articles/PMC9330886/. Accessed 2 Jun 2025

Smart Nanotechnology for Tissue Engineering

Pawan Kumar, Jitender Sharma, Mohini Goyal, Meenu Saini, Anil Sindhu, and Brijnandan Singh Dehiya

Abstract Smart nanotechnology has transformed tissue engineering by creating adaptive, responsive, and multifunctional nanosystems capable of sensing, decision-making, and self-regulation within biological environments. This chapter provides a thorough overview of the principles, design strategies, and biomedical uses of smart nanomaterials in regenerative medicine. It highlights the importance of the interaction between cells, scaffolds, and signaling cues, and how nanotechnology improves their integration to support biomimetic, dynamic tissue regeneration. Advanced nanomaterials, including stimuli-responsive polymers, hybrid nanocomposites, and self-assembling, programmable nanostructures, enable controlled drug delivery, targeted cell guidance, and real-time feedback—bridging the gap between structural and functional tissue engineering. The discussion also covers 3D bioprinting, where nanoparticle-containing bioinks and AI-assisted manufacturing produce intelligent scaffolds capable of dynamic remodeling and cellular adaptation. Additionally, tissue-specific applications are examined across bone, cartilage, neural, cardiac, vascular, and skin regeneration, demonstrating smart nanotechnology's ability to replicate native extracellular matrix (ECM) properties, promote new blood vessel growth, and facilitate electrochemical and mechanical signaling. Finally, the chapter discusses future directions, including AI-driven design, digital twins, 4D bioprinting, and nanorobotics, positioning smart nanotechnology as a key element of next-generation, patient-specific regenerative medicine.

P. Kumar (✉) · J. Sharma
Department of Biotechnology, Kurukshetra University, Kurukshetra, Haryana, India

M. Goyal
Department of Botany, Kurukshetra University, Kurukshetra, India

M. Saini
Department of Physics, Capital University, Koderma, Jharkhand, India

A. Sindhu
Department of Biotechnology, Deenbandhu Chhotu Ram University of Science and Technology, Murthal, Haryana, India

B. S. Dehiya
Department of Chemical Engineering, Deenbandhu Chhotu Ram University of Science and Technology, Murthal, Haryana, India

J. Chen (ed.), *AI-Driven Nanotechnology in Precision Medicine and Healthcare*, Smart Nanomaterials Technology, https://doi.org/10.1007/978-981-92-1690-1_13

Keywords Smart nanotechnology · Tissue engineering · Biomimetic scaffolds · Stimuli-responsive nanomaterials · 3D bioprinting · Regenerative medicine · Artificial intelligence

1 Introduction

Over the past two decades, nanotechnology has emerged as one of the most transformative tools in modern biomedical science, enabling the engineering of materials and devices at the nanoscale for targeted, responsive, and efficient therapeutic and diagnostic applications. The convergence of nanotechnology with precision medicine, a paradigm emphasizing individualized treatment based on genetic, environmental, and lifestyle factors, has catalyzed the emergence of smart nanotechnology, an adaptive and intelligent approach to disease management that integrates nanoscale engineering, artificial intelligence (AI), and systems biology [1, 2].

Smart nanotechnology refers to nanosystems capable of sensing, decision-making, and adaptive responses within biological environments. These systems are often designed to be stimuli-responsive, releasing drugs or imaging agents in response to internal (pH, redox potential, enzyme activity) or external (light, temperature, magnetic field) cues [3, 4]. This responsiveness enables precise spatiotemporal control of therapeutic actions, minimizing off-target effects and maximizing therapeutic efficacy, a hallmark of precision medicine [5]. For instance, smart polymeric micelles and metal-organic frameworks (MOFs) have shown potential for programmable drug delivery and real-time feedback [6]. Beyond therapeutics, smart nanoplatforms have also revolutionized diagnostics and monitoring. Nanobiosensors and nanoparticle-based imaging probes enable real-time disease tracking with molecular specificity, a crucial feature for early detection and personalized intervention [7]. The integration of AI and machine learning (ML) algorithms further enhances smart nanotechnology's capabilities—allowing predictive modeling of nanomaterial behavior, optimization of nanoparticle design, and real-time interpretation of complex biomedical data [8]. However, despite remarkable advances, translational challenges persist. Regulatory, ethical, and manufacturing barriers slow the path from laboratory innovation to clinical deployment. Concerns regarding nanotoxicity, immune interactions, and long-term biosafety underscore the need for standardized frameworks for evaluation [9]. Furthermore, equitable access to nanomedicine technologies remains an ongoing concern in global healthcare systems [10].

In essence, smart nanotechnology embodies the fusion of precision, adaptability, and intelligence in medicine, enabling a shift from reactive to predictive and preventive healthcare. The ongoing integration of AI-driven analytics, stimuli-responsive materials, and real-time feedback mechanisms marks a new era of nanomedicine poised to redefine disease management and patient care.

2 Fundamentals of Smart Nanotechnology

2.1 *"Smart" Nanomaterials*

Smart nanomaterials represent a transformative class of engineered nanosystems that can dynamically respond to specific biological or environmental stimuli. Unlike conventional nanomaterials that passively interact with their surroundings, smart nanomaterials are "intelligent" constructs, designed to sense, adapt, and act based on physiological cues such as pH, redox potential, enzyme concentration, temperature, or magnetic fields [11, 12]. These materials exhibit programmable responsiveness, enabling controlled drug release, targeted delivery, and selfreporting diagnostic feedback, making them indispensable in precision medicine [10, 13].

Structurally, smart nanomaterials can be broadly classified into stimuli-responsive, selfassembling, and hybrid multifunctional systems (Table 1). Stimuli-responsive nanomaterials, such as pH- or temperature-sensitive liposomes and polymeric micelles, alter their configuration to release therapeutic payloads only under specific conditions, thus minimizing off-target effects [14, 15]. Meanwhile, self-assembling nanostructures can form in situ in response to physiological triggers, thereby enhancing bioavailability and tissue-specific localization. Hybrid nanosystems, which combine metallic, polymeric, and biomolecular components, integrate multiple functionalities, such as imaging, therapy, and biosensing, into a single platform [16, 17]. Recent studies emphasize that "smartness" arises not only from material composition but also from design logic—the deliberate integration of feedback mechanisms and sensing capabilities into nanoscale architectures. For instance, polymer–gold hybrid nanostructures have demonstrated autonomous decision-making, releasing drugs in response to intracellular ATP levels while simultaneously enabling fluorescence imaging [18, 19]. Such systems exemplify the convergence of nanotechnology, materials science, and bioinformatics in the development of next-generation adaptive nanomedicine platforms.

2.2 *Functionalization Strategies: Surface Chemistry and Targeting*

Functionalization is the cornerstone of smart nanotechnology, enabling nanomaterials to achieve targeted, responsive, and biocompatible performance in precision medicine. The surface chemistry of nanomaterials determines their biodistribution, interaction with biomolecules, and therapeutic selectivity. Tailoring these surfaces with biomimetic, polymeric, or ligand-based modifications enhances both biological recognition and stability within physiological environments [39].

Table 1 Classification and functional features of smart nanomaterials in precision medicine

Type of smart nanomaterial	Stimulus/ mechanism of action	Biomedical applications	Representative examples	Refs.
pH-responsive nanocarriers	Structural change or polymer swelling in the acidic tumor microenvironment triggers drug release	Targeted chemotherapy and tumor-specific delivery	Poly(L-histidine)PEG micelles, pHsensitive liposomes	[20, 21]
Temperature-responsive nanomaterials	Polymer phase transition (LCST/ UCST) regulates release upon hyperthermia	Hyperthermia assisted drug delivery and thermo-triggered release	PNIPAM-based nanogels, thermosensitive liposomes	[22–24]
Enzyme-responsive nanoparticles	Enzymatic cleavage of linkers or degradation of the coating releases the payload	Cancer therapy, inflammation targeting	MMP-sensitive polymeric micelles, peptide-crosslinked nanocarriers	[25, 26]
Redox-responsive nanostructures	Reduction of disulfide bonds in high-GSH environments enables release	Intracellular targeted therapy and controlled gene delivery	Disulfide-linked polymeric nanoparticles, GSH-sensitive dendrimers	[27–29]
Magnetic responsive nanomaterials	External magnetic field guides localization or induces hyperthermia	Magnetic drug targeting, MRI guided therapy	Fe_3O_4-based hybrid nanoparticles	[30–32]
Light-responsive nanoplatforms	Photothermal or photochemical activation triggers the release	Photodynamic therapy and image-guided treatment	Gold nanorods, upconversion nanoparticles	[33, 34]
Self-assembling smart nanostructures	Spontaneous formation under physiological conditions enables structural adaptation	Personalized nanomedicine and tissue-specific delivery	Peptide-based nanofibers, DNA origami systems	[34–36]
Hybrid multifunctional nanosystems	Integration of multiple stimuli responses with imaging and therapeutic features	Theranostics, personalized drug release	Polymer–gold hybrids, magnetic luminescent nanocomposites	[37, 38]

2.2.1 Covalent and Non-covalent Functionalization Approaches

Two primary strategies dominate nanomaterial functionalization: covalent conjugation (through amide, ester, or thiol–maleimide linkages) and non-covalent interactions (such as electrostatic or π–π stacking). Covalent methods ensure stability and reproducibility, making them ideal for applications requiring long-term circulation, whereas non-covalent techniques preserve the native structure of biomolecules and enable reversible drug attachment [40, 41]. For instance, gold and silica nanoparticles functionalized with aptamers or peptides demonstrate high-affinity targeting in tumor microenvironments while maintaining biocompatibility [42, 43].

2.2.2 Targeting Functionalization and Molecular Recognition

Surface functionalization with ligands, antibodies, or aptamers enhances specific binding to overexpressed receptors in diseased cells, such as folate, transferrin, or HER2 receptors in cancer. Smart lipid nanoparticles (SLNs), for example, are engineered with antibody mediated targeting to improve intracellular delivery of gene-editing payloads and mRNA therapeutics [2, 44]. Similarly, programmable DNAzyme-functionalized gold nanoplatforms exhibit sequence-specific recognition and catalytic precision in biosensing applications [45].

2.2.3 Stimuli-Responsive and Multifunctional Designs

Emerging "smart" nanocarriers integrate stimuli-responsive linkers that respond to pH, enzymatic activity, redox conditions, or light to enable controlled, site-specific release. MOFs and nanodiamonds (NDs) exemplify multifunctional scaffolds, in which hybrid organic–inorganic chemistry enables dual functionality: imaging and therapy [46, 47]. The trend toward *multi-ligand co-functionalization* provides synergistic effects, such as combining cell-penetrating peptides with hydrophilic polymers to improve both uptake and circulation time [48].

2.3 Intelligent Nanostructures: Responsive, Programmable, and Self-assembling Systems

The emergence of intelligent nanostructures marks a transformative phase in precision medicine, enabling nanoscale materials to sense, adapt, and respond dynamically to biological environments. These "smart" systems integrate stimuli-responsiveness, programmable molecular logic, and self-assembly mechanisms to achieve unprecedented control in biomedical applications such as targeted therapy, diagnostics, and tissue regeneration [49, 50]. Self-assembling systems harness non-covalent interactions, including hydrogen bonding, hydrophobic effects, π–π stacking, and electrostatic forces, to organize functional nanostructures in situ [51, 52]. Lei et al. reported the use of short peptides and DNA-based nanostructures that self-assemble into programmable, stimuli-responsive scaffolds for precision biomedical applications. These systems demonstrate adaptive behavior, such as structural rearrangement in response to pH or enzyme activity, thereby optimizing drug release and biomolecular recognition [53].

Programmability in nanomaterials often relies on nucleic acid-based designs (DNA origami, RNA scaffolds) or synthetic polymer architectures that encode responsiveness. DNA nanotechnology offers precise structural control and biocompatibility, while synthetic polymers such as poly(Nisopropylacrylamide) and poly(ethylene glycol) can be functionalized with responsive moieties to trigger conformational changes or degradation in response to specific physiological cues [54, 55]. These features make them ideal for *on-demand* drug delivery and biosensing applications. Meanwhile, stimuli-responsive nanostructures can be externally or internally activated. Externally triggered systems, such as magnetic, light, or ultrasound-responsive nanoparticles, offer spatiotemporal control over therapeutic delivery, while internally responsive systems, sensitive to pH gradients, oxidative stress, or enzymatic environments, align closely with tumor microenvironments and inflammatory tissues [56, 57]. Hybrid designs combining both modalities, often termed dual- or multi-stimuli-responsive systems, are increasingly used to achieve finer control over nanosystem activation [58].

Another frontier in intelligent nanostructures involves integration with computational logic and AI. AI-assisted design tools can model molecular interactions to predict the stability and behavior of self-assembling nanostructures, thereby reducing the number of experimental trials. These computationally guided nanostructures represent the transition from *passive carriers* to *adaptive nanosystems* capable of decision-making and environmental feedback [59, 60]. However, challenges remain in scaling up self-assembled nanostructures and maintaining reproducibility in biological environments. The dynamic and complex nature of self-assembly often leads to variability in morphology and functionality. Addressing these limitations requires advances in in situ imaging, machine-learning-guided fabrication, and bioinspired materials that mimic the selforganizing properties of living systems [61–63].

3 Nanotechnology in Tissue Engineering: The Conceptual Framework

3.1 Triad of Tissue Engineering: Cells, Scaffolds, and Signals

The classical paradigm of tissue engineering rests on the synergistic interaction between three fundamental components, cells, scaffolds, and signaling cues that collectively govern tissue formation, integration, and regeneration. This triad forms the conceptual and functional foundation upon which smart nanotechnology introduces transformative precision and responsiveness.

Cells act as the living functional units of tissue regeneration, contributing to matrix synthesis, remodeling, and homeostasis. Stem cells, particularly mesenchymal stem cells (MSCs) and induced pluripotent stem cells (iPSCs), have emerged as central players due to their multilineage differentiation potential and immunomodulatory capabilities [64]. Smart nanotechnology enhances cell functionality by providing nanoscale topographical and biochemical cues that mimic the native extracellular matrix (ECM). Nanoscale features influence cell adhesion, migration, and fate decisions through mechanotransduction pathways such as YAP/TAZ and integrin-mediated signaling [65].

Scaffolds serve as the structural framework for cellular organization and tissue ingrowth. Smart nanotechnology allows for the creation of nanostructured or nanocomposite scaffolds that exhibit tunable mechanical strength, biodegradability, and bioactivity. Incorporating nanoparticles such as hydroxyapatite, graphene oxide, or silica nanospheres within polymer matrices can replicate ECM stiffness gradients and biochemical functionality [66]. Furthermore, stimuli-responsive nanomaterials, such as thermoresponsive hydrogels or magnetic nanocomposites, enable dynamic modulation of scaffold properties in response to external or internal triggers [67]. This adaptability allows scaffolds to act not merely as passive supports but as *intelligent bioactive interfaces* that guide tissue formation through real-time feedback and responsiveness to biological signals.

Signals, encompassing biochemical, mechanical, and electrical cues, regulate cellular behavior and tissue maturation. In native tissues, these signals are orchestrated in a tightly regulated spatiotemporal manner. Smart nanotechnology introduces precision-controlled signaling platforms, where nanocarriers deliver growth factors, morphogens, or nucleic acids in a programmable fashion [68]. For instance, nanoparticle-mediated delivery of bone morphogenetic protein-2 (BMP-2) or vascular endothelial growth factor (VEGF) allows localized and sustained signaling for osteogenic or angiogenic differentiation, respectively [69]. Magnetic and piezoelectric nanomaterials further enable mechanical or electrical stimulation at the cellular level, enhancing differentiation pathways relevant to bone, muscle, and neural tissues [70]. Integration of AI-driven data analytics has also begun to optimize the dosing and temporal release of these cues, enabling adaptive and personalized tissue regeneration strategies [71]. The convergence of cells, scaffolds, and signals within a nanoscale framework establishes a dynamic and self-regulating regenerative system. Unlike conventional tissue engineering approaches, smart nanotechnology enables bidirectional communication between materials and biological systems, allowing real-time sensing and modulation of the microenvironment.

3.2 Role of Nanotechnology in Mimicking Native ECM

ECM is a dynamic and hierarchically organized network that provides both structural and biochemical cues essential for cellular adhesion, proliferation, differentiation, and tissue homeostasis. Mimicking this intricate milieu remains a central challenge in tissue engineering. Nanotechnology offers an unprecedented ability to replicate ECM architecture and functionality by controlling material properties at the molecular and nanoscale levels, enabling precise imitation of native tissue environments. At its core, the ECM is composed of nanoscale fibrous proteins such as collagen, elastin, and laminin, interspersed with proteoglycans and glycosaminoglycans that regulate hydration and signaling [72]. Smart nanotechnology replicates this complexity through nanofibrous scaffolds, nanocomposites, and bioactive coatings that mimic the topographical and biochemical features of native ECM. Electrospinning and self-assembly techniques have been widely used to fabricate nanofibers with diameters ranging from 50 to 500 nm, comparable to collagen fibrils, allowing for the recreation of ECM-like architectures that guide cellular alignment and migration [66, 73]. Incorporation of bioactive nanoparticles such as hydroxyapatite, silica, or carbon nanotubes enhances mechanical properties while providing sites for biomolecule adsorption and controlled release. These multifunctional nanostructures bridge the gap between mechanical integrity and biological performance, facilitating improved cell–matrix interactions and tissue integration [74].

Beyond physical mimicry, chemical and biochemical functionality is equally vital for ECM emulation. Nanomaterial functionalization with adhesion peptides (e.g., RGD, IKVAV) and growth factors enables selective interaction with integrin receptors, thereby directing cellular behavior and lineage specification [75]. Smart nanocarriers embedded within scaffolds can release bioactive molecules in response to environmental stimuli, simulating the natural dynamic turnover of ECM components. For instance, pH- or enzyme-responsive nanocarriers can deliver transforming growth factor-β (TGF-β) or fibroblast growth factor (FGF) in localized regions undergoing tissue remodeling [76]. Such spatiotemporal control of bioactive signaling is essential for orchestrating complex regeneration processes.

Moreover, nanotechnology provides tools to emulate the mechanical and viscoelastic properties of native ECM, which play crucial roles in mechanotransduction and cellular differentiation. Nanoscale reinforcement using graphene oxide, cellulose nanocrystals, or peptide amphiphiles can tune stiffness and elasticity to match specific tissue requirements—from the rigidity of bone to the softness of neural or cardiac tissues [77]. Smart hydrogels incorporating dynamic crosslinkers or reversible supramolecular bonds offer self-healing properties, reflecting the adaptability of biological ECM under physiological stresses. These materials not only provide mechanical support but also participate actively in signal transduction, influencing cellular responses through deformation and stress feedback [78].

3.3 Design Considerations for Nano-Enabled Scaffolds

The design of nano-enabled scaffolds requires integrating biological functionality, mechanical stability, and intelligent responsiveness. These scaffolds serve as bioactive matrices that guide cellular behavior while supporting tissue regeneration. Structurally, mechanical and architectural optimization is critical to match native tissue characteristics. Nanocomposite scaffolds combining polymers with nanoparticles like hydroxyapatite or silica improve strength, toughness, and elasticity, enabling application-specific tailoring from soft neural to rigid bone tissues. Nanoscale reinforcement and computational modeling together ensure mechanical precision and long-term stability [79, 80].

Surface nanoarchitecture significantly influences cell adhesion and differentiation. Nanopatterned or electrospun fibers (50–200 nm) mimic ECM fibrils and promote organized tissue regeneration. Hierarchical pore structures enhance oxygen and nutrient diffusion, facilitating vascularization and tissue integration [81, 82]. Incorporating biochemical functionality, via immobilized peptides, proteins, or growth factors, provides localized biochemical signaling. Stimuli-responsive nanocarriers embedded within scaffolds enable controlled release in response to pH, enzyme, or temperature variations, achieving precise spatiotemporal control over tissue morphogenesis [60]. Ensuring biocompatibility and biodegradability remains a key design priority. Smart coatings such as PEG or zwitterionic layers enhance hemocompatibility and reduce immune recognition. Emerging AI-driven models are increasingly used to predict scaffold behavior, optimize pore geometry, and simulate degradation kinetics [83, 84]. Ultimately, nano-enabled scaffolds represent a convergence of materials science and regenerative biology, designed to be adaptive, interactive, and biomimetic.

3.4 Smart Nanotechnology in 3D Bioprinting

The integration of smart nanotechnology into 3D bioprinting has redefined the precision and functionality of tissue fabrication. By embedding nanoscale materials into printable bioinks, researchers can now produce scaffolds that not only replicate the structural complexity of native tissues but also respond actively to physiological cues. Smart nanomaterials, such as graphene, gold nanoparticles, carbon nanotubes, and magnetic nanoparticles, enhance the mechanical strength, electrical conductivity, and bioactivity of bioprinted constructs [85, 86]. These features are especially valuable for cardiac, neural, and musculoskeletal applications, where electrical or mechanical signaling is critical for cellular function. Moreover, nanofillers improve print fidelity and layer adhesion, ensuring structural stability during and after printing [87].

Stimuli-responsive nanocomponents incorporated within bioinks introduce a level of adaptability previously unattainable. pH-, temperature-, or enzyme-sensitive nanocarriers can release bioactive agents post-printing in response to the local microenvironment, supporting dynamic tissue maturation [88]. Magnetic nanoparticles further enable remote control over cell alignment and scaffold architecture under applied magnetic fields, improving tissue anisotropy and integration [89]. Additionally, AI-assisted bioprinting platforms enable real-time monitoring and adjustment of printing parameters, thereby optimizing cell viability and spatial organization. Machine learning algorithms can predict bioink flow dynamics, curing behavior, and nutrient diffusion profiles, streamlining the fabrication of complex tissues [90].

4 Smart Nanomaterials for Specific Tissue Applications

Smart nanomaterials are nanoscale (1–1000 nm) systems whose properties or behavior can be reversibly tuned by external or internal stimuli. In tissue engineering and biomedicine, they are designed to perform targeted actions with high precision, such as targeting a tissue, releasing a drug on demand, or signaling the tissue through diagnostic readouts [3]. Figure 1 illustrates the diverse applications of smart nanotechnology across four key tissue types: bone and cartilage, neural, cardiac and vascular, and skin and soft tissue. Each domain benefits from nanoscale innovations such as bioactive nanocomposites, conductive and stimuli-responsive materials, and AI-assisted scaffold design, all of which enhance tissue-specific regeneration, integration, and functional recovery.

4.1 Bone and Cartilage Tissue Engineering

Bone and cartilage tissues possess distinct yet interrelated regenerative requirements, both of which rely on the precise coordination of structural support, biochemical signaling, and mechanical stimulation. Smart nanotechnology provides an advanced toolkit to address these challenges by engineering biomimetic scaffolds that promote osteogenesis and chondrogenesis while enabling controlled therapeutic modulation [79, 91].

Bone tissue engineering benefits greatly from nanostructured materials that replicate the hierarchical organization of native bone, which consists of collagen fibrils mineralized with hydroxyapatite nanocrystals. Nano-hydroxyapatite (nHA), bioactive glass nanoparticles, and calcium phosphate nanocomposites, when incorporated into polymeric scaffolds, enhance osteoconductivity and mechanical stiffness, promoting matrix mineralization and vascular infiltration [92, 93]. Stimuli-responsive nanocarriers further refine bone regeneration by enabling localized and

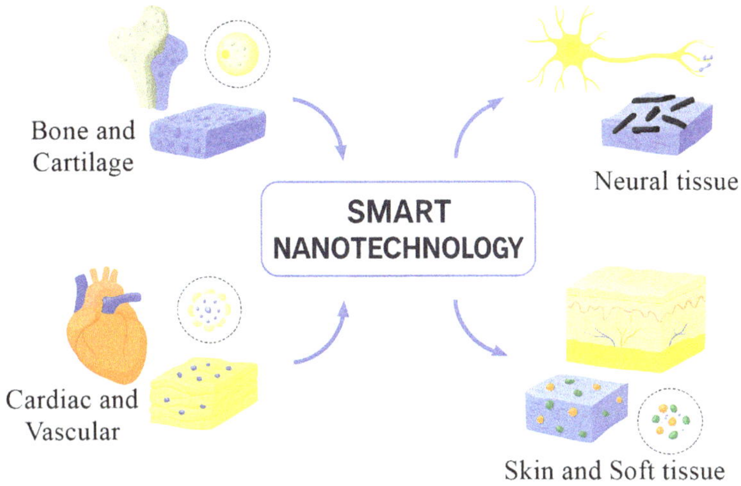

Fig. 1 Applications of smart nanotechnology in tissue engineering

sustained delivery of osteoinductive factors such as bone morphogenetic protein-2 (BMP-2) or vascular endothelial growth factor (VEGF) in response to pH or enzyme activity [94]. Moreover, magnetic and piezoelectric nanoparticles integrated within scaffolds can convert external magnetic or mechanical fields into biochemical cues, activating mechanotransduction pathways (e.g., MAPK, Wnt/β-catenin) that stimulate bone formation [95].

Cartilage regeneration, on the other hand, demands scaffolds with high elasticity, low friction, and sustained chondrogenic signaling. Nanocomposite hydrogels reinforced with graphene oxide, cellulose nanofibers, or silica nanoparticles replicate the viscoelastic nature of cartilage while enhancing mechanical resilience and nutrient permeability [96]. Smart nanocarriers embedded within these matrices can release transforming growth factor-β (TGF-β) or insulin-like growth factor-1 (IGF-1) in a temporally controlled manner to sustain chondrocyte differentiation and extracellular matrix synthesis [97]. Additionally, thermoresponsive and self-healing nanohydrogels offer dynamic adaptability under cyclic loading, mimicking the mechanical responsiveness of native cartilage [96].

Recent advances in AI-driven scaffold optimization and 3D bioprinting with nanoparticle-laden bioinks have further improved the structural precision and spatial control of osteochondral constructs. These hybrid systems enable gradient deposition of osteogenic and chondrogenic zones, closely resembling the natural bone–cartilage interface [98]. Such integration of nanotechnology, artificial intelligence, and biofabrication allows simultaneous enhancement of mechanical functionality and biological signaling, accelerating functional tissue restoration.

4.2 Neural Tissue Engineering

Neural tissue engineering presents unique challenges due to the complex architecture and limited regenerative capacity of the central and peripheral nervous systems. Smart nanotechnology offers innovative strategies to overcome these barriers by creating biomimetic, conductive, and responsive platforms that facilitate neuronal growth, synaptic connectivity, and functional recovery [99]. Nanomaterials such as graphene oxide, carbon nanotubes (CNTs), gold nanoparticles, and conducting polymers have been widely employed to enhance electrical conductivity and neural interface performance [100]. These materials promote neuron adhesion, axonal elongation, and electrophysiological signaling by mimicking the electroactive nature of native neural tissue. Their nanoscale roughness and high surface area improve protein adsorption and integrin-mediated attachment, supporting the formation of neural networks. Moreover, hybrid nanocomposites combining CNTs or graphene with biopolymers (e.g., chitosan, gelatin, or silk fibroin) yield flexible, biocompatible scaffolds suitable for peripheral nerve regeneration [101].

Stimuli-responsive nanostructures play an essential role in controlled neurotrophic factor delivery. Encapsulation of nerve growth factor (NGF), brain-derived neurotrophic factor (BDNF), or glialderived neurotrophic factor (GDNF) within pH- or redox-responsive nanoparticles enables sustained and localized release, maintaining a pro-regenerative microenvironment [102]. Additionally, magnetic and piezoelectric nanoparticles can transduce external physical stimuli into electrical signals, enhancing neuronal differentiation through voltage-gated ion channel activation and calcium influx [103]. Recent advances in 3D nanofiber scaffolds and microchannel-guided conduits have facilitated spatial alignment of neurons and Schwann cells, crucial for directional nerve regeneration. Incorporating nanosensors within these constructs enables real-time monitoring of electrical activity, ion exchange, and metabolic states, bridging the gap between regenerative scaffolds and neural prosthetics [104]. Furthermore, AI-driven modeling assists in optimizing scaffold geometry, conductivity, and release kinetics, enabling personalized neuroregenerative designs based on patient-specific injury profiles [105]. In essence, smart nanotechnology transforms neural tissue engineering from a static reconstruction process into a dynamic, feedback-driven system capable of electrical communication and biological adaptability.

4.3 Cardiac and Vascular Tissue Regeneration

Cardiac and vascular tissue engineering requires highly specialized materials that can synchronize with the body's electrical and mechanical dynamics while supporting angiogenesis and tissue remodeling. Smart nanotechnology enables the fabrication of responsive, conductive, and bioactive scaffolds that promote cardiomyocyte alignment, vascular growth, and synchronized contraction [106, 107].

Nanostructured conductive materials, such as graphene, carbon nanotubes (CNTs), gold nanowires, and polypyrrole, have shown exceptional potential in cardiac tissue regeneration by restoring electrical coupling between damaged myocardial cells. These materials facilitate cell–cell communication through nanoscale conduction pathways and improve synchronous beating in engineered cardiac patches. When integrated into biodegradable polymers such as polycaprolactone (PCL) or gelatin methacryloyl (GelMA), they provide mechanical elasticity and electrical conductivity comparable to native myocardium [101, 107].

In **vascular tissue engineering**, nanotechnology plays a vital role in replicating the hierarchical structure and functionality of blood vessels. Electrospun nanofibrous scaffolds with nanoscale porosity enhance endothelial cell adhesion and migration, promoting rapid endothelialization and vessel maturation [108]. The inclusion of bioactive nanoparticles, such as nitric oxide (NO)releasing nanocarriers or VEGF-loaded nanoliposomes, stimulates angiogenesis and maintains antithrombotic activity [109]. Additionally, smart, stimuli-responsive nanocarriers can release proangiogenic factors in response to local hypoxia or mechanical stress, thereby mimicking the adaptive signaling of native vascular tissues [110].

Mechanical and electrical stimulation mediated by magnetic or piezoelectric nanoparticles further enhances cardiac and vascular regeneration. Magnetoelectric nanocomposites can translate applied magnetic fields into localized electrical impulses, promoting cardiomyocyte contractility and vascular smooth muscle proliferation [111]. Similarly, AI-assisted modeling and simulation tools are being employed to design nanoscaffold architectures that optimize fluid shear stress, oxygen diffusion, and electrical propagation patterns within engineered tissues [112]. Overall, the integration of smart nanotechnology within cardiac and vascular tissue engineering provides multifunctional platforms capable of sensing, adapting, and stimulating biological processes in real time.

4.4 Skin and Soft Tissue Reconstruction

Skin and soft tissue regeneration demand biomaterials that integrate structural strength, elasticity, antimicrobial protection, and bioactivity while supporting angiogenesis and re-epithelialization. Smart nanotechnology provides an adaptive framework for achieving these multifaceted goals by designing bioactive, stimuli-responsive, and nanostructured systems that closely mimic native dermal architecture [113]. Nanocomposite scaffolds integrating polymers (such as chitosan, collagen, or polycaprolactone) with nanoparticles like silver, zinc oxide, or cerium oxide exhibit enhanced mechanical resilience and antimicrobial properties, crucial for wound healing and infection control [114]. These nanosystems promote fibroblast proliferation, collagen synthesis, and angiogenesis while reducing oxidative stress and bacterial colonization. Moreover, electrospun nanofibers and self-assembling peptide hydrogels provide ECM-like topography that guides keratinocyte migration and tissue remodeling [115].

Stimuli-responsive nanocarriers embedded within wound dressings or hydrogels can release growth factors (e.g., EGF, VEGF, and PDGF) or anti-inflammatory agents in response to environmental triggers such as pH, temperature, or enzyme activity [113]. This on-demand release enables spatiotemporal control of healing stages, hemostasis, proliferation, and remodeling, leading to accelerated wound closure and scar reduction. Magnetic and photothermal nanoparticles also support controlled hyperthermia and oxygen generation, improving perfusion and antibacterial efficacy in chronic wounds [116]. Recent developments in AI-guided nanomaterial design and 3D bioprinting have enabled the fabrication of gradient nanoscaffolds that replicate the multilayered structure of skin, incorporating dermal and epidermal compartments with embedded vasculature. These intelligent constructs can integrate nanosensors to monitor pH, temperature, or oxygen concentration, providing real-time feedback for adaptive therapeutic modulation [117].

5 Future Directions and Outlook

Smart nanotechnology stands at the threshold of transforming tissue engineering from a structural science into an intelligent, adaptive, and self-regulating discipline. The next decade will likely witness the convergence of nanomaterials, AI, and biofabrication to create dynamic, personalized, and feedback-controlled regenerative systems. Future research will focus on developing multifunctional and self-healing nanomaterials capable of sensing, responding, and adapting to the physiological microenvironment. These materials will integrate biosensing, drug delivery, and mechanical reinforcement into a single platform, enabling real-time monitoring and regulation of tissue regeneration [60]. For instance, hybrid nanostructures combining conductive polymers, magnetic nanoparticles, and biodegradable matrices can both stimulate and track cellular behavior through optical, electrical, or magnetic feedback. The integration of AI and machine learning (ML) will further accelerate the design and optimization of nanomaterials. Predictive modeling of nanostructure–cell interactions and in silico simulation of degradation and release kinetics will reduce experimental uncertainty and cost. Digital twins, virtual replicas of biological systems, may soon be used to simulate scaffold performance and optimize regenerative therapies in a patient-specific manner [60, 118]. Another major direction is the emergence of 4D bioprinting, where smart nanocomposite bioinks enable constructs that can evolve in response to stimuli such as temperature, pH, or mechanical stress. This approach holds promise for producing dynamic tissues that adapt to the host environment, closely mimicking developmental biology [119]. Furthermore, the integration of nanorobotics and microactuators could enable in vivo manipulation, repair, and remote-controlled regeneration at the cellular level [120]. Despite these advances, several challenges persist. Long-term biosafety, immunogenicity, and nanoparticle clearance remain key regulatory hurdles. Standardized protocols for evaluating nanotoxicity, degradation, and functional performance are essential for clinical translation. Ethical considerations related to AI-driven design and equitable access to smart biomaterials must also be addressed to ensure responsible innovation.

6 Conclusion

In conclusion, smart nanotechnology is redefining the frontiers of tissue engineering—ushering in a new era of intelligent regenerative medicine characterized by self-adaptation, predictive control, and patient-specific functionality. The integration of responsive nanomaterials, computational design, and real-time biofeedback promises to shift the field from static biomaterials toward *living, learning, and evolving systems* that can autonomously guide tissue repair and regeneration.

References

1. Bi Y, Xie S, Li Z, Dong S, Teng L (2025) Precise nanoscale fabrication technologies, the "last mile" of medicinal development. Acta Pharm Sin B 15:2372–2401
2. Chehelgerdi M, Chehelgerdi M, Allela OQB et al (2023) Progressing nanotechnology to improve targeted cancer treatment: overcoming hurdles in its clinical implementation. Mol Cancer 22:169
3. Sun L, Liu H, Ye Y, Lei Y, Islam R, Tan S, Tong R, Miao Y-B, Cai L (2023) Smart nanoparticles for cancer therapy. Signal Transduct Target Ther 8:418
4. Parvin N, Joo SW, Mandal TK (2025) Biodegradable and stimuli-responsive nanomaterials for targeted drug delivery in autoimmune diseases. J Funct Biomater. https://doi.org/10.3390/jfb16010024
5. Taneja A, Panda HS, Panda JJ, Singh TG, Kour A (2025) Revolutionizing precision medicine: unveiling smart stimuli-responsive nanomedicine. Adv Ther 8:e00073
6. Guo Z, Xiao Y, Wu W, Zhe M, Yu P, Shakya S, Li Z, Xing F (2025) Metal–organic framework-based smart stimuli-responsive drug delivery systems for cancer therapy: advances, challenges, and future perspectives. J Nanobiotechnology 23:157
7. Huang Y, Guo X, Wu Y, Chen X, Feng L, Xie N, Shen G (2024) Nanotechnology's frontier in combatting infectious and inflammatory diseases: prevention and treatment. Signal Transduct Target Ther 9:34
8. Arya A, Garg D, Parashar A, Joshi HC, Ahmad W (2025) Nano-AI synergy: emerging trends, applications, challenges, and future prospects. Next Nanotechnol 8:100309
9. Younis MA, Tawfeek HM, Abdellatif AAH, Abdel-Aleem JA, Harashima H (2022) Clinical translation of nanomedicines: challenges, opportunities, and keys. Adv Drug Deliv Rev 181:114083
10. Kazi RN, Hasani IW, Khafaga DSR, Kabba S, Farhan M, Aatif M, Muteeb G, Fahim YA (2025) Nanomedicine: the effective role of nanomaterials in healthcare from diagnosis to therapy. Pharmaceutics 17:987
11. Singh R, Sharma A, Saji J, Umapathi A, Kumar S, Daima HK (2022) Smart nanomaterials for cancer diagnosis and treatment. Nano Converg 9:21
12. Kashyap BK, Singh VV, Solanki MK, Kumar A, Ruokolainen J, Kesari KK (2023) Smart nanomaterials in cancer theranostics: challenges and opportunities. ACS Omega 8:14290–14320

13. Srivastava R (2025) Stimuli-responsive nanomaterials for the precision delivery of mRNA cancer vaccines. Nano Trends 11:100147
14. Chen X, Wu D, Chen Z (2024) Biomedical applications of stimuli-responsive nanomaterials. MedComm 5:e643
15. Handa M, Singh A, Flora SJS, Shukla R (2022) Stimuli-responsive polymeric nanosystems for therapeutic applications. Curr Pharm Des 28:910–921
16. Yaraki MT, Zahed Nasab S, Zare I, Dahri M, Moein Sadeghi M, Koohi M, Tan YN (2022) Biomimetic metallic nanostructures for biomedical applications, catalysis, and beyond. Ind Eng Chem Res 61:7547–7593
17. Seaberg J, Montazerian H, Hossen MN, Bhattacharya R, Khademhosseini A, Mukherjee P (2021) Hybrid Nanosystems for biomedical applications. ACS Nano 15:2099–2142
18. Harun-Ur-Rashid M, Jahan I, Foyez T, Imran AB (2023) Bio-inspired nanomaterials for micro/nanodevices: a new era in biomedical applications. Micromachines 14:1786
19. Xia L, Liu Y, Wang L, Yao X (2025) The architecture of nanorobots: from advanced materials to structural visualization. Small n/a:e09588
20. Wu H, Zhu L, Torchilin VP (2013) pH-sensitive poly(histidine)-PEG/DSPE-PEG copolymer micelles for cytosolic drug delivery. Biomaterials 34:1213–1222
21. Kale AA, Torchilin VP (2010) Environment-responsive multifunctional liposomes BT - liposomes: methods and protocols, volume 1: pharmaceutical nanocarriers. In: Weissig V (ed) Humana Press. Totowa, NJ, pp 213–242
22. Karimi M, Sahandi Zangabad P, Ghasemi A et al (2016) Temperature-responsive smart Nanocarriers for delivery of therapeutic agents: applications and recent advances. ACS Appl Mater Interfaces 8:21107–21133
23. Deshpande S, Sharma S, Koul V, Singh N (2017) Core–shell nanoparticles as an efficient, sustained, and triggered drug-delivery system. ACS Omega 2:6455–6463
24. Guo Y, Sun L, Wang Y, Wang Q, Jing D, Liu S (2022) Nanomaterials based on thermosensitive polymer in biomedical field. Front Chem. https://doi.org/10.3389/fchem.2022.946183
25. Nagel G, Sousa-Herves A, Wedepohl S, Calderón M (2020) Matrix metalloproteinase sensitive multistage nanogels promote drug transport in 3D tumor model. Theranostics 10:91–108
26. Massi L, Najer A, Chapman R, Spicer CD, Nele V, Che J, Booth MA, Doutch JJ, Stevens MM (2020) Tuneable peptide cross-linked nanogels for enzyme-triggered protein delivery. J Mater Chem B 8:8894–8907
27. Cheng R, Feng F, Meng F, Deng C, Feijen J, Zhong Z (2011) Glutathione-responsive nano-vehicles as a promising platform for targeted intracellular drug and gene delivery. J Control Release 152:2–12
28. Chibh S, Kour A, Yadav N, Kumar P, Yadav P, Chauhan VS, Panda JJ (2020) Redox-responsive dipeptide nanostructures toward targeted cancer therapy. ACS Omega 5:3365–3375
29. Meng X, Shen Y, Zhao H, Lu X, Wang Z, Zhao Y (2024) Redox-manipulating nanocarriers for anticancer drug delivery: a systematic review. J Nanobiotechnology 22:587
30. Kumar CSSR, Mohammad F (2011) Magnetic nanomaterials for hyperthermia-based therapy and controlled drug delivery. Adv Drug Deliv Rev 63:789–808
31. Dutta B, Shetake NG, Patra S, Chakravarty R, Vimalnath KV, Chakraborty A, Chakraborty S, Pandey BN, Hassan PA, Barick KC (2024) pH-responsive magnetic nanocarriers for chelator-free bimodal (MRI/SPECT-CT) image-guided chemohyperthermia therapy in human breast carcinoma. J Mater Chem B 12:11759–11777
32. Nori ZZ, Bahadori M, Moghadam M, Tangestaninejad S, Mirkhani V, MohammadpoorBaltork I, Jafari SS, Emamzadeh R, Alem H (2023) Synthesis and characterization of a new gold-coated magnetic nanoparticle decorated with a thiol-containing dendrimer for targeted drug delivery, hyperthermia treatment and enhancement of MRI contrast agent. J Drug Deliv Sci Technol 81:104216
33. Luo L, Sun W, Feng Y, Qin R, Zhang J, Ding D, Shi T, Liu X, Chen X, Chen H (2020) Conjugation of a scintillator complex and gold nanorods for dual-modal image-guided photothermal and X-ray-induced photodynamic therapy of tumors. ACS Appl Mater Interfaces 12:12591–12599

34. Chan M-H, Chen S-P, Chen C-W, Chan Y-C, Lin RJ, Tsai DP, Hsiao M, Chung R-J, Chen X, Liu R-S (2018) Single 808 nm laser treatment comprising photothermal and photodynamic therapies by using gold nanorods hybrid upconversion particles. J Phys Chem C 122:2402–2412
35. Delfi M, Sartorius R, Ashrafizadeh M et al (2021) Self-assembled peptide and protein nanostructures for anti-cancer therapy: targeted delivery, stimuli-responsive devices and immunotherapy. Nano Today 38:101119
36. Habibi N, Kamaly N, Memic A, Shafiee H (2016) Self-assembled peptide-based nanostructures: smart nanomaterials toward targeted drug delivery. Nano Today 11:41–60
37. Nan X, Zhang X, Liu Y, Zhou M, Chen X, Zhang X (2017) Dual-targeted multifunctional nanoparticles for magnetic resonance imaging guided cancer diagnosis and therapy. ACS Appl Mater Interfaces 9:9986–9995
38. Govindan B, Sabri MA, Hai A, Banat F, Haija MA (2023) A review of advanced multifunctional magnetic nanostructures for cancer diagnosis and therapy integrated into an artificial intelligence approach. Pharmaceutics 15:868
39. Mout R, Moyano DF, Rana S, Rotello VM (2012) Surface functionalization of nanoparticles for nanomedicine. Chem Soc Rev 41:2539–2544
40. Su Y-C, Chen G, Lai Y-J, Song G-Z, Wu T-L, Yeh Y-C (2025) Tailoring the dynamic nanocomposite hydrogels through surface-functionalized nanomaterials and interfacial crosslinking chemistry toward multifunctional biomedical and engineering applications. Chem Soc Rev. https://doi.org/10.1039/D5CS00975H
41. Percoco RM, Schirinzi S, Mandriota G, Cortese B, Gigli G, Palamà IE, Maiorano G (2025) Smart inorganic nanoparticles in nanomedicine: strategies for synthesis and functionalization. Int J Pharm 684:126192
42. Kumar A, Ma H, Zhang X, Huang K, Jin S, Liu J, Wei T, Cao W, Zou G, Liang X-J (2012) Gold nanoparticles functionalized with therapeutic and targeted peptides for cancer treatment. Biomaterials 33:1180–1189
43. Aravind A, Veeranarayanan S, Poulose AC, Nair R, Nagaoka Y, Yoshida Y, Maekawa T, Kumar DS (2012) Aptamer-functionalized silica nanoparticles for targeted cancer therapy. BioNanoScience 2:1–8
44. Nabih NW, Hassan HAFM, Preis E, Schaefer J, Babker A, Abbas AM, Amin MU, Bakowsky U, Fahmy SA (2025) Antibody-functionalized lipid nanocarriers for RNA based cancer gene therapy: advances and challenges in targeted delivery. Nanoscale Adv 7:5905–5931
45. Wang K, Zhu W, Li X, Ling G, Zhang P (2025) Programmable DNAzyme walkers on gold nanoplatforms: from rational design to bioimaging/biosensing breakthroughs. Acta Biomater 206:74–87
46. Mi P (2020) Stimuli-responsive nanocarriers for drug delivery, tumor imaging, therapy and theranostics. Theranostics 10:4557–4588
47. Pattnaik AK, Priyadarshini N, Priyadarshini P, Behera GC, Parida K (2024) Recent advancements in metal organic framework-modified multifunctional materials for photodynamic therapy. Mater Adv 5:6030–6051
48. Guo S, Wang J, Wang Q, Wang J, Qin S, Li W (2024) Advances in peptide-based drug delivery systems. Heliyon 10:e26009
49. Rabiee N (2025) Programmable biointerfaces and adaptive functionality in NextGeneration green nanomaterials. ACS Synth Biol 14:3862–3879
50. Amoozadeh M, Hariri A, Zarepour A, Khosravi A, Iravani S, Zarrabi A (2025) Biophotonic (nano)structures: from fundamentals to emerging applications. RSC Adv 15:26138–26172
51. Niu J, Wang C, Liao M, Liu Y, Ding Y, Yao H, Zhang H, Cao H (2025) Supramolecular self-assembly of emamectin benzoate with β-cyclodextrin and polylactic acid: enhancing pesticide efficacy and environmental safety. Chem Eng J 509:160834
52. Sánchez-Iglesias A, Grzelczak M, Altantzis T et al (2012) Hydrophobic interactions modulate self-assembly of nanoparticles. ACS Nano 6:11059–11065
53. Lei G, Hanyan X, Xin J, Jing Y (2023) Oligopeptide self-assembly: mechanisms, stimuli-responsiveness, and biomedical applications. CCS Chem 6:69–90

54. Aye SL, Sato Y (2022) Therapeutic applications of programmable DNA nanostructures. Micromachines 13:315
55. Speed SK, Gupta K, Peng Y-H, Hsiao SK, Krieg E (2023) Programmable polymer materials empowered by DNA nanotechnology. J Polym Sci 61:1713–1729
56. Blum AP, Kammeyer JK, Rush AM, Callmann CE, Hahn ME, Gianneschi NC (2015) Stimuli-responsive nanomaterials for biomedical applications. J Am Chem Soc 137:2140–2154
57. Liu H, Xing F, Yu P, Shakya S, Peng K, Liu M, Xiang Z, Ritz U (2024) Integrated design and application of stimuli-responsive metal–organic frameworks in biomedicine: current status and future perspectives. J Mater Chem B 12:8235–8266
58. Yoon MS, Lee JM, Jo MJ et al (2025) Dual-drug delivery systems using hydrogel– nanoparticle composites: recent advances and key applications. Gels 11:520
59. T A, Narayan R, Shenoy PA, Nayak UY (2022) Computational modeling for the design and development of nano based drug delivery systems. J Mol Liq 368:120596
60. Parvin N, Aslam M, Alam MN, Mandal TK (2025) Nanotechnology driven innovations in modern pharmaceutics: therapeutics, imaging, and regeneration. Nano 15:1733
61. Abdullah M, Obayedullah M, Shariful Islam Shuvo M, Abul Khair M, Hossain D, Nahidul Islam M (2025) A review on multifunctional applications of nanoparticles: analyzing their multi-physical properties. Results Surf Interfaces 21:100635
62. Chen R, Jia X, Pang W, Yang Y, Zhou H, Gao Z (2025) Applications of DNA origami in biomedicine: advances, challenges, and prospects. Adv Compos Hybrid Mater 8:375
63. Guo L, Zhu Q, Slater AG (2025) Self-assembly under continuous flow conditions. Chem Commun 61:10236–10251
64. Hsiung N, Ju Y, Yang K, Yang P, Zeng W, Zhao H, Zou P, Ye J, Yi K, Wang X (2025) Organoid-based tissue engineering for advanced tissue repair and reconstruction. Mater Today Bio 33:102093
65. Mishra A, Modi U, Sharma R, Bhatia D, Solanki R (2025) Biochemical and biophysical cues of the extracellular matrix modulates stem cell fate: Progress and prospect in extracellular matrix mimicking biomaterials. Biomed Eng Adv 9:100143
66. Narayanan KB (2025) Nanotopographical features of polymeric nanocomposite scaffolds for tissue engineering and regenerative medicine: a review. Biomimetics 10:317
67. Teixeira do Nascimento A, Stoddart PR, Goris T, Kael M, Manasseh R, Alt K, Tashkandi J, Kim BC, Moulton SE (2025) Stimuli-responsive materials for biomedical applications. Adv Mater 37:e07559
68. O'Grady B, Balikov DA, Wang JX, Neal EK, Ou Y-C, Bardhan R, Lippmann ES, Bellan LM (2019) Spatiotemporal control and modeling of morphogen delivery to induce gradient patterning of stem cell differentiation using fluidic channels. Biomater Sci 7:1358–1371
69. Zhu L, Liu Y, Wang A, Zhu Z, Li Y, Zhu C, Che Z (2022) Application of BMP in bone. Tissue Eng 10:1–25
70. Liu L, Li Z, Zhu T, Sun Y, Xu J (2025) Advances in applications of low-dimensional piezoelectric materials in musculoskeletal system. Mater Today Bio 33:102065
71. Singh H, Balusamy SR, Sukweenadhi J, Saravanan M, Aruchamy M, Mijakovic I, Singh P (2025) Smart hybrid nanomaterials for chronic infections: microbiome-responsive and sustainable therapeutic platforms. J Nanobiotechnology 23:698
72. Kelleher CM, Vacanti JP (2010) Engineering extracellular matrix through nanotechnology. J R Soc Interface 7:S717–S729
73. Zhang Y, Su B, Venugopal J, Lim CT (2007) Biomimetic and bioactive nanofibrous scaffolds from electrospun composite nanofibers. Int J Nanomed 2:623–638
74. Nikolova MP, Apostolova MD (2023) Advances in multifunctional bioactive coatings for metallic bone implants. Materials (Basel) 16:183
75. Aazmi A, Zhang D, Mazzaglia C, Yu M, Wang Z, Yang H, Huang YYS, Ma L (2024) Biofabrication methods for reconstructing extracellular matrix mimetics. Bioact Mater 31:475–496
76. Naghib SM, Amiri S, Mozafari MR (2024) Stimuli-responsive chitosan-based nanocarriers for drug delivery in wound dressing applications: a review. Carbohydr Polym Technol Appl 7:100497

77. Huang G, Li F, Zhao X, Ma Y, Li Y, Lin M, Jin G, Lu TJ, Genin GM, Xu F (2017) Functional and biomimetic materials for engineering of the three-dimensional cell microenvironment. Chem Rev 117:12764–12850
78. Xue L, An R, Zhao J, Qiu M, Wang Z, Ren H, Yu D, Zhu X (2025) Self-healing hydrogels: mechanisms and biomedical applications. MedComm 6:e70181
79. Nikolova MP, Chavali MS (2019) Recent advances in biomaterials for 3D scaffolds: a review. Bioact Mater 4:271–292
80. Gartia AK, Chakraverty S (2025) Advanced computational Modeling and mechanical behavior analysis of multi-directional functionally graded nanostructures: a comprehensive review. Comput Model Eng Sci 142:2405–2455
81. Litowczenko J, Woźniak-Budych MJ, Staszak K, Wieszczycka K, Jurga S, Tylkowski B (2021) Milestones and current achievements in development of multifunctional bioscaffolds for medical application. Bioact Mater 6:2412–2438
82. Musílková J, Beran M, Sedlář A et al (2025) Composite polylactide/polycaprolactone foams with hierarchical porous structure for pre-vascularized tissue engineering. Int J Mol Sci 26:2974
83. Cheng Y, Lu Y (2025) Physical stimuli-responsive polymeric patches for healthcare. Bioact Mater 43:342–375
84. Sutradhar SC, Shin H, Kim W, Jang H (2025) Hydrogel films in biomedical applications: fabrication. In: Properties and therapeutic potential gels, vol 11, p 918
85. Rahman M, Dip TM, Nur MG, Hossain MH, Snow F, Hossain NB, Mirabedini A, Quigley A, Padhye R, Houshyar S (2025) Nanomaterial-integrated 3D biofabricated structures for advanced biomedical applications. Macromol Mater Eng 310:2500083
86. Agarwal P, Mathur V, Kasturi M, Srinivasan V, Seetharam RN, S Vasanthan K (2024) A futuristic development in 3D printing technique using nanomaterials with a step toward 4D printing. ACS Omega 9:37445–37458
87. Jiang Y, Latif M, Kim J (2024) Three-dimensional printing of lignocellulose structures: improving mechanical properties and shape fidelity. ACS Omega 9:23442–23450
88. Cook AB, Decuzzi P (2021) Harnessing endogenous stimuli for responsive materials in theranostics. ACS Nano 15:2068–2098
89. Santhosh M, Choi J-H, Choi J-W (2019) Magnetic-assisted cell alignment within a magnetic nanoparticle-decorated reduced graphene oxide/collagen 3D nanocomposite hydrogel. Nano 9:1293
90. Budharaju H, Singh RK, Kim H-W (2025) Bioprinting for drug screening: a path toward reducing animal testing or redefining preclinical research? Bioact Mater 51:993–1017
91. Zhou J, Li Q, Tian Z, Yao Q, Zhang M (2023) Recent advances in 3D bioprinted cartilage-mimicking constructs for applications in tissue engineering. Mater Today Bio 23:100870
92. Kumar P, Dehiya BS, Sindhu A (2019) Synthesis and characterization of nHA-PEG and nBG-PEG scaffolds for hard tissue engineering applications. Ceram Int 45:8370–8379
93. Kumar P, Saini M, Dehiya BS, Umar A, Sindhu A, Mohammed H, Al-Hadeethi Y, Guo Z (2020) Fabrication and in-vitro biocompatibility of freeze-dried CTS-nHA and CTS-nBG scaffolds for bone regeneration applications. Int J Biol Macromol 149:1–10
94. Tong J, Wang Y, Cao Y, Liang B, Yu K (2025) Advanced bio-polymers for bone regeneration: harnessing anti-inflammatory, oxidative stress, and pro-angiogenic strategies. Mater Des 260:114986
95. Gu J, Gui L, Yan D, Xia X, Xie Z, Xue L (2025) Magnetic-responsive material-mediated magnetic stimulation for tissue engineering. Magnetochemistry 11:82
96. Hashemi-Afzal F, Fallahi H, Bagheri F, Collins MN, Eslaminejad MB, Seitz H (2025) Advancements in hydrogel design for articular cartilage regeneration: a comprehensive review. Bioact Mater 43:1–31
97. Wang M, Wang J, Xu X, Li E, Xu P (2024) Engineering gene-activated bioprinted scaffolds for enhancing articular cartilage repair. Mater Today Bio 29:101351
98. Gao D, Li R, Pan J, Li C, Zhang W, Qin L, Lai Y (2025) 3D bioprinting bone/cartilage organoids: construction, applications, and challenges. J Orthop Transl 55:75–93

99. Huang W-H, Ding S-L, Zhao X-Y, Li K, Guo H-T, Zhang M-Z, Gu Q (2023) Collagen for neural tissue engineering: materials, strategies, and challenges. Mater Today Bio 20:100639
100. Şak B, Sousa HBA, Prior JAV (2025) Carbon nanomaterial-based electrochemical biosensors for Alzheimer's disease biomarkers: Progress, challenges, and future perspectives. Biosensors 15:684
101. Sancar E, Oktay B, Ahlatcıoğlu Özerol E (2024) Carbon based materials in conducting tissue engineering. Mater Res Express 11:122001
102. Bathina S, Das UN (2015) Brain-derived neurotrophic factor and its clinical implications. Arch Med Sci 11:1164–1178
103. Wang X, Sun Y, Han C, Meng X, Wen K, Wu J, Min P, Li K, Zhang Y (2025) Research trends of piezoelectric materials in neurodegenerative disease applications. Bioact Mater 52:366–392
104. Wang J, Liu Y, Lv M, Zhao X, So KF, Li H, EL-Newehy M, EL-Hamshary H, Morsi Y, Mo X (2023) Regulation of nerve cells using conductive nanofibrous scaffolds for controlled release of *Lycium barbarum* polysaccharides and nerve growth factor. Regener Biomater 10:rbad038
105. Haramshahi SMA, Hamblin MR, Ravesh RK, Sadr H, Ahmadirad N, Mehrabi F, Taherian Z, Hosseingolipour S, Barzegar Z, Mehrabi S (2025) Biomaterials for CNS disorders: a review of development from traditional methods to AI-assisted optimization. J Mater Sci Mater Med 36:85
106. Tisha ZT, Alam KT, Ahmed T (2025) Cardiac tissue engineering using stimuli responsive biomaterials for the targeted therapy of myocardial infarction. MedComm—Biomater Appl 4:e70009
107. Ashtari K, Nazari H, Ko H et al (2019) Electrically conductive nanomaterials for cardiac tissue engineering. Adv Drug Deliv Rev 144:162–179
108. Rickel AP, Deng X, Engebretson D, Hong Z (2021) Electrospun nanofiber scaffold for vascular tissue engineering. Mater Sci Eng C 129:112373
109. Lee KH, Song M-Y, Lee S et al (2024) Nitric oxide releasing nanofiber stimulates revascularization in response to ischemia via cGMP-dependent protein kinase. PLoS One 19:e0303758
110. Zhuang J, Zhang X, Liu Q, Zhu M, Huang X (2022) Targeted delivery of nanomedicines for promoting vascular regeneration in ischemic diseases. Theranostics 12:6223–6241
111. Faricelli I, Lenzuni M, Giannoni P, Ravazzani P, Marrella A (2025) Magnetoelectric/piezoelectric-based materials for coupled electrical and mechanical stimulation for bone repair: an in silico study. Nanoscale Adv 7:7768–7779
112. Bermejillo Barrera MD, Franco-Martínez F, Díaz Lantada A (2021) Artificial intelligence aided design of tissue engineering scaffolds employing virtual tomography and 3D convolutional neural networks. Materials (Basel) 14:5278
113. Jiang Y, Zhou Y, Tian Y, Nabavi N, Ashrafizadeh M, Conde J, Li Z, Guo L (2025) Conductive polymers in smart wound healing: from bioelectric stimulation to regenerative therapies. Mater Today Bio 34:102114
114. Jangra N, Singla A, Puri V, Dheer D, Chopra H, Malik T, Sharma A (2025) Herbal bioactive-loaded biopolymeric formulations for wound healing applications. RSC Adv 15:12402–12442
115. Nazarnezhad S, Baino F, Kim H-W, Webster TJ, Kargozar S (2020) Electrospun nanofibers for improved angiogenesis: promises for tissue engineering applications. Nano 10:1609
116. Fahimirad S, Fattahi F, Hatami M, Shabani S, Ghorbanpour M (2025) Nanotechnology based biotherapeutics for physiological wound healing phases. Ind Crop Prod 226:120608
117. Di Stefano AB, Urrata V, Schilders K, Franza M, Di Leo S, Moschella F, Cordova A, Toia F (2025) Three-dimensional bioprinting techniques in skin regeneration: current insights and future perspectives. Life 15:787
118. Bamisaye A, Adekola MA, Abati SM, Etafo NO, Ademola OS, Joseph PT, Samuel O, Ogunlaja OO, Langmi H, Idowu MA (2025) Recent advances in metal/metal-oxide nanoparticle-polymer nanohybrid for biomedical applications. Mater Today Chem 49:103086

119. Arif ZU, Khalid MY, Zolfagharian A, Bodaghi M (2022) 4D bioprinting of smart polymers for biomedical applications: recent progress, challenges, and future perspectives. React Funct Polym 179:105374
120. Liu S, Gao C, Peng F (2022) Micro/nanomotors in regenerative medicine. Mater Today Adv 16:100281

Smart Nanotechnology for Nucleic Acid Delivery

Bancha Yingngam

Abstract Nucleic acid-based drugs, such as DNA, messenger RNA (mRNA), small interfering RNA (siRNA), microRNA (miRNA), and gene-editing constructs, such as CRISPR/Cas, have transformed biological research and enabled new therapeutic approaches to treat genetic diseases, cancer, and infectious diseases. Nevertheless, the real use of these nucleic acids in the clinic is restricted by several biological barriers, such as enzymatic degradation, inefficient cellular uptake, fast clearance from the blood, and possible off-target effects. To address these challenges, intelligent nanotechnology has become an attractive and innovative strategy for developing well-defined, smart, and responsive delivery systems that are designed in a tailor-made fashion for the unique physicochemical characteristics of nucleic acids. This chapter provides a comprehensive summary of recent advanced smart nanocarrier platforms for nucleic acid delivery. The chapter discusses the design considerations of lipid, polymeric, inorganic, and hybrid nanocarriers with an emphasis on stimuli-responsive systems. These systems are designed to respond to endogenous stimulators (pH, redox conditions, and enzymatic activity) as well as exogenous triggers, such as light, magnetic fields, and ultrasound, for controlled and localized therapeutic delivery. This chapter includes a discussion of the essential steps of nanocarrier synthesis, physicochemical analysis, biological performance (cellular uptake), and endosomal escape. The chapter also discusses translational aspects, safety perspectives, and new therapeutic opportunities as well as the fusion of nanotechnology with gene-editing approaches and design optimization via artificial intelligence. Finally, future aspects and regulatory considerations to offer an important reference for the successful clinical translation of smart nanotechnology-based nucleic acid therapeutics are summarized.

Keywords Gene therapy · Biocompatibility · Endosomal escape · Targeted therapeutics · Controlled release · Stimuli-responsive nanocarriers · CRISPR/Cas systems · Molecular diagnostics

B. Yingngam (✉)
Department of Pharmaceutical Chemistry and Technology, Faculty of Pharmaceutical Sciences, Ubon Ratchathani University, Ubon Ratchathani, Thailand
e-mail: bancha.y@ubu.ac.th

J. Chen (ed.), *AI-Driven Nanotechnology in Precision Medicine and Healthcare*, Smart Nanomaterials Technology, https://doi.org/10.1007/978-981-92-1690-1_14

Abbreviations

AI	Artificial intelligence
LNPs	Lipid-based nanoparticles
LNs	Lipid-based nanocarriers
miRNA	microRNA
ML	Machine learning
mRNA	Messenger RNA
siRNA	Small interfering RNA

1 Introduction

The therapeutic application of nucleic acids has rapidly emerged as a class of therapeutics with profound potential to modulate gene expression, treat genetic defects, and manage complex diseases at the molecular level [1]. Nucleic acids, such as plasmid DNA, messenger RNA (mRNA), small interfering RNA (siRNA), microRNA (miRNA) or antisense oligonucleotides, and gene-editing constructs, provide the opportunity for targeted intervention at the level of transcription, translation, or genomic sequence [2, 3]. Over the past 20 years, progress in molecular biology, genomics, and synthetic biology has accelerated the creation of these therapeutics, resulting in several key clinical achievements, including mRNA vaccines and RNA interference-based drugs [4]. Notwithstanding these developments, the broad clinical application of nucleic acid drugs is largely dependent on the adaptation of safe, effective, and tunable delivery technology [5].

Unlike traditional small molecules, nucleic acids have intrinsic physical properties that prevent their use via direct application. These large, hydrophilic, and negatively charged macromolecules exhibit limited membrane permeability [6]. Nucleic acids are very susceptible to degradation, often resulting in off-target effects and dose-limiting toxicity [7]. Furthermore, once cells internalize nucleic acid cargoes, they often become trapped in endosomal and lysosomal vesicles, where they can be degraded prior to accessing their active intracellular sites [8]. The principal biological barriers limiting the clinical translation of nucleic acid therapeutics are illustrated in Fig. 1. Together, these complex challenges constitute one of the greatest impediments in gene and nucleic acid therapy.

Viral vectors are predominantly used in early delivery strategies because of their ability to transfect efficiently and enable sustained gene expression to maintain gene expression [9]. However, immunogenicity, insertional mutagenesis, cargo titer capacity, complex manufacturing, and regulatory issues have triggered the search for nonviral alternatives [10]. Owing to the various advantages of nanotechnology, it is considered an effective platform for nucleic acid delivery. Nanocarriers can be

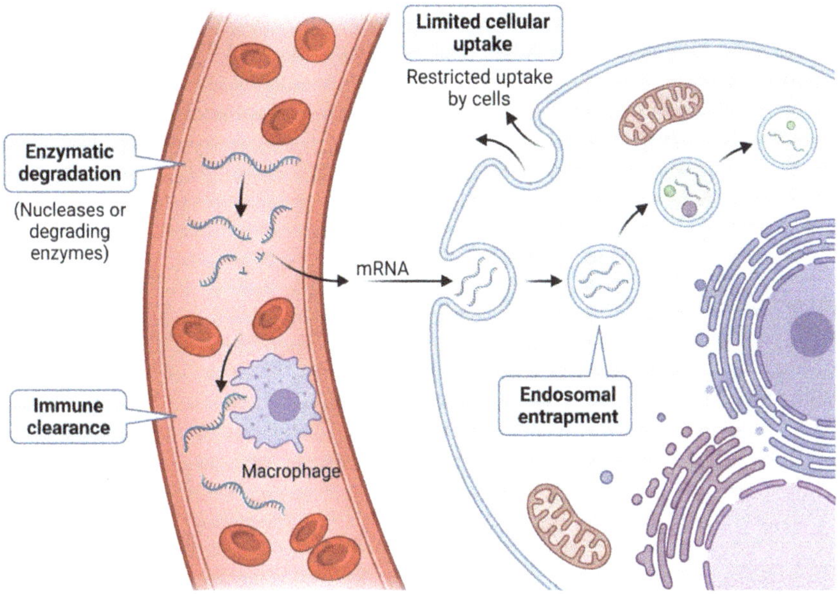

Fig. 1 Schematic representation of the cellular barriers to nucleic acid therapeutics after systemic administration, including enzymatic degradation, immunogenic clearance, restricted uptake by cells and entrapment within endosomes. Created with MindtheGraph (www.mindthegraph.com)

designed to protect nucleic acids against degradation and improve their pharmacokinetics [11]. Furthermore, they can enhance cellular uptake and endosomal escape as well as provide targeted release at the tissue or cellular level [12].

In the broader context of nanomedicine, the development of smart nanotechnology is a transition from passive delivery systems to active, responsive systems that can respond to biological environments [13]. Intelligent nanocarriers are intended to detect or react with particular outside conditions or inside environments, resulting in structural conversion or release of cargos in a regulated way [14]. These stimuli can be endogenous (pH gradients, redox potential, enzymatic activity, or metabolite concentration) or exogenous (light, magnetic fields, ultrasound, and electrical signals). By utilizing these triggers, smart nanocarriers are expected to achieve ideal therapeutic efficacy with low systemic exposure and side effects [15].

The importance of smart nanotechnology has become much more evident with the rapid development of nucleic acid-based applications. RNA interference platforms require effective cytoplasmic delivery to induce sequence-specific gene suppression, whereas the modality of mRNA drugs requires intact cytosolic bioavailability for protein translation [16]. Gene-editing tools such as complexes based on CRISPR–Cas systems also have an additional level of complexity in that separate components—DNA and/or RNA and/or proteins—must be effectively codelivered into the same cells [17]. Smart nanocarriers can be designed to meet such diverse delivery needs via a modular design with a multifunctional architecture [18].

In addition to therapeutic uses, nucleic acid delivery systems are also highly important for molecular diagnostics, regenerative medicine, and personalized patient care. For diagnostics, RNA probes and sensors are used for early detection of diseases as well as real-time monitoring; gene manipulations aid in tissue engineering and cell therapy [19]. The function of the delivery system affects both the clinical feasibility and performance under all these scenarios [20]. Therefore, in nanocarrier design, in addition to delivery efficiency, biocompatibility, scalability, reproducibility, and regulation are needed.

In recent years, considerable progress has been made in materials science, branching off a wide range of possibilities to develop smart nanocarriers. Lipid-based nanoparticles, polymeric particles, inorganic nanomaterials, and hybrids can all be tailored to the exact specifications of size, surface charge, morphology, and functionalization [4]. These characteristics dictate important biointeractions, such as protein adsorption, biodistribution, the mechanism of cellular uptake pathways, and intracellular transport. Importantly, surface functionalization techniques (the presence of targeting ligands, polyethylene glycol, and biomimetic coatings) have facilitated the ability of nanocarriers to escape immune detection and preferentially accumulate at sites where disease is present [21].

Notably, increasing our understanding of cellular biology and intracellular trafficking processes has provided a more rational background for delivery. Understanding the cellular uptake of nanoparticles, endosomal release, and interaction with the intracellular machinery has driven the design of stimuli-responsive and multifunctional systems [22]. For example, pH-labile materials utilize the acidity of the endosome to drive membrane perturbation and cargo release, whereas redox-responsive linkages can capitalize on increased glutathione in cells for cytosolic decomplexation of nucleic acids [23, 24]. Such approaches demonstrate the principles of smart nanotechnology between materials engineering and biological response.

However, even with these significant advances, the translation of smart nanocarriers from the bench to the clinic is still challenging. Challenges related to upscaling production, long-term safety, immunogenicity, and interpatient variability remain obstacles [25]. In addition, regulation guidelines for advanced nanomedicines are far from unified, and further development of reliable characterization and standardized criteria remains the main challenge [26]. Overcoming these challenges will demand interdisciplinary efforts among chemists, engineers, biologists, clinicians, and regulatory scientists.

The purpose of this chapter is to present a systematic and critical review of the use of smart nanotechnology for the delivery of nucleic acids. This chapter starts with a summary of the basic properties and drug delivery design criteria for various classes of nucleic acids, followed by a detailed review of materials and designs for nanocarriers. In the following sections, stimuli-responsive delivery systems, targeting strategies, and cellular uptake mechanisms are described, indicating that intelligent design can help circumvent biological obstacles. Translation hurdles and recently emerged clinical applications are discussed to set the ongoing work from a real-world perspective. Finally, future prospects are highlighted, especially with respect to the growing need to incorporate artificial intelligence, machine learning, and gene-editing technologies during the development of next-generation smart nanocarriers. The aim of this chapter is to be used as a practical reference point for the academic, clinical, and industry sectors that are currently active in the nuclease-based field working at the nanotechnological nucleic acid intersection by consolidating recent progress and highlighting opportunities and challenges.

2 Nucleic Acid Type and Delivery Requirements

Nucleic acid-based therapeutics present a large family of biopolymers with very different structures, sizes, stability properties, intracellular targets, and modes of action [27]. These differences require unique delivery systems and ultimately dictate design criteria for intelligent nanocarriers. To rationally design efficient delivery systems, a complete picture of both the biological roles and physicochemical properties of each class of nucleic acid is needed [28]. This section reviews major classes of therapeutic nucleic acids, including DNA, messenger RNA (mRNA), siRNA, miRNA, and CRISPR/Cas-based gene editing, with a focus on their unique delivery challenges and requirements. An overview of the major classes of therapeutic nucleic acids and their intracellular targets is summarized in Fig. 2.

2.1 DNA-Based Therapeutics

Plasmid DNA is one of the earliest classes and best characterized modalities of nucleic acid-based therapeutics. Genetic approaches for transforming cells with DNA are designed to deliver foreign genetic material into target cells [29]. Therefore, therapeutic proteins such as enzymes, growth factors, or antigens can be expressed over a longer duration. Unlike RNA therapeutics, plasmid DNA must be transported into the nucleus to exert transcriptional activity, which is highly desirable for effective delivery [30]. Plasmid DNA is a large, highly negatively charged and structurally rigid molecule. These characteristics restrict passive diffusion through cell membranes and make DNA highly vulnerable to degradation by intracellular and extracellular nucleases. In addition, DNA delivery involves multiple biological barriers, such as opsonization within the blood stream, uptake by the reticuloendothelial system, inefficient endosomal release, and restricted nuclear entry in quiescent cells [31]. As such, nanocarriers for DNA delivery should serve several critical functions: condensing of the DNA to compact structures to permit cellular uptake, protecting against nucleases/enzymes, encouraging endosomal release, and optimally facilitating nuclear entry. To achieve this, cationic polymers and lipids have been used to complex DNA electrostatically. The issues of cytotoxicity and immunogenic activity have prompted the design of more advanced systems that can disassemble in response to a stimulus and undergo biodegradation [32].

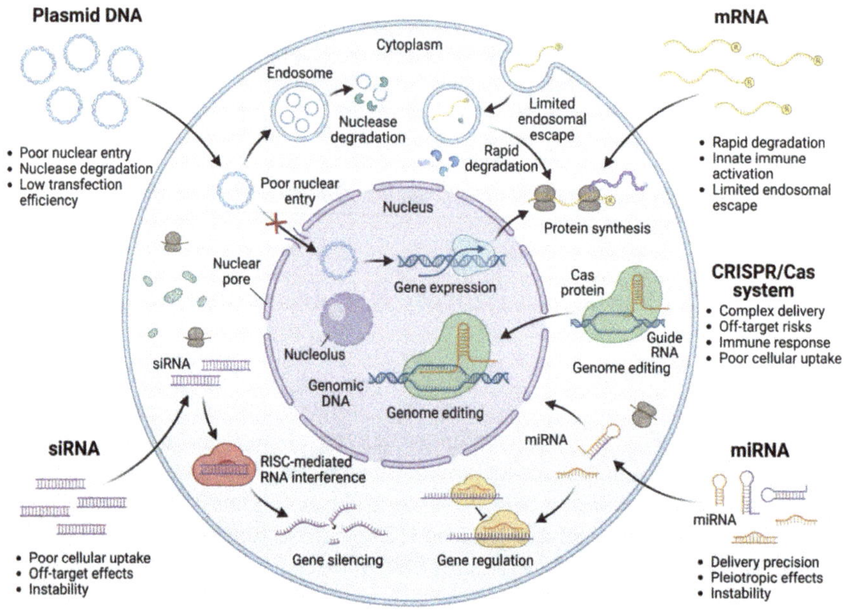

Fig. 2 Overview of the main classes of therapeutic nucleic acids, such as the DNA, mRNA, siRNA, miRNA, and CRISPR/Cas systems, with descriptions of their intracellular targets and some relevant delivery hurdles. Created with MindtheGraph (www.mindthegraph.com)

2.2 *Messenger RNA*

Messenger RNA has recently been highlighted for its successful use in vaccines and protein replacement therapies. Because mRNA works in the cytoplasm and does not need to be taken into the nucleus, it poses no threat of genomic integration and can allow for very fast protein expression [33]. These features make mRNAs attractive vehicles for transient therapeutic use, such as in vaccination, cancer immunotherapy, and regenerative medicine [4]. However, the delivery of mRNA is challenging. It is intrinsically unstable, extremely sensitive to ubiquitous RNases and likely activates the innate immune response by stimulating pattern recognition receptors [34]. Moreover, mRNAs are negatively charged large molecules and have a disabled ability to pass through cell membranes without any assistance [35]. Thus, successful mRNA delivery systems must have strong shieldability in circulation, enable efficient cytoplasmic release, and avoid unintended immune activation. For mRNA delivery, intelligent nanocarriers often involve ionizable compounds that are neutral at physiologic pH to reduce off-target effects and that yield a positive charge that is elastic within slightly acidic environments (for example, endosomes), promoting membrane lysis and release [36]. This is especially true for protein translation efficiency, including controlled release and rapid escape from endosomes [37].

2.3 Small Interfering RNAs

Small interfering RNAs are a class of double-stranded RNA molecules that are 20–25 base pairs in length and play a variety of roles in biology. Upon entry into the cellular cytoplasm, siRNA is incorporated into an RNAi-induced silencing complex where it cleaves complementary messenger RNA (mRNA), resulting in the inhibition of target gene expression [38]. Compared with the delivery of DNA or mRNA, the delivery of siRNAs, which are simple and small molecules, is limited. It is quickly degraded in biological fluids, has low cellular uptake and can cause off-target effects or immunogenic responses if not delivered adequately. In addition, the success of RNAi requires that the drug be well targeted in the cytoplasm so that endosomal escape becomes an issue of critical importance to therapeutic effectiveness [39]. Nanocarriers intended for siRNA delivery need to strike a delicate balance of strong binding and efficient release in the intracellular environment. Overstabilization can impede dissociation and reduce the ability of gene silencing provided by the complex, whereas insufficient stabilization may cause degradation to be premature [40]. Novel nanotechnology approaches, including redox-responsive or enzyme-cleavable linkages, have been designed to address these contradictory requirements and increase the potential for the intracellular-trigger-mediated release of siRNA [41].

2.4 MicroRNA

MicroRNAs are endogenous small noncoding RNAs that interact with target mRNAs, modulating their expression posttranscriptionally. Deregulation of miRNA expression is a feature of many diseases, including cancer, cardiovascular diseases, and neurodegenerative disorders [42]. Therapeutic strategies for miRNAs have generally been divided into two categories: replacement, which involves the introduction of a miRNA mimic, and inhibition, which involves the use of anti-miRNA oligonucleotides to inhibit dysregulated miRNAs [43]. Delivery obstacles to miRNA therapeutics are similar to those of siRNAs, with pleiotropic effects. One single aspect that miRNA-mediated therapies must account for is that one miRNA can regulate many genes, which increases the possibility of noncontrolled and off-target biological effects if its delivery is not appropriately attempted [44]. In addition, miRNA mimics and inhibitors have different chemical compositions and modifications that impact their stability, binding efficiency, and intracellular activity [45]. Therefore, smart nanocarriers for miRNA delivery must pursue precision and targeting, which are generally established with tissue-specific ligands or stimuli-responsive components to restrict activity at the target sites. Controlled release profiles are of particular importance to prevent overmodulation of gene networks and to reach a therapeutically effective but safe level of regulation [46].

2.5 CRISPR/Cas Systems

CRISPR/Cas-mediated gene-editing technologies are widely considered among the most potent tools in modern biomedical research. This technology also allows for the targeted editing of genomic DNA through site-specific cleavage and repair mechanisms. The clinical uses of CRISPR/Cas systems include gene repair, knockout of disease-related genes, and insertion of functional genetic elements [47]. Compared with other nucleic acid systems, CRISPR/Cas systems have greater delivery barrier constraints. Depending on the method of delivery, components can comprise plasmid DNA that encodes the Cas protein and gRNA, mRNA that encodes the Cas protein along with a guide RNA or preassembled ribonucleoprotein complexes. The mode of delivery and the associated sensitizing or desensitizing effects, as well as the power, duration of action and safety, are compared in each case for special advantages and disadvantages [48, 49]. Intelligent nanocarriers engineered for CRISPR/Cas delivery require the delivery of several cargos of different sizes, charges, and stabilities. In addition to efficient cellular internalization and endosomal escape, these systems need to offer such fine-tuned temporal control of component release that off-target editing and immune responses are minimized [50]. In this context, the emergence of stimuli-responsive and multifunctional nanocarriers, which allow for the coordinated administration and triggering of gene-editing tools, has been particularly encouraging [51].

2.6 Comparative Delivery Considerations

Although no single mechanism of action is shared by the DNA, mRNA, siRNA, miRNA, or CRISPR/Cas systems, there are various common delivery obstacles for all nucleic acid modalities. These effects are related to protection from degradation, a lack of nonspecific interactions, cell uptake, and intracellular trafficking. Moreover, each category of nucleic acid has specific demands for subcellular distribution, action time, and biosafety [52]. Smart nanoscale technology provides a versatile platform to meet these requirements through modular design and responsive properties. Nanocarriers can be designed to carry specific nucleic acid cargos and achieve therapeutic goals on the basis of the composition of materials, surface functionality, and stimulus responsiveness [53]. The specific biological and physicochemical behavior of each nucleic acid form is well understood. Thus, the intelligent design of next-generation delivery systems is a basic requirement. A comparative overview of the major classes of therapeutic nucleic acids and their delivery requirements is summarized in Table 1.

In the next section, the materials and design principles involved in the delivery of smart nanocarriers to nucleic acids are reviewed. This section highlights how some of these designs have benefited from advances in the field of nanotechnology because of the increasing complexity and efficacy of therapeutic strategies.

Table 1 Comparison of the major classes of therapeutic nucleic acids and their delivery requirements

Nucleic acid type	Primary mechanism of action	Intracellular target	Key advantages	Major delivery challenges	Critical delivery requirements	References
Plasmid DNA	Transcription of therapeutic genes into mRNA	Nucleus	Long-term gene expression; suitable for gene replacement	Large size; nuclease degradation; poor nuclear entry; risk of low transfection efficiency	DNA condensation; protection from nucleases; efficient endosomal escape; nuclear localization	[54]
mRNA	Translation into therapeutic proteins	Cytoplasm	No risk of genomic integration; rapid protein expression	High instability; RNase degradation; innate immune activation	Strong extracellular protection; efficient cytoplasmic release; minimized immunogenicity	[55]
siRNA	RNA interference-mediated gene silencing	Cytoplasm	High specificity; transient and reversible effects	Rapid degradation; poor cellular uptake; off-target effects	Stable complexation; efficient endosomal escape; controlled intracellular release	[56]
miRNA (mimics/ inhibitors)	Post-transcriptional gene regulation	Cytoplasm	Ability to modulate gene networks; disease-relevant targets	Pleiotropic effects; delivery precision needed; instability	Targeted delivery; dose control; tissue specificity	[57]
CRISPR/ Cas systems	Genome editing via site-specific DNA cleavage	Nucleus (DNA editing) or cytoplasm (RNP assembly)	Permanent gene modification; high versatility	Multicomponent delivery; off-target editing; immune response	Coordinated delivery of components; temporal control; precise intracellular trafficking	[58]

3 Smart Nanocarriers: Design and Materials

The performance of delivery vehicles is critical for accessing nucleic acid-based therapeutics. Smart nanocarriers are advanced nonviral delivery systems specifically designed to shield nucleic acids, travel through complicated physiological barriers, and be safely activated to unload the cargo in a tempered, targeted response. The development of these nanocarriers integrates advances in materials science, molecular biology, and biomedical engineering. Important considerations are biocompatibility, biodegradability, efficacy of cargo loading, retention in physiological environments, cellular uptake, and endosomal escape. This section outlines the main types of smart nanocarriers—lipid-based, polymeric, inorganic, and hybrid systems—through highlighting their design secrets as well as advantages/disadvantages in nucleic acid delivery.

3.1 Design Principles of Smart Nanocarriers

Compared with passive nanoparticles, smart nanocarriers have an active response in that they are dynamic in response to internal biological signals or extrinsic stimuli. Central to these designs, they address multiple delivery barriers in a tandem manner. First, the nanocarrier is required to protect nucleic acids from degradation by enzymes and nonspecific binding during bloodstream circulation. It should then accumulate at the target site, promote cellular internalization, escape from endosomal compartments, and then release nucleic acid cargo unless it is targeted to an ideal intracellular location [59].

Biological behaviors are largely determined by physicochemical properties, including size, surface charge, shape, and surface chemistry. In general, nanoparticles in the size range of ~20–200 nm are desired for systemic delivery, as they have a long circulation time and good tissue permeation [60, 61]. The surface charge needs to be carefully tailored, as cationic surfaces can promote nucleic acid binding and cellular interactions, but too much positive charge can increase cytotoxicity and reduce immuno-evasion. Smart nanocarriers typically use charge-shifting or ionizable materials whose charge changes are dependent on the local pH to improve their performance in different biological environments [4]. The key design principles underlying smart nanocarriers for nucleic acid delivery are schematically presented in Fig. 3. The key material classes and representative examples of smart nanocarriers used for nucleic acid delivery are summarized in Table 2.

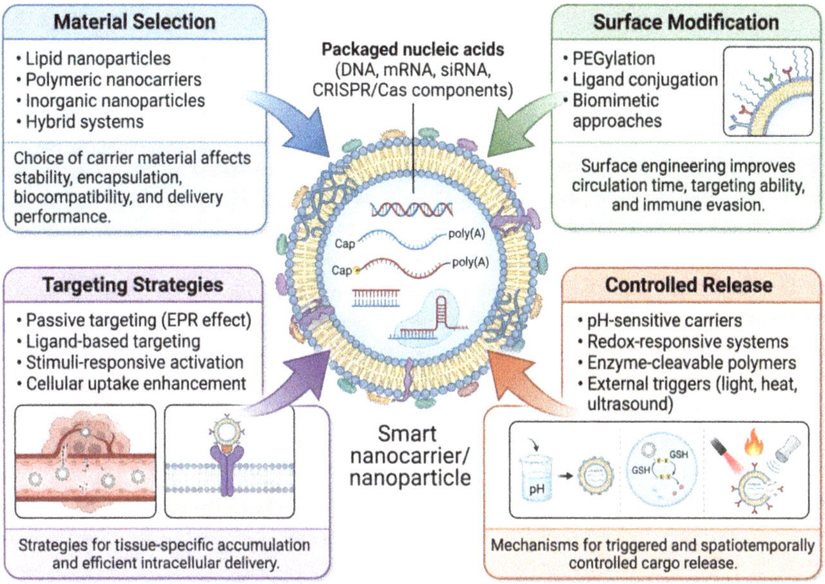

Fig. 3 Design principles of smart nanocarriers for nucleic acid delivery, emphasizing material selection, surface modification, targeting strategies, and controlled release mechanisms. Created with MindtheGraph (www.mindthegraph.com)

3.2 Lipid-Based Nanocarriers

Lipid-based nanocarriers (LNs) represent among the most advanced nucleic acid carriers in clinical development. These systems generally consist of lipid nanoparticles (LNPs), liposomes, or lipid–nucleic acid complexes via electrostatic interactions between cationic or ionizable lipids and negatively charged nucleic acids. Their similarity to the natural structure of biological membranes enables efficient cellular internalization and membrane fusion, which is highly suited for cytoplasmic delivery [4]. Ionizable lipid nanoparticles have become a platform technology for mRNA and siRNA delivery. The lipids remain nearly uncharged at physiological pH, making them relatively nontoxic in vivo but becoming cationic under acidic conditions (experienced during nanoparticle formulation or within endosomes). This pH-responsive property facilitates the formation of pores in the membrane and endosomal release. In addition to ionizable lipids, LNPs typically include helper lipids, cholesterol for structural support, and polyethylene glycol-lipids for enhanced colloidal stability and circulation half-life [66]. Although lipid-based systems offer attractive benefits, lipid-based delivery systems have several limitations, including immunogenicity, nonspecific organ localization (particularly in the liver), and less flexibility in accommodating specific nucleic acid payloads. Current studies are aimed at fine-tuning lipid structures, developing biodegradable lipids, and including targeting

Table 2 Representative materials used in smart nanocarriers for nucleic acid delivery

Nanocarrier material class	Representative materials	Key advantages	Major limitations	Typical applications	References
Lipid-based systems	Ionizable lipids, cationic lipids, lipid nanoparticles, liposomes	High transfection efficiency; biocompatibility; efficient endosomal escape	Potential immunogenicity; liver accumulation; formulation sensitivity	mRNA vaccines, siRNA therapeutics	[62]
Polymeric systems	Polyethylenimine, PLGA, chitosan, block copolymers	Structural tunability; multifunctionality; controlled release	Cytotoxicity (cationic polymers); batch variability	DNA, siRNA, miRNA, CRISPR delivery	[63]
Inorganic systems	Gold nanoparticles, silica nanoparticles, magnetic nanoparticles	Unique optical/magnetic properties; theranostic capability	Limited biodegradability; long-term accumulation	Imaging-guided delivery; externally triggered release	[64]
Hybrid systems	Lipid–polymer hybrids, polymer–inorganic composites	Combined advantages of multiple materials; enhanced stability	Increased complexity; manufacturing challenges	Combination therapies; gene editing	[65]

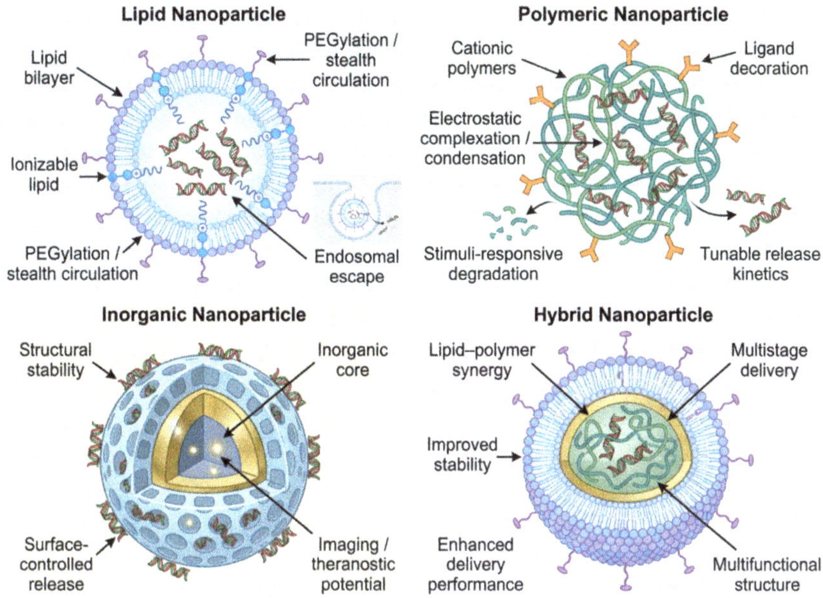

Fig. 4 Representative structures of lipid-based, polymeric, inorganic, and hybrid nanocarriers used for nucleic acid delivery. Created with MindtheGraph (www.mindthegraph.com)

ligands to increase their potential [4, 60]. Representative architectures of lipid-based, polymeric, inorganic, and hybrid nanocarriers are shown in Fig. 4.

Previous experimental research by Maeki et al. [67] demonstrated how the creative design of a nanocarrier can overcome the intrinsic barriers in plasmid DNA delivery. In their study, they synthesized polymer–lipid hybrid nanoparticles that were specifically designed for efficient delivery of large plasmid DNA (15 kbp) by encapsulating a positively charged pDNA–polyethylenimine core inside lipid nanoparticles. In the optimized formulation, a mean particle size of approximately 35 nm with a slightly positive zeta potential (approximately + 10 mV) was observed compared with conventional pDNA-loaded lipid nanoparticles of approximately 75 nm and an almost neutral charge. *An* in vitro transfection assay with HeLa cells demonstrated that the optimized hybrid nanoparticles mediated 40% of the gene expression of the 15 kbp pDNA, and this value was greater than that of commercial lipid nanoparticles encapsulating negatively charged pDNA (eightfold). Interestingly, in these experiments, the cell viability was still equivalent to that of the control groups (> 85–90%). These results demonstrate that enhanced gene expression was achieved without compromising cell viability. The transmission electron microscopy and small-angle X-ray scattering results revealed that the high transfection efficiency was related to the reduced size of the nanoparticles and mostly their unilamellar interior, which was beneficial for achieving enhanced cellular uptake and endosomal escape of large plasmid DNA.

3.3 Polymeric Nanocarriers

Polymeric nanocarriers offer unprecedented flexibility with respect to chemical composition and structural design. Both natural and synthetic polymers, such as polyethylenimine, poly(lactic-co-glycolic acid), chitosan, dextran, and different block copolymers, have been studied as nucleic acid delivery carriers. These can be designed with specific molecular weights, functional groups, or architectures for designed delivery properties [68]. Cationic polymers have been used to condense nucleic acids into nanometer-sized complexes through electrostatic interactions. Nevertheless, a high cationic charge density is frequently associated with cytotoxicity and inflammation. To overcome this obstacle, intelligent polymeric systems feature the use of biodegradable backbones, cleavable linkages, or stimuli-responsive groups that reduce toxicity after internalization into cells [69]. For example, redox-active polymers with disulfide cross-links are stable in the extracellular space and degrade rapidly in the intracellular reductive environment for nucleic acid delivery. Polymeric micelles/nanoparticles may also be programmed to serve as multifunctional platforms containing targeting ligands, imaging agents, or therapeutic cargos. The modular design of polymeric nanocarriers makes them particularly attractive for more sophisticated delivery applications, such as combination therapies or CRISPR/Cas systems, although issues, notably scalability and batch-to-batch reproducibility, still remain [70].

3.4 Inorganic Nanocarriers

Inorganic nanocarriers are composed of multiple types, such as gold nanoparticles, silica nanoparticles, magnetic nanoparticles, and quantum dots. These systems possess novel physicochemical properties that are difficult to obtain with organic materials only, e.g., tunable optical, magnetic, or electronic properties. Therefore, inorganic nanocarriers are commonly studied for theranostic purposes that combine the delivery of nucleic acids with imaging or external stimuli-triggered release [71]. These nanomaterials include gold nanoparticles, which are readily functionalized via thiol chemistry and possess excellent biocompatibility and surface tunability properties. The photothermal features of these materials can achieve light-induced release or endosomal escape under irradiation. Magnetic nanoparticles are capable of remote magnetic guidance and magnetic field targeting, which allows site-specific delivery. The choice is mesoporous silica nanoparticles, which provide the excellent surface area and pore volume available for nucleic acid loading as well as the ability to modify pore gates sensitive to user-defined stimulants [72]. However, long-term biocompatibility, biodegradability, and tissue retention remain challenges for the clinical translation of many inorganic carriers. The development of novel design strategies particularly highlights surface modifications, biodegradable coatings, and organic–inorganic hybrid materials to seek solutions for these problems [73].

3.5 Hybrid Nanocarrier Systems

Hybrid nanovectors combine the synergistic benefits of distinct material classes in one tool. By combining lipid, polymeric, and inorganic components, hybrid systems aim to offer improved stability, delivery efficiency, and multifunctionality [74, 75]. For example, lipid–polymer hybrid nanoparticles are often designed with a polymeric core for mechanical support and cargo encapsulation and are surrounded by a lipid corona that enhances biocompatibility and interaction with cells [76]. Hybrid systems are particularly useful for complex delivery scenarios, e.g., co-delivery of nucleic acids and small-molecule drugs or proteins or concurrent delivery of more than one CRISPR/Cas component. The modular design of hybrid nanocarriers provides the ability to tailor each functional layer finely to address its own challenges for delivery [77]. However, the added structural complexity may significantly intensify the manufacturing, characterization, and regulatory assessment.

3.6 Materials Selection and Safety Considerations

The choice of materials is a critical factor that significantly affects the performance and translational potential of a nanocarrier. In addition to their efficient delivery properties, smart nanocarriers should have acceptable safety profiles with low cytotoxicity, weak immunogenicity, and predictable biodegradation behaviors. From a regulatory perspective, particular emphasis is placed on comprehensive physicochemical characterization and standardized biological testing to ensure comparability, reproducibility, and safety for large-scale and clinical applications. The development of materials science has broadened the repertoire of smart nanocarriers for nucleic acid delivery. Through the application of rational design principles and a more comprehensive understanding of the properties governing biological interactions, increasingly complex systems are being fabricated that can respond to ever finer or more stringent demands for modern nucleic acid therapeutics. How these materials are modified to yield even more control over nucleic acid release and activity via stimuli-responsive processes is then discussed in the subsequent section.

4 Stimuli-Responsive Delivery Systems

Stimuli-responsive delivery systems play an important role in the field of smart nanotechnology for nucleic acid delivery. Unlike passive nanocarriers, which generally rely on either an enhanced permeability and retention effect or nonselective cellular uptake to target pathological sites for site-selective drug release, responsive nanocarriers are carefully designed to have specific changes in their physicochemical properties and structure in response specifically to selected triggers [78]. These fast advances enable the controlled, site-specific, and temporally chorus release of nucleic acids, leading to enhanced therapeutic effectiveness while avoiding off-target effects and systemic toxicity. The stimuli employed for promoting this behavior can be divided into two general categories: endogenous, in which the activity is elicited from within the biological environment, and exogenous, where it is extrinsically applied [79].

4.1 Endogenous Stimuli-Responsive Systems

Stimuli-activated endogenous nanocarriers take advantage of the natural differences between healthy and diseased tissues or between the extracellular and intracellular environments. These systems also have the advantage of not requiring exogenous intervention to induce their activation and are therefore highly attractive for systemic delivery [80]. The mechanisms of endogenous stimuli-responsive nucleic acid release, including pH-, redox-, and enzyme-triggered pathways, are depicted in Fig. 5.

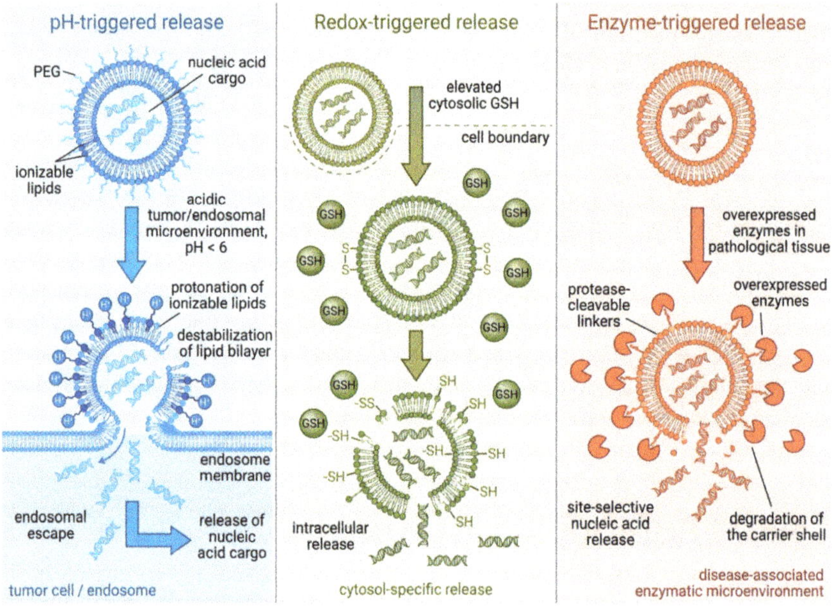

Fig. 5 Mechanisms of endogenous stimuli-responsive nucleic acid release, including pH-, redox-, and enzyme-triggered activation within pathological or intracellular microenvironments. Created with MindtheGraph (www.mindthegraph.com). *GSH* glutathione (γ-glutamylcysteinyl glycine); *PEG* polyethylene glycol; *-SH* sulfhydryl; *-SS* disulfide bond

4.1.1 pH-Responsive Systems

The pH gradient is one of the most widely used endogenous triggers for the specific design of nanocarriers. The extracellular pH in tumor tissues and inflammatory sites is usually slightly acidic relative to that in normal tissues, whereas endosomes and lysosomes have much lower pH values. pH-responsive nanocarriers are stable at physiological pH and swell or degrade in response to acidic conditions as a result of protonation [81]. In nucleic acid delivery, pH-sensitive materials are often used to facilitate endosomal release. Ionizing lipids and polymers become positively charged at acidic pH values, leading to electrostatic interactions with endosome membranes and membrane destabilization. This provides a means for the release of nucleic acids into the cytoplasm before lysosomal degradation commences [80, 82]. In addition, acid-cleavable linkages (e.g., hydrazone or acetal bonds) can be introduced into the structure of nanocarriers to cause preferential release of cargos in acidic environments [15].

4.1.2 Redox-Responsive Systems

Redox-responsive nanocarriers take advantage of the large difference in redox potential values between extracellular fluids and the intracellular environment. Intracellular spaces, especially the cytosol, contain a high abundance of reducing agents (e.g., glutathione). Redox-sensitive materials are usually formed through disulfide bonds, which remain inert in the bloodstream but deteriorate quickly following cellular internalization [83, 84]. With respect to nucleic acid delivery, redox-responsive systems effectively increase the cytosolic release of nucleic acid–nanocarrier complexes. This particularly benefits siRNA, miRNA, and CRISPR/Cas delivery, where fast release within the cytoplasm or nucleus is essential for therapeutic efficacy. Redox-sensitive nanoparticles used to deliver siRNAs for the suppression of oxidatively stressed or age-associated osteoblastogenesis redox-triggered degradation of polymeric nanocarriers combine efficient intracellular release with improved delivery efficacy and reduced off-target interactions [85, 86].

4.1.3 Enzyme-Responsive Systems

Enzyme-sensitive nanocarriers are smart enough to be responsible for disease-related enzymes that are overexpressed in certain tissues, such as tumors or inflamed areas. Enzymes that are often the target of such a strategy include proteases, esterases, and matrix metalloproteinases. Nanocarriers with enzyme-cleavable linkers or polymer coatings can be degraded or undergo structural changes upon exposure to such enzymes, which leads to the release of nucleic acids [87]. This method provides increased specificity because the expression of enzymes can differ widely among tissues. In the delivery of nucleic acids, enzyme-sensitive systems have been frequently integrated with targeting ligands for better selectivity [88]. However, the

Table 3 Endogenous and exogenous stimuli employed in stimuli-responsive nanocarrier systems

Stimulus type	Specific stimulus	Triggering environment	Activation mechanism	Delivery advantage	References
Endogenous	pH	Tumor microenvironment; endosomes	Protonation; acid-labile bond cleavage	Enhanced endosomal escape	[81]
	Redox potential	Cytosol (high glutathione)	Disulfide bond cleavage	Intracellular nucleic acid release	[83]
	Enzymes	Disease-specific tissues	Enzyme-mediated degradation	Tissue-specific activation	[87]
Exogenous	Light	Targeted irradiation sites	Photothermal or photochemical effects	Spatial and temporal control	[89]
	Magnetic field	External application	Magnetic guidance or heating	Noninvasive targeting	[90]
	Ultrasound	Localized acoustic exposure	Cavitation or membrane disruption	Deep tissue penetration	

differences in enzyme expression between patients need to be addressed through careful design and patient selection. The endogenous and exogenous stimuli exploited in stimuli-responsive nanocarrier systems, along with their activation mechanisms, are summarized in Table 3.

4.2 Exogenous Stimulus-Responsive Systems

Exogenous stimuli-responsive nanocarriers can release nucleic acids in response to externally exerted triggers with high spatial and temporal resolution. Such systems provide clinicians with the ability to finely tune therapeutic response at will, offering benefits in applications requiring localized or repeated activation. Exogenous stimuli-responsive delivery strategies activated by light, magnetic fields, or ultrasound are illustrated in Fig. 6.

4.2.1 Light-Responsive Systems

Photoresponsive nanoparticles employ photoactive substances that undergo changes in their structure or chemistry under the action of certain wavelengths. Near-infrared light is especially preferable, as it can penetrate tissue to a greater extent with less phototoxicity. Photoresponsive systems offer the opportunity for controlled release of nucleic acids at a specific site and not into systemic circulation, thus preventing

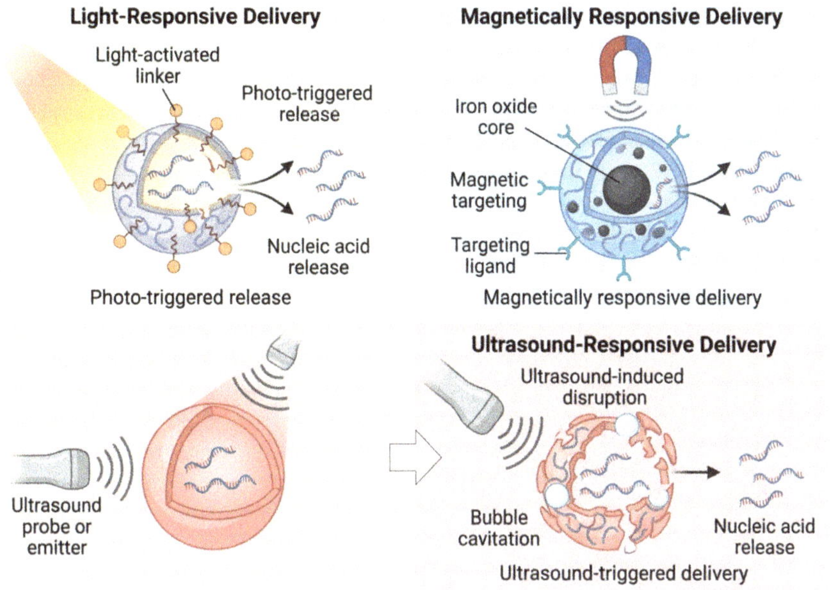

Fig. 6 Exogenous stimulus-responsive nanocarriers for the controlled release of nucleic acids, with light-, magnetic-, or ultrasonic-triggered activation for on-demand spatial and temporal control. Created with MindtheGraph (www.mindthegraph.com)

off-site effects [29]. Photothermal or photochemical effects are utilized to disrupt nanocarrier systems and/or endosomal membranes to facilitate nucleic acid release. Light-responsive modalities have great potential for treating surface tumors or tissues that can be accessed by light through optical fibers, but limited tissue penetration still limits deeper targets [91].

4.2.2 Magnetic Field-Responsive Systems

Magnetic field-responsive nanocarriers are usually based on magnetic particles that are moved (or triggered) by an external magnetic field. This approach can provide focused accumulation at specific anatomical sites of nucleic acid-loaded nanocarriers, with on-command magnetothermal activation for release [92, 93]. With magnetic targeting noninvasively, repeatable control over delivery may be possible, but it is equipment-dependent and requires careful tuning of equipment to achieve sufficient field strength and fidelity [94]. The safety concerns of high-temperature and long-term accumulation of magnetic materials are also issues that need to be solved [90].

4.2.3 Ultrasound-Responsive Systems

Acoustic stimuli containing ultrasound can be harnessed to trigger mechanical or thermal effects to improve nucleic acid loading. Ultrasound can increase vascular permeability, stimulate cellular uptake, and induce payload release via cavitation or nanoparticle abrasion. This strategy is very attractive given the noninvasive nature of light, good deep penetration in tissues, and wide applicability. Ultrasound-triggered delivery has demonstrated potential in cases of gene therapy for cardiovascular and neurological diseases; however, detailed control of exposure is needed to prevent tissue damage [95].

For example, Li and colleagues introduced an innovative ultrasonication-responsive folic acid (FA)-conjugated nanobubble system established for the effective and secure administration of nucleic acids. To the best of the authors' knowledge, the coupling of a targeted nanocarrier with low-intensity focused ultrasound (LIFU) is a major step forward. The SIRPα siRNA- and Fe_3O_4 nanoparticle-encapsulating nanobubbles also showed remarkable encapsulation, with an encapsulation capacity of approximately 73.3%, and presented outstanding biocompatibility, with over 85% cell survival at a concentration of 200 μg/mL. Upon exposure to LIFU at the modified position of 1 W/cm for 2 min and a 50% duty cycle, the perfluoropentane nanobubbles undergo acoustic droplet vaporization, causing cavitation and sonoporation, promoting cellular uptake, and furthering endosomal breaks. Under these circumstances, the siRNA delivery performance in M2-like macrophages increased from approximately 49.8% in the absence of ultrasound to 75.8% with LIFU, and with the FA-targeted nanobubbles, it increased to more than 95%, surpassing that of Lipofectamine. To the best of our knowledge, the ultrasonography-formed group increased the intracellular transcriptional efficiency of siRNAs in all three systems. The LIFU-enabled system resulted in substantial SIRPα downregulation and total M1 macrophage polarization and increased antitumor immune actions, generating unwavering responses to the developing tumor and presenting a decent restraining influence in vivo without systemic toxicity, highlighting the ability of ultrasonication-responsive systems to safely dispense nucleic acids [96].

4.3 Multi-stimuli and Adaptive Systems

To improve sensitivity and stability, advanced smart nanocarriers usually use various stimuli-responsive moieties. Multiple stimulus couples can act sequentially and synergistically with different switches, such as pH and redox status, providing precise control of nucleic acid release. Adaptive systems could also include feedback loops that modify the carrier's behavior to follow changes in the biological environment. The integration of multiple responsive features raises complexity in design, yet it also offers great advantages in addressing diverse and dynamic biological barriers.

These systems would be of particular importance for the delivery of more sophisticated cargo, for example, CRISPR/Cas components, where controlled activation is as important for safety as it is for efficacy.

4.4 Advantages and Limitations

Stimuli-responsive delivery systems offer significant benefits, such as targeted drug release, low toxicity, and improved therapeutic efficiency. However, their resolution depends on tailored responsiveness and a good understanding of their biological variability. Challenges include the tissue specificity of off-target activation, differences in stimulus intensity across tissues, and additional manufacturing complexity. However, despite these barriers, stimuli-responsive nanocarriers represent an important breakthrough in the field of nucleic acid delivery. Through controlled and intelligent release mechanisms, they reconcile material design with biological function. In the following section, the targeting strategy and cellular uptake of the smart nanocarrier-delivery system are discussed to ensure the specificity and efficacy of lead nucleic acid delivery.

5 Delivery Strategies and Cellular Uptake Mechanisms

The efficient delivery of nucleic acid-based therapeutics requires more than just protecting and releasing the cargo in a controlled manner. Moreover, proper targeting to cells or tissues, as well as successful internalization into target cells, is needed. Thus, targeting approaches and cellular internalization mechanisms are among the key factors for therapy relevance and safety. Intelligent nanocarriers combine physicochemical constructions with biological targeting designs to improve the selective accumulation, receptor-mediated uptake, and intracellular release of nucleic acids. This section reviews the main targeting strategies and cellular uptake pathways associated with smart nanotechnology-facilitated nucleic acid delivery. The targeting strategies, cellular uptake pathways, and endosomal escape mechanisms of smart nanocarriers are summarized in Fig. 7. The major targeting strategies and cellular uptake pathways relevant to smart nanocarrier-mediated nucleic acid delivery are outlined in Table 4.

5.1 Passive Targeting

Passive targeting is based on the physiological and pathological properties of tissues where nanoparticle accumulation increases without requiring specific targeting ligands. One typical passive targeting mechanism involves enhanced permeability

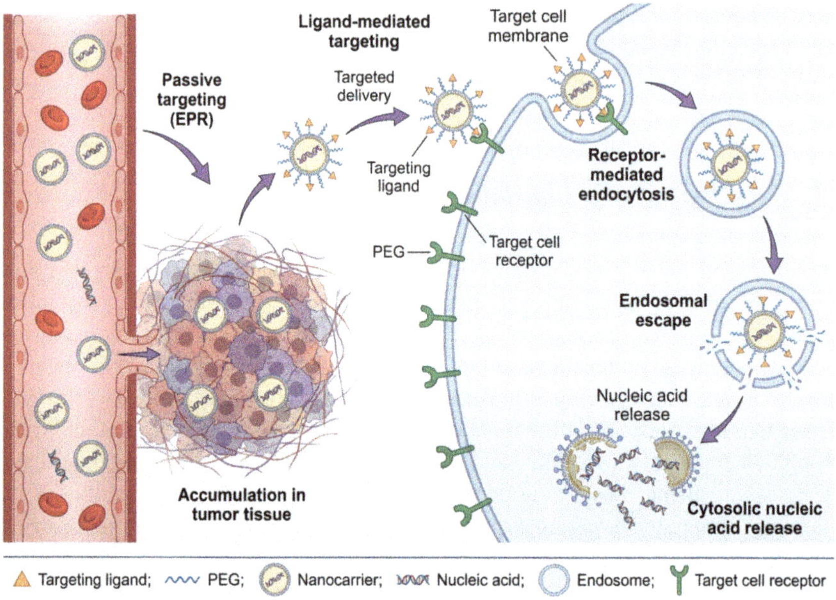

Fig. 7 Targeting strategies and cellular uptake pathways of smart nanocarriers, including ligand-mediated targeting, receptor-mediated endocytosis, and endosomal escape processes. Created with MindtheGraph (www.mindthegraph.com)

Table 4 Targeting strategies and cellular uptake pathways of smart nanocarriers

Targeting strategy	Ligand or feature	Target receptor or mechanism	Uptake pathway	Key benefit	References
Passive targeting	Optimized size and surface charge	Enhanced permeability and retention	Nonspecific endocytosis	Increased accumulation in diseased tissue	[12]
Active targeting	Peptides, antibodies, aptamers, small molecules	Cell-specific surface receptors	Receptor-mediated endocytosis	Improved specificity and uptake	[97]
Biomimetic targeting	Cell membrane coatings	Immune evasion; homotypic binding	Natural cellular interaction	Prolonged circulation	[98]
Endosomal escape strategies	pH-responsive or membrane-disruptive materials	Endosomal membrane destabilization	Cytosolic release	Enhanced biological activity	[99]

and retention, which is due to the leaky vasculature and impaired lymphatic drainage that are common in tumors and inflamed tissues [100]. Nanocarriers of the right size are able to accumulate preferentially in dim areas, leading to higher local concentrations of nucleic acids [12]. Despite the success of passive targeting in preclinical models, its efficacy may differ substantially among patients and disease types. Vascular heterogeneity, interstitial pressure, and disease stage affect nanoparticle distribution. Without additional assistance, passive targeting is generally not enough to achieve accurate delivery of nucleic acids in either a non-oncological context; therefore, active targeting methods are needed [6].

5.2 Active Targeting by Ligand Functionalization

Active targeting is achieved by modifying the surface of nanocarriers with ligands that are selectively bound to the target cell. These ligands could include small molecules, peptides, antibodies, antibody fragments, aptamers, or carbohydrates. The ligands and their native receptors promote discrimination, binding to/enhancing receptor-mediated endocytosis of nucleic acids, which are involved in cellular uptake/intracellular release [12]. The predominant receptors that have been targeted are transferrin, folate, integrins, epidermal growth factor, and different types of cell type-specific surface markers. For nucleic acid delivery, however, proper selection of ligand density, orientation, and affinity is important to achieve the right trade-off between targeting efficacy and blood circulation [101]. The overpresentation of ligands can exceed immunogenicity levels or lead to rapid first-pass clearance; conversely, the lack of targeting may not facilitate uptake [97].

5.3 Biomimetic and Cell Membrane-Coated Systems

Recently, interest in biomimetic targeting strategies to improve delivery specificity and overcome immune recognition has increased. These strategies include the surface functionalization of nanocarriers by utilizing biological membranes from cells, such as red blood cells, platelets, immune cells, or cancer cells. The resulting membrane-camouflaged nanoparticles retain surface proteins and antigens that are advantageous for immune evasion and homology targeting [98]. By mimicking biomicroneeds for the delivery of nucleic acids, active agents also have advantages in long-term circulation and target-tissue targeting [14]. For example, nanoparticles coated with cancer cell membranes can use homotypic binding to increase the target efficiency of tumors [102]. However, membrane sourcing, scalability, and batch heterogeneity issues need to be overcome before these platforms can support clinical implementation on a large scale [103].

5.4 Cellular Uptake Pathways

After accumulation in target tissues, cellular internalization is predominantly attributed to endocytic processes. The dominant uptake mode depends on numerous factors, such as nanoparticle size, shape, charge, chemistry, or cell type. Clathrin-mediated endocytosis, caveolae-mediated endocytosis, macropinocytosis, and phagocytosis are the major endocytic pathways [12, 104]. Receptor-mediated endocytosis is more than relevant for active targeted nanocarriers since ligand–receptor binding triggers the process of internalization. Caveolae-mediated routes may have advantages for nucleic acid delivery owing to lysosomal bypass; clathrin-mediated routes often lead to endosomal transit [105]. Nanosystem design informed by such pathways facilitates the development of nanocarriers that target desired intracellular paths [106].

5.5 Endosomal Escape and Intracellular Trafficking

Endosomal release is a fundamental problem for efficient nucleic acid transfer. After being internalized, nanocarriers are commonly trapped in endosome delivery systems, which evolve into lysosomes. In these compartments, the acidic conditions and activity of degradative enzymes are highly damaging to the integrity of nucleic acids. In the context of these limitations, more advanced nanocarriers use different tactics to break down the endosomal membrane and facilitate the release of their payload in the cytoplasm [107]. Endosomal escape can be modulated by the proton sponge effect, membrane fusion, pore formation, and disruption induced by photothermal or ultrasound stimuli. The efficiency of such escape mechanisms is central because it has direct implications for the biological in vivo activity of nucleic acids, particularly therapeutic agents such as siRNAs, miRNAs, and mRNAs, which exert their effects inside the cytoplasm [108]. In addition, with the use of gene-editing tools such as DNA and CRISPR/Cas systems, subsequent nuclear shuttling is an additional layer of complexity that needs to be addressed before therapeutic success can be achieved [99, 109].

5.6 Subcellular Targeting Considerations

In addition to cellular uptake, efficient nucleic acid delivery sometimes also requires specific subcellular targeting. For DNA-based therapeutics and some components of the gene-editing system, access to the nucleus is needed, with RNA-based therapeutics remaining in the cytoplasm. Intelligent nanocarriers with a nuclear localization signal, pH-sensitive material, or sequential release should be designed for the targeted intracellular location. The targeting moieties and stimuli-responsive features incorporated within smart nanocarriers allow better navigation through the intracellular space. By regulating tissue targeting, cellular uptake, endosomal release, and subcellular trafficking, these systems optimize therapeutic efficacy and avoid side effects [81].

In conclusion, strategies for targeting and mechanisms of cellular entry are critical determinants in achieving successful smart nanotechnology-mediated delivery of nucleic acids. Progress in ligand design, biomimicry, and a mechanistic description of intracellular transport will refine these strategies. The next section addresses translational hurdles and clinical potential, emphasizing how these advanced delivery systems behave in actual therapeutic settings.

6 Translational Challenges and Clinical Applications

Although great progress has been made in the development of intelligent nanocarriers for nucleic acid delivery, enormous challenges remain during the process from the laboratory to the clinic. Preclinical studies have recapitulated efficacy across many platforms, but few of the thousands identified have progressed to clinical trials or become approved drugs [110]. This discrepancy highlights the complexity of biological systems, the tough safety and production standards naive to clinical development, and the dynamic regulatory framework surrounding nanomedicine [111]. This section will focus on the central translation challenges and mention some particular clinical directions of smart-nanoenabled nucleic acid therapeutics. Key translational challenges and considerations for the clinical application of smart nanotechnology-based nucleic acid delivery systems are summarized in Table 5.

6.1 In Vivo Biological Barriers

After entering systemic circulation, nanocarriers encounter multiple biological barriers that can severely limit delivery efficiency. When in the bloodstream, nanoparticles are rapidly coated with plasma proteins, which stick to their surface, creating a so-called protein corona. This modification of their physicochemical characteristics may substantially change this biological behavior, in most cases resulting in recognition and elimination by the mononuclear phagocyte system, especially that of the liver and spleen. As a result, fewer nucleic acids enter the target tissue [118]. Vascular obstacles also complicate delivery because the vascular covering and permeability of endothelial cells are quite different in various tissues and disease conditions. Even when nucleic acid-loaded nanocarriers can extravasate into a targeted tissue, dense extracellular matrices and increased interstitial fluid hamper penetration deep within

Table 5 Translational considerations for smart nanotechnology-based nucleic acid delivery systems

Translational aspect	Key issues	Impact on clinical development	Mitigation strategies	References
Safety and toxicity	Cytotoxicity; material accumulation	Dose limitation; adverse effects	Biodegradable materials; dose optimization	[112]
Immunogenicity	Innate immune activation; hypersensitivity	Reduced efficacy; patient variability	Surface modification; nucleic acid modification	[113]
Manufacturing and scalability	Complex formulations; reproducibility	Delayed translation; high cost	Microfluidic production; quality-by-design	[114]
Regulatory challenges	Lack of standardization; complex characterization	Approval delays	Early regulatory engagement	[115, 116]
Clinical efficacy	Biological variability; delivery efficiency	Variable patient response	Personalized delivery strategies	[117]

the tissue [97]. Intelligent micro/nanocarriers capable of responding to local microenvironmental cues may help mitigate some of these problems; however, intrinsic patient and disease heterogeneity are still one of the major difficulties [103].

6.2 *Toxicity and Safety Aspects*

Safety is also a chief challenge for the clinical translation of nanocarrier-mediated nucleic acid delivery systems. Possible toxicities could be related to the components of the nanocarrier, their degradation products, or secondary biological effects. Cationic lipids and polymers, although suitable for the complexation of nucleic acids, are associated with membrane lysis, inflammation, and organ toxicity at relatively high doses [112]. Even greater importance is given to long-term safety concerns, especially for chronic or repeated dosing schedules. The presence of nonbiodegradable materials in tissues may lead to long-term effects that the short duration of a study has not shown. In this context, increasing attention is being given to the design of biodegradable elements in smart nanocarriers to degrade into nontoxic metabolites [119]. For regulatory submission, a thorough toxicological assessment, including assessment of immunogenicity, genotoxicity, and reproductive toxicity, is needed.

6.3 Immune Response and Immunogenicity

Nucleic acid therapeutics and delivery systems can trigger innate and adaptive immunity. Nonmodified RNAs can be recognized by pattern recognition receptors, thus initiating an inflammation cascade [120]. In the same way, on the one hand, nanoparticle surfaces can interact with immune cells with subsequent cytokine release or hypersensitivity reactions [116, 121]. The strategies to overcome immunogenicity are chemical modification of nucleic acids, the use of stealth coatings, including polyethylene glycol, or in-depth nanocarrier formulation optimization [122, 123]. However, it is also important to note that immune responses are not eliminated completely and may differ between individuals. These responses are especially important for understanding and controlling applications such as gene editing, which may require prolonged or repeated exposure [124].

6.4 Manufacturing and Scalability

Intelligent nanocarriers are often multicomponent systems, and proper designs in terms of structure and composition should be provided. It is also difficult to maintain uniformity, scalability, and reproducibility in batches for platforms with stimuli-responsive or multifunctional properties [125]. The production procedure must be in accordance with the general principles of good manufacturing practices and allow strict quality control. The particle size distribution, surface chemistry, and nucleic acid content/loading capability/stability must be carefully controlled [115, 126]. Although microfluidic fabrication and continuous processing technology have been developed to handle such problems, more work is needed for their large-scale use [114].

6.5 Regulatory Challenges

Regulatory nanomedicine frameworks are still under development, especially in the case of smart nanocarriers for nucleic acid-free delivery, and a specific regulatory approach is needed because of their intrinsic complexity. Regulatory bodies require a full characterization of the nanocarrier and its nucleic acid load as well as comprehensive data regarding the pharmacokinetic, biodistribution, and safety profiles [115, 116]. Evaluating and comparing different platforms are extremely difficult because of the lack of standardized testing protocols and reference materials. Developers should therefore seek to actively engage with regulators, developing a mutual understanding of expectations. Such discussion is necessary to support that preclinical and clinical studies sufficiently consider appropriate safety and efficacy end points. Ensuring that there is no ambiguity around regulatory pathways is especially important in the context of gene editing, where ethical considerations and safety concerns are more pronounced [127].

6.6 *Clinical Applications and Case Examples*

Despite these limitations, several intelligent nanotechnology-mediated nucleic acid delivery systems have been successful in the clinic. LNP-encapsulated mRNA delivery has enabled accelerated discovery and deployment of vaccines and therapeutic proteins. Furthermore, RNA interference drugs packaged in nanoparticle delivery systems have already been marketed as treatments for rare genetic and metabolic diseases [4, 126]. In oncology, smart nanocarriers are also designed for siRNAs and miRNAs to inhibit oncogene expression or modify the tumor microenvironment. Responsive systems that release nucleic acids in response to tumor-specific signals have shown increased efficacy and decreased toxicity in early clinical trials. CRISPR/Cas-based gene-editing therapies are progressing toward clinical investigation, and nanocarrier-mediated delivery represents a nonviral delivery strategy that provides an improved safety and temporal control interface [128, 129].

6.7 *Bridging the Gap from Bench to Bedside*

Rational design, comprehensive preclinical validation, and rigorous clinical development are essential for the effective translation of smart nanocarriers. Interdisciplinary efforts, including those of material scientists, biologists, clinicians, and regulatory scientists, are needed to detect and resolve translational bottlenecks at an early stage of biomaterial development. Patient-related factors, such as disease heterogeneity and genetic background, should also be considered when evaluating delivery efficacy. Adaptable nanocarrier systems that can be designed on a patient-specific basis offer significant potential for personalized and precision-medicine-based strategies. Finally, smart nanotechnology developments are slowly pushing forward nucleic acid-based therapeutics toward the clinic, although translational hurdles remain evident. In the last part of this chapter, the author discusses the outlook on the next generation of nucleic acid carriers, such as new technologies, artificial intelligence-based designs, and changing regulations.

Progress in this area is indicative of the rapid development of smart nanotechnology for nucleic acid delivery. Table 6 summarizes representative works that describe various nanocarrier systems, nucleic acid types, and therapeutic applications. This table summarizes 14 original studies released in the last two years, demonstrating the cutting edge of smart nanocarrier systems for nucleic acid delivery. The work covered here employs diverse therapeutic modalities: siRNA, mRNA, CRISPR/Cas systems, and DNA constructions in combination with lipid-based (liposomes), polymeric (polyplexes), inorganic, and hybrid nanomaterials for delivery. Together, these studies address in vitro, in vivo, and nascent clinical applications for disorders ranging from cancer to genetic disease and neurodegeneration. The heterogeneity in strategies and the incremental advancements in delivery modalities seen across these publications reflect a maturing field ready for further translation into the clinic.

7 Future Perspectives

In the author's opinion, the convergence of artificial intelligence (AI) and machine learning (ML) algorithms will lead to critical breakthroughs in smart nanotechnology for nucleic acid delivery. In general, formulations have been based foremost on empirical optimization, which is currently not sufficient to address complex in vivo dynamics for nanocarrier–biological interactions. AI-based approaches make it possible to predict delivery efficacy, toxicity profiles, and intracellular processes via extensive multidimensional data [144]. This feature not only speeds up the rational design but also reduces the high burden of experiments in these procedures. It is anticipated that AI and ML will evolve from supportive analytical tools to core design engines driving nanocarrier selection, formulation optimization, and treatment customization for individual patients, ultimately leading toward more reproducible and personalized nucleic acid therapeutic delivery [145].

Emerging trends and future directions in smart nanotechnology for nucleic acid delivery are conceptually outlined in Fig. 8. The demand for smart delivery systems is expected to increase in response to progress toward gene-editing technologies and combination therapies [146]. Upcoming gene-editing tools, specifically engineered CRISPR-based instruments, require stringent temporal and spatial control—features that can be marshaled solely by high-technology, multifunctional nanocarriers [147]. Moreover, the incorporation of bioinspired and sustainable materials is emerging as an essential approach to improve biocompatibility, strengthen immune tolerance, and ensure long-term safety [148]. Materials derived from biological membranes, peptides and systems that are naturally degradable are also consistent with clinical safety issues as well as future sustainability interests [149]. Thus, the junction of AI-enabled nanocarrier design, the latest development of gene-editing technology, and bioinspired material engineering is expected to define a new era for smart nanotechnology, guiding research toward more clinically relevant and environmentally friendly solutions.

8 Conclusion

Smart nanotechnology has been identified as a key facilitator in the development of nucleic acid-based therapeutics, effectively resolving many critical challenges that have impeded their clinical translatability. With the driving forces emerging from material science, molecular biology, and biomedicine engineering fields, smart

Table 6 Selected smart nanocarriers for nucleic acid delivery

Nucleic acid	Carrier (material/type)	Stage	Target disease/application	Key findings/outcomes	References
mRNA (IL-21, IL-7, 4-1BBL)	Lipid nanoparticles (LNPs)	In vivo (mouse)	Solid tumors (colon carcinoma, melanoma)	Triple-cytokine mRNA LNPs synergistically activated tumor-specific T cells and memory immunity. Intratumoral LNPs encoding IL-21 + IL-7 + 4-1BBL elicited robust $CD8^+$ T-cell infiltration, eradicated established tumors, generated long-term immunity, and outperformed checkpoint inhibitors	[130]
CRISPR–Cas9 RNP (iGeoCas9 + sgRNA)	Lipid nanoparticles (LNPs with thermostable Cas9)	In vivo (mouse)	Genetic diseases (lung/liver gene editing; surfactant deficiency model)	Engineered thermostable Cas9 (iGeoCas9) ribonucleoprotein, delivered in LNPs, achieved high-efficiency editing in lung (up to 37%) and liver (30%) cells. LNPs corrected a human SFTPC mutation in neonatal mice (≈19% editing), restoring protein; showed robust HDR using codelivered ssDNA. This demonstrates potent nonviral in vivo gene editing (lung, liver)	[131]

(continued)

Table 6 (continued)

Nucleic acid	Carrier (material/type)	Stage	Target disease/ application	Key findings/outcomes	References
CRISPR components (Cas9 mRNA + sgRNA)	Lipid nanoparticles (RNA LNPs)	In vitro and in vivo (GBM model)	Glioblastoma (cancer stem cells)	LNPs coencapsulating Cas9 mRNA and sgRNA achieved potent gene editing of GFP and therapy genes. In vitro, up to 67% GFP knockdown in stem-like GBM cells. In vivo (orthotopic GBM model), intratumoral LNPs produced ~ 80% GFP knockout and ~ 70% target gene editing, validating therapeutic targets with minimal toxicity	[132]
Plasmid DNA (large gene)	LNPs with PEGylation and self-protein corona	In vitro and in vivo (mouse)	Gene therapy (generic; proof-of-concept)	Designed PEGylated LNPs loaded with large (~5.5 kb) plasmid DNA and coated with endogenous protein corona. Achieved efficient DNA transfection in vitro and in vivo (rodent liver), low immunogenicity, and stable particle integrity. Novel morphology ("nanobioarchitectures") overcame size constraints, enabling sustained gene expression in vivo	[133]

(continued)

Table 6 (continued)

Nucleic acid	Carrier (material/type)	Stage	Target disease/application	Key findings/outcomes	References
Antisense oligonucleotides (anti-miR-21, anti-miR-210)	Polyethylenimine-coated ultrasmall mesoporous silica (USMP-PEI)	In vitro (2D/3D), BBB model	Glioblastoma (brain tumor)	Sub40 nm silica NPs (USMP) modified with PEI achieved efficient codelivery of anti-miR21 and anti-miR210. Optimized PEI ratio reduced toxicity while preserving uptake. In GBM spheroids, combined anti-miRs reduced spheroid size (~30%) and colony formation (~40%). Crucially, NPs showed transient BBB penetration in vitro. Thus, USMP-PEI effectively delivered anti-miRs to GBM cells	[134]

(continued)

Table 6 (continued)

Nucleic acid	Carrier (material/type)	Stage	Target disease/ application	Key findings/outcomes	References
siRNA (anti-YTHDF2)	pH-responsive ionizable lipid nanoparticles (Lip@si-YTHDF2)	In vivo (mouse HCC model)	Hepatocellular carcinoma (HCC)	Developed acid-triggered LNPs (Lip@si-YTHDF2) for targeted siRNA delivery to liver. Lip@si-YTHDF2 downregulated YTHDF2 in myeloid-derived suppressor cells (MDSCs), leading to reduced MYC expression (via m6A pathway), reprogrammed immune microenvironment, and strong synergy with PD-1 blockade. In mice, this combination suppressed tumor growth and reinvigorated T-cell immunity	[135]
DNA (self-assembled DNA origami carrying gene)	DNA origami nanostructures (folded genetic elements)	In vitro (human cells)	Gene expression platform (vaccine/ therapy)	Constructed DNA origami nanotubes encoding a GFP reporter. Demonstrated that human cells can transcribe and express genes embedded within highly folded DNA nanostructures. Gene expression level depended on the density/ placement of ssDNA domains in the origami. This shows DNA nanocarriers themselves can serve as genetic cargo, with tunable expression	[136]

(continued)

Table 6 (continued)

Nucleic acid	Carrier (material/type)	Stage	Target disease/ application	Key findings/outcomes	References
Self-amplifying RNA (saRNA)	Bioreducible poly(ethylenimine-ran-propylenimine) copolymer	In vitro and in vivo (mice, reporter study)	Gene vaccination/ long-term expression	Novel polymeric carriers (random PEI–PPI copolymers) effectively complexed saRNA in < 100 nm polyplexes. In vitro, polymer/saRNA polyplexes matched or exceeded lipofection efficiency across cell lines with high viability. In vivo, polymer-saRNA complexes yielded stronger and more sustained bioluminescence in mice ($\approx$35 days) than naked or PEI controls, indicating prolonged protein expression (higher and longer than LNP-like carriers)	[137]

(continued)

Table 6 (continued)

Nucleic acid	Carrier (material/type)	Stage	Target disease/application	Key findings/outcomes	References
Plasmid DNA (EGFP reporter)	PLGA/PVA nanoparticles, anti-CD19 antibody-coated	In vitro and in vivo (human/zebrafish lymphoma model)	B-cell lymphoma (targeted gene therapy)	Developed antibody-targeted PLGA/PVA NPs loaded with therapeutic pDNA (EGFP). Anti-CD19 coating enabled selective B-cell targeting. In vitro, targeted NPs safely transfected lymphoma cells, producing the protein without toxicity. In a zebrafish xenograft, targeted NPs increased localized EGFP expression in B-cell tumors without systemic toxicity. This proves a safe targeted platform for tumor-cell-specific DNA expression	[138]

(continued)

Table 6 (continued)

Nucleic acid	Carrier (material/type)	Stage	Target disease/ application	Key findings/outcomes	References
mRNA (GFP, β2-AMPK) and RNA aptamers	Glutathione-responsive polycationic nanogels (disulfide-crosslinked)	In vitro (mammalian cells)	Gene therapy, mRNA vaccine development	Designed GSH-sensitive nanogels incorporating new cationic monomers (polyarginine, polyhistidine, guanidine). Nanogels showed ~ 90% mRNA encapsulation and triggered release in reductive conditions. They outperformed commercial reagents in hard-to-transfect cells, enabling strong GFP and β2-AMPK expression with controlled release. Mechanistic studies (TIRF) confirmed high mRNA loading per particle. The platform is highly tunable, biocompatible, and effective for diverse mRNA therapeutics	[139]

(continued)

Table 6 (continued)

Nucleic acid	Carrier (material/type)	Stage	Target disease/application	Key findings/outcomes	References
siRNA (anti-SNCA)	PEI/PPI dendrimer-liposome hybrid nanoparticles (Tyrosine-modified PEI/PPI)	In vivo (mouse, intranasal)	Parkinson's disease (neurological)	Developed cationic PEI and PPI dendrimer nanocomplexes with siRNA targeting α-synuclein. Intranasal administration in Thy1-α-synuclein mice bypassed BBB, delivering siRNA broadly in brain regions (olfactory bulb, cortex, SN). After 4 days, SNCA mRNA and α-syn protein levels were significantly reduced without toxicity or inflammation. This noninvasive NP system markedly lowered pathogenic α-synuclein in vivo, suggesting a novel PD therapy	[140]

(continued)

Table 6 (continued)

Nucleic acid	Carrier (material/type)	Stage	Target disease/ application	Key findings/outcomes	References
siRNA (anti-survivin)	ZnO-chitosan core nanoparticles, coated with cancer cell membrane	In vitro and in vivo (mouse 4T1 breast cancer)	Breast cancer (chemotherapy + RNAi)	Designed ZnO NP/chitosan cores that dissolve in acidic lysosomes, releasing siRNA and disrupting endosomes. Coating with homologous cancer-cell membrane enhanced tumor targeting. In 4T1 tumor models, pH-triggered ZnO dissolution synergized with si-survivin to induce apoptosis and inhibit tumor growth. ZnO-mediated lysosomal escape of siRNA, amplifying gene silencing. This biomimetic NP system enabled effective tumor-specific siRNA delivery and anticancer synergy	[141]

(continued)

Table 6 (continued)

Nucleic acid	Carrier (material/type)	Stage	Target disease/ application	Key findings/outcomes	References
CRISPR/Cas9 plasmid DNA	Lipidated arginine-rich peptide dendrimers (D-enantiomer)	In vitro (HEK293, HeLa cells)	Gene editing (reporter/GFP)	Optimized a D-peptide dendrimer (Z22) for plasmid delivery. Z22 (arginine-rich, lipidated) formed stable ~ 200 nm complexes with 9 kb CRISPR-GFP plasmids. It achieved > 83% GFP transfection in HEK cells (no toxicity) and functional Cas9 editing, outperforming Lipofectamine. Critical features included D-amino acids (protease resistance) and lipid anchors. This single-component dendrimer enabled highly efficient CRISPR pDNA delivery in vitro	[142]

(continued)

Table 6 (continued)

Nucleic acid	Carrier (material/type)	Stage	Target disease/ application	Key findings/outcomes	References
mRNA and CRISPR RNP	Ionizable lipids with branched tails (BEND lipids) in LNPs	In vivo (mouse liver and ex vivo T cells)	Liver gene editing; T-cell engineering	Developed a new class of branched ionizable lipids ("BEND") that enhance endosomal escape. Compared to straight-chain lipids, BEND LNPs greatly improved hepatic delivery: showing higher mRNA expression and Cas9 editing in mouse liver. BEND LNPs also efficiently transfected primary human T cells ex vivo (>70%). Mechanistic studies confirmed that branched chains promote lipid-phase disruption, facilitating endosomal release. Thus, BEND lipids enable potent mRNA and RNP delivery for both in vivo gene editing and cell engineering	[143]

LNPs lipid nanoparticles; *RNP* ribonucleoprotein; *sgRNA* single-guide RNA; *pDNA* plasmid DNA; *saRNA* self-amplifying RNA; *siRNA* small interfering RNA; *miRNA* microRNA; *ASO* antisense oligonucleotide; *PEG* polyethylene glycol; *PEI* polyethylenimine; *PPI* poly(propylenimine); *PLGA* poly(lactic–coglycolic acid); *PVA* poly(vinyl alcohol); *USMP* ultrasmall mesoporous silica nanoparticles; BBB, blood–brain barrier; *FA* folic acid; Fe_3O_4 magnetite nanoparticles; *LIFU* low-intensity focused ultrasound; *GSH* glutathione; *ZnO* zinc oxide; *EGFP* enhanced green fluorescent protein; *GFP* green fluorescent protein; *IL* interleukin; *PD-1* programmed cell death protein-1; *HCC* hepatocellular carcinoma; *GBM* glioblastoma multiforme; *SNCA* α-synuclein gene; *RBC* red blood cell; *HDR* homology-directed repair

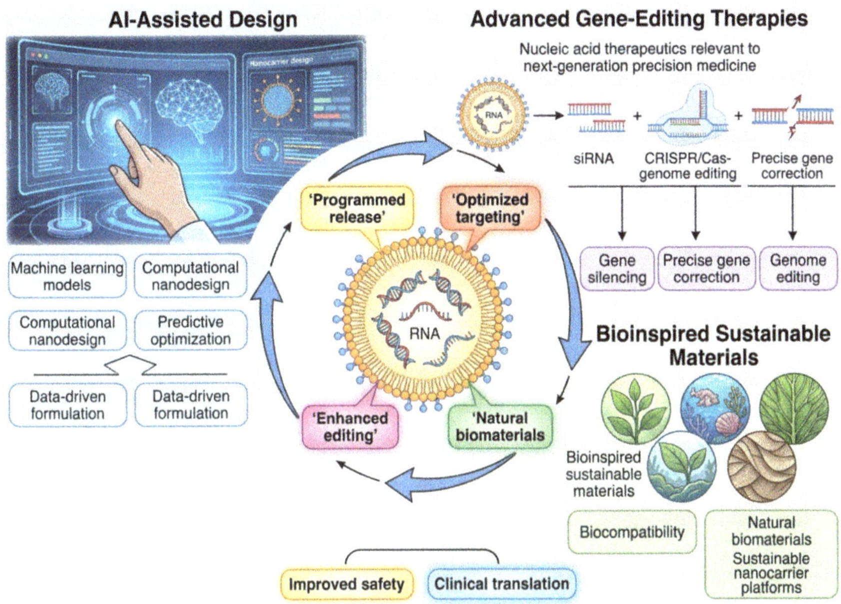

Fig. 8 Outlook on smart nanotechnology for nucleic acid delivery, focusing on AI-assisted design, advanced gene-editing therapies and bioinspired sustainable materials for clinical translation. Created with MindtheGraph (www.mindthegraph.com)

nanocarriers enable the effective protection of, targeted release of and on-demand control over various nucleic acid cargos, including DNA, mRNA, siRNA, miRNA, and gene-editing systems. The dynamic response to biological and external triggers endows these platforms with superior delivery accuracy, intracellular trafficking, and therapeutic activities.

This chapter discusses the design patterns, material choices, and mechanisms of function involved in smart nanocarriers, particularly their ability to avoid stumbling blocks such as enzymatic digestion, low cellular uptake, or entrapment within endosomes. Emerging trends in the development of stimuli-responsive systems, targeting approaches, and multifunctional designs indicate the growing complexity of delivery platforms via nanotechnology. However, a number of important translational challenges remain to be addressed, including periodicity and concerns related to safety (i.e., manufacturing, immunogenicity and regulatory approval), but these challenges need to be resolved for successful translation into the clinic.

In the future, the synergy of smart nanotechnology with advancing approaches, including artificial intelligence, precision medicine, and new generation gene-editing technology, will promote development in this field. Continued interdisciplinary cooperation and early inclusion of translation and regulatory perspectives will be essential to closing the gap between bench-side work and clinical application. Together, these advances place smart nanotechnology at the forefront of future nucleic acid therapeutics, which hold significant potential to transform the practice of preventing, diagnosing, and treating a variety of diseases that cannot be overstated.

References

1. Rafique MG, Laurent Q, Dore MD et al (2025) Sequence-defined DNA polymers: new tools for DNA nanotechnology and nucleic acid therapy. Acc Chem Res 58:177–188. https://doi.org/10.1021/acs.accounts.4c00580
2. Vosoughi P, Naghib SM, Kangarshahi BM et al (2025) A review of RNA nanoparticles for drug/gene/protein delivery in advanced therapies: current state and future prospects. Int J Biol Macromol 295:139532. https://doi.org/10.1016/j.ijbiomac.2025.139532
3. Wang Y, Xuan W, Mao C et al (2025) Inhalable nucleic acid therapeutics for chronic pulmonary disease: progress, challenges, and prospects. Acta Biomater 204:187–204. https://doi.org/10.1016/j.actbio.2025.07.068
4. Yingngam B (2026) Chapter 8—the development of lipid nanoparticle technology for the management of COVID-19. In: Chen J-T (ed) Nanotechnology and antivirus nanomaterials for combating COVID-19. Elsevier, Amsterdam, pp 201–250. https://doi.org/10.1016/B978-0-443-33002-5.00006-3
5. Van Der Meel R, Wender PA, Merkel OM et al (2025) Next-generation materials for nucleic acid delivery. Nat Rev Mater 10:490–499. https://doi.org/10.1038/s41578-025-00814-1
6. Bian X, Zhou L, Luo Z et al (2025) Emerging delivery systems for enabling precision nucleic acid therapeutics. ACS Nano 19:4039–4083. https://doi.org/10.1021/acsnano.4c11858
7. Ullah A, Khan M, Zhang Y et al (2025) Advancing therapeutic strategies with polymeric drug conjugates for nucleic acid delivery and treatment. Int J Nanomed 20:25–52. https://doi.org/10.2147/IJN.S429279
8. Pal S, Cannata JN, Rouge JL (2025) Nucleic acid nanocapsules as a new platform to deliver therapeutic nucleic acids for gene regulation. Acc Chem Res 58:1951–1962. https://doi.org/10.1021/acs.accounts.5c00126
9. Xu L, Shao Z, Fang X et al (2025) Exploring precision treatments in immune-mediated inflammatory diseases: harnessing the infinite potential of nucleic acid delivery. Exploration 5:20230165. https://doi.org/10.1002/EXP.20230165
10. Geng G, Xu Y, Hu Z et al (2025) Viral and nonviral vectors in gene therapy: current state and clinical perspectives. eBioMedicine 118:105834. https://doi.org/10.1016/j.ebiom.2025.105834
11. Gonzalez JC, Park KW, Evans DB et al (2025) Nano approaches to nucleic acid delivery: barriers, solutions, and current landscape. Wiley Interdiscip Rev Nanomed Nanobiotechnol 17:e70010. https://doi.org/10.1002/wnan.70010
12. Chen X, Luo X, Yin W et al (2025) Framework nucleic acid nanomaterials for central nervous system therapies: design for barrier penetration, targeted delivery, cellular uptake, and endosomal escape. ACS Nano 19:24335–24376. https://doi.org/10.1021/acsnano.5c03945
13. Abaza T, Mohamed EE, Zaky MY (2025) Lipid nanoparticles: a promising tool for nucleic acid delivery in cancer immunotherapy. Med Oncol 42:409. https://doi.org/10.1007/s12032-025-02939-3
14. Esmaeilpour D, Ghomi M, Zare EN et al (2025) Nanotechnology-enhanced siRNA delivery: revolutionizing cancer therapy. ACS Appl Bio Mater 8:4549–4579. https://doi.org/10.1021/acsabm.5c00489
15. Hao J, Li Y, Huang L et al (2025) Smart nanoarchitectures for precision RNA delivery: harnessing endogenous and exogenous stimuli in cancer treatment. Theranostics 15:7747–7778. https://doi.org/10.7150/thno.112492

16. You X, Zeng Q, Xia T et al (2025) Rolling circle amplification/transcription-based nanotechnology for efficient delivery of nucleic acid drugs. Nanoscale Horiz 10:2285–2303. https://doi.org/10.1039/d5nh00364d
17. Baig MMFA, Chien WT, Chair SY (2025) Nanotechnological approaches for the targeted delivery of CRISPR-Cas systems for genomic modifications, biomolecular sensing, and precision medicine. Biomater Sci 13:4597–4638. https://doi.org/10.1039/D5BM00711A
18. Peng W, Shi M, Hu B et al (2025) Nanotechnology-leveraged CRISPR/Cas systems: icebreaking in trace cancer-related nucleic acids biosensing. Mol Cancer 24:78. https://doi.org/10.1186/s12943-024-02222-5
19. Kansara K, Mansuri A, Kumar A et al (2025) DNA nanobiomaterials based futuristic technologies for tissue engineering and regenerative therapeutics. Small 21:2504361. https://doi.org/10.1002/smll.202504361
20. Chattopadhyay S, Goswami A, Sil M (2025) Nanobiotechnology: traditional reinterpreting personalized medicine through targeted therapies and regenerative solutions. Naunyn Schmiedebergs Arch Pharmacol 398:10025–10048. https://doi.org/10.1007/s00210-025-04038-6
21. Kuran SA, Adıyıl R, Zıkşahna K et al (2025) Nanotechnology-enabled miRNA delivery systems next-generation molecular strategies in cancer therapy. Biochem Biophys Res Commun 792:152923. https://doi.org/10.1016/j.bbrc.2025.152923
22. Kavita, Singh B, Sharma T (2025) Riding the wave of innovation: nanotechnology in nucleic acid-based cancer therapy. 3 Biotech 15:226. https://doi.org/10.1007/s13205-025-04397-0
23. Fu Y, Jiang W, Jang M-S et al (2025) Bioresponsive nucleic acid base pair-inspired nanogels for codelivery of chemo-/protein drugs in targeted cancer therapy. Biomacromolecule 26:7177–7189. https://doi.org/10.1021/acs.biomac.5c01632
24. Mohamed RGA, Ali SM, Ahmed IS et al (2025) Next-generation nanocarriers for colorectal cancer: passive, active, and stimuli-responsive strategies for precision therapy. Biomater Sci 13:5626–5664. https://doi.org/10.1039/D5BM01176K
25. Sarma L, Singh S, Anjali et al (2025) Challenges and future directions for next-generation drug delivery systems. In: Pathak A, Singh SP (eds) Next-generation drug delivery systems. Springer US, New York, NY, pp 571–596. https://doi.org/10.1007/978-1-0716-4554-3_25
26. Mangla B, Kumar P, Javed S et al (2025) Regulating nanomedicines: challenges, opportunities, and the path forward. Nanomedicine 20:1911–1927. https://doi.org/10.1080/17435889.2025.2533107
27. Liu M, Wang Y, Zhang Y et al (2025) Landscape of small nucleic acid therapeutics: moving from the bench to the clinic as next-generation medicines. Signal Transduct Target Ther 10:73. https://doi.org/10.1038/s41392-024-02112-8
28. Chaturvedi DP, Ghangal R, Bansal N et al (2025) Nucleic acid therapeutics. Resonance 30:209–228. https://doi.org/10.1007/s12045-025-1744-y
29. Huang B, De Smedt SC, De Vos WH et al (2025) Light-triggered nanocarriers for nucleic acid delivery. Drug Deliv 32:2502346. https://doi.org/10.1080/10717544.2025.2502346
30. Singh AK, Goel K, Dhanawat M (2025) Chapter three—plasmid DNA and mRNA delivery: approaches and challenges. In: Singh AK, Mishra N, Ashique S, Prabhakar PK (eds) Advances in immunology. Academic Press, London, pp 63–87. https://doi.org/10.1016/bs.ai.2024.12.001
31. Xu Z, Xie Y, Chen W et al (2025) Nanocarrier-based systems for targeted delivery: current challenges and future directions. MedComm 6:e70337. https://doi.org/10.1002/mco2.70337
32. Anderluzzi G, Mohamed T, Moschetti G et al (2025) The impact of using different cationic polymers on the formation of efficient lipopolyplexes for pDNA delivery. Int J Nanomed 20:10021–10041. https://doi.org/10.2147/IJN.S513568
33. Yao T, Dong X, Wang X et al (2025) Engineering exosomes for mRNA delivery: a review. Int J Biol Macromol 316:144662. https://doi.org/10.1016/j.ijbiomac.2025.144662
34. Zhang T, Qin J, Zhu B et al (2025) Topical VEGF-A mRNA therapy promotes angiogenesis and accelerates wound healing. Cell Investig 1:100043. https://doi.org/10.1016/j.clnves.2025.100043

35. Sanadgol N, Abedi M, Hashemzaei M et al (2025) Exosomes as nanocarriers for brain-targeted delivery of therapeutic nucleic acids: advances and challenges. J Nanobiotechnology 23:453. https://doi.org/10.1186/s12951-025-03528-2
36. Lian Z, Zheng L, Liu S et al (2025) Breaking endosomal barriers: thiol-mediated uptake lipid nanoparticles for efficient mRNA vaccine delivery. J Am Chem Soc 147:31530–31540. https://doi.org/10.1021/jacs.5c05367
37. Wang J, Cai L, Li N et al (2025) Developing mRNA nanomedicines with advanced targeting functions. Nano-Micro Lett 17:155. https://doi.org/10.1007/s40820-025-01665-9
38. Fan J, Xiao Z, Dong Y et al (2025) Nanocarrier-mediated RNA delivery platform as a frontier strategy for hepatic disease treatment: challenges and opportunities. Adv Healthc Mater 14:2402933. https://doi.org/10.1002/adhm.202402933
39. Askarizadeh A, Vahdat-Lasemi F, Karav S et al (2025) Lipid nanoparticle-based delivery of small interfering RNAs: new possibilities in the treatment of diverse diseases. Eur Polym J 223:113624. https://doi.org/10.1016/j.eurpolymj.2024.113624
40. Sharma H, Kumar V, Bahal R (2025) Functional bioconjugates for precision delivery of nucleic acid therapeutics. Trends Chem 7:742–755. https://doi.org/10.1016/j.trechm.2025.10.002
41. Zhang Y, Zhang M, Song H et al (2025) Tumor microenvironment-responsive polymer-based RNA delivery systems for cancer treatment. Small Methods 9:2400278. https://doi.org/10.1002/smtd.202400278
42. Lou Y, Wang Y, Lu J et al (2025) MicroRNA-targeted nanoparticle delivery systems for cancer therapy: current status and future prospects. Nanomedicine 20:1181–1194. https://doi.org/10.1080/17435889.2025.2492542
43. Xu L, Liu X, Chen J et al (2025) Use of mesenchymal stem cell-targeting 3WJ nanoparticles and reported specific delivery of anti-miRNA 138 to treat osteoporosis. ACS Omega 10:10633–10641. https://doi.org/10.1021/acsomega.4c11505
44. Valizadeh-Otaghsara M, Haji Molla Hoseini M, Karima S et al (2025) MiRNA delivery systems in cancer stem cell therapy: exosomes versus chitosan nanoparticles. Stem Cell Res Ther 16:576. https://doi.org/10.1186/s13287-025-04667-x
45. Cheng Y-H, Chen H-Y, Kato K et al (2025) Chitosan nanoparticles as a targeted delivery system for anti-fibrotic microRNAs for oral submucosal fibrosis treatment. Colloids Surf B Biointerfaces 252:114657. https://doi.org/10.1016/j.colsurfb.2025.114657
46. Subramony A, Raj D, Fairooz F et al (2025) Genosomes: an introspection into transfection, future perspectives and applications. Adv Pharm Bull 15:482–498. https://doi.org/10.34172/apb.025.43191
47. Kaur G, Arora J, Sodhi AS et al (2025) Nanotechnology and CRISPR/Cas-mediated gene therapy strategies: potential role for treating genetic disorders. Mol Biotechnol 67:3800–3822. https://doi.org/10.1007/s12033-024-01301-8
48. Aussel C, Cathomen T, Fuster-García C (2025) The hidden risks of CRISPR/Cas: structural variations and genome integrity. Nat Commun 16:7208. https://doi.org/10.1038/s41467-025-62606-z
49. Kalter N, Fuster-García C, Silva A et al (2025) Off-target effects in CRISPR-Cas genome editing for human therapeutics: progress and challenges. Mol Ther Nucleic Acids 36:102636. https://doi.org/10.1016/j.omtn.2025.102636
50. Bolideei M, Barzigar R, Gahrouei RB et al (2025) Applications of gene editing and nanotechnology in stem cell-based therapies for human diseases. Stem Cell Rev Rep 21:905–934. https://doi.org/10.1007/s12015-025-10857-0
51. Gholamian S, Baghaee P, Doroudian M (2025) Nanotechnology in gene editing: pioneering CRISPR-Cas delivery systems to tackle antibiotic resistance. Adv Therap 8:2400412. https://doi.org/10.1002/adtp.202400412
52. Ferreira D, Rodrigues LR (2025) RNA-based therapeutics: cutting-edge advances in clinical research. Adv Therap 8:e00082. https://doi.org/10.1002/adtp.202500082
53. Chauhan P, Tiwari I, Sharma NK (2025) Challenges and opportunities used in RNA-based therapeutics. In: Sharma NK, Chand Dakal T, Maurya PK, Wadhwa G (eds) RNA-based cancer therapeutics: techniques, applications and challenges in prognosis. Springer, Singapore, pp 231–264. https://doi.org/10.1007/978-981-96-3456-9_10

54. Tebas P, Patel A, Agnes JT et al (2025) Safety and pharmacokinetics of SARS-CoV-2 DNA-encoded monoclonal antibodies in healthy adults: a phase 1 trial. Nat Med 31:4150–4159. https://doi.org/10.1038/s41591-025-03969-0
55. Anand P, Zhang Y, Patil S et al (2025) Metabolic stability and targeted delivery of oligonucleotides: advancing RNA therapeutics beyond the liver. J Med Chem 68:6870–6896. https://doi.org/10.1021/acs.jmedchem.4c02528
56. Chen L, Bosmajian C, Woo S (2025) Mechanistic intracellular PK/PD modeling to inform development strategies for small interfering RNA therapeutics. Mol Ther Nucleic Acids 36:102516. https://doi.org/10.1016/j.omtn.2025.102516
57. Martino MTD, Tagliaferri P, Tassone P (2025) MicroRNA in cancer therapy: breakthroughs and challenges in early clinical applications. J Exp Clin Cancer Res 44:126. https://doi.org/10.1186/s13046-025-03391-x
58. Barber HM, Pater AA, Gagnon KT et al (2025) Chemical engineering of CRISPR–Cas systems for therapeutic application. Nat Rev Drug Discov 24:209–230. https://doi.org/10.1038/s41573-024-01086-0
59. Zhang J, Gai T, Wang J et al (2025) Nucleic acid-locked smart carrier for photothermal/chemotherapy-amplified immunogenic cell death to enhance systemic antitumor efficacy. Adv Sci 12:2503299. https://doi.org/10.1002/advs.202503299
60. Yingngam B (2024) Advances in nanomaterials for drug delivery. In: Prabhakar PK, Prakash A (eds) Cutting-edge applications of nanomaterials in biomedical sciences. IGI Global Scientific Publishing, Hershey, PA, pp 22–85. https://doi.org/10.4018/979-8-3693-0448-8.ch002
61. Sabit H, Pawlik TM, Radwan F et al (2025) Precision nanomedicine: navigating the tumor microenvironment for enhanced cancer immunotherapy and targeted drug delivery. Mol Cancer 24:160. https://doi.org/10.1186/s12943-025-02357-z
62. Yingngam B, Sethabouppha B, Yingngam K et al. (2024) Biomedical applications of hybrid nanomaterials. In: Khanna V, Sharma P, Mahajan P (eds) Technological applications of nano-hybrid composites. IGI Global Scientific Publishing, Hershey, PA, pp 170–225. https://doi.org/10.4018/979-8-3693-1261-2.ch007
63. Li Y, Tian R, Xu J et al (2024) Recent developments of polymeric delivery systems in gene therapeutics. Polym Chem 15:1908–1931. https://doi.org/10.1039/d4py00124a
64. Zhou Q, Liu Q, Wang Y et al (2024) Bridging smart nanosystems with clinically relevant models and advanced imaging for precision drug delivery. Adv Sci 11:2308659. https://doi.org/10.1002/advs.202308659
65. Ahmed T, Liu F-CF, Wu XY (2024) An update on strategies for optimizing polymer-lipid hybrid nanoparticle-mediated drug delivery: exploiting transformability and bioactivity of PLN and harnessing intracellular lipid transport mechanism. Expert Opin Drug Deliv 21:245–278. https://doi.org/10.1080/17425247.2024.2318459
66. Lu Z-R, Sun D (2025) Mechanism of pH-sensitive amphiphilic endosomal escape of ionizable lipid nanoparticles for cytosolic nucleic acid delivery. Pharm Res 42:1065–1077. https://doi.org/10.1007/s11095-025-03890-8
67. Maeki M, Uno S, Sugiura K et al (2024) Development of polymer–lipid hybrid nanoparticles for large-sized plasmid DNA transfection. ACS Appl Mater Interfaces 16:2110–2119. https://doi.org/10.1021/acsami.3c14714
68. Koirala S, Cheng K (2025) Lipid and polymeric nanocarriers for siRNA delivery to the brain. J Control Release 388:114286. https://doi.org/10.1016/j.jconrel.2025.114286
69. Zhang J, Yang X, Chang Z et al (2025) Polymeric nanocarriers for therapeutic gene delivery. Asian J Pharm Sci 20:101015. https://doi.org/10.1016/j.ajps.2025.101015
70. Thakur A, Sharma R, Sharma R et al (2026) CRISPR/Cas9-integrated nanocarriers for cancer gene therapy: mechanistic insights, challenges, and clinical translation. J Drug Deliv Sci Technol 115:107705. https://doi.org/10.1016/j.jddst.2025.107705
71. Alahverdi M, Dadmehr M, Sahebkar A (2025) Nanocarriers for microRNA delivery: a review of applied platforms and perspectives. Int J Biol Macromol 319:145463. https://doi.org/10.1016/j.ijbiomac.2025.145463

72. Hoshino R, Sugawara K, Terashita K et al (2026) Acid-triggered nucleic acid release from gold nanoparticles via Schiff base linkages: *in vitro* validation of endosomal escape and gene silencing. Biomater Adv 178:214477. https://doi.org/10.1016/j.bioadv.2025.214477
73. Dutour R, Bruylants G (2025) Gold nanoparticles coated with nucleic acids: an overview of the different bioconjugation pathways. Bioconjug Chem 36:1133–1156. https://doi.org/10.1021/acs.bioconjchem.5c00098
74. Chen Y, Guo Z, Li J et al (2025) Smart nucleic acid nanodrug delivery system for precision therapeutics. Coord Chem Rev 535:216673. https://doi.org/10.1016/j.ccr.2025.216673
75. Karati D, Mukherjee S, Prajapati B et al (2024) A review on lipid-polymer hybrid nanocarriers in cancer. J Drug Deliv Sci Technol 97:105827. https://doi.org/10.1016/j.jddst.2024.105827
76. Liu X, Sun M, Dai L et al (2025) Functional nucleic acid-powered hybrid nanocarriers for synchronized targeted delivery of dual-solubility therapeutics. J Biotechnol 406:127–135. https://doi.org/10.1016/j.jbiotec.2025.07.002
77. Spataru I, Mahoro F, Badea I (2025) Development of inorganic and hybrid nanosystems for delivery of CRISPR-based gene editors. Mol Pharm 22:4373–4393. https://doi.org/10.1021/acs.molpharmaceut.5c00251
78. Cheung TH, Shoichet MS (2025) The interplay of endosomal escape and RNA release from polymeric nanoparticles. Langmuir 41:7174–7190. https://doi.org/10.1021/acs.langmuir.4c05176
79. Xiao X, Gao Y, Jia Y et al (2025) Impact of stimuli-responsiveness on the mRNA delivery efficiency of low-generation dendrimer nanogels. Biomacromolecule 26:6962–6972. https://doi.org/10.1021/acs.biomac.5c01272
80. Zeng H, Zhang Y, Liu N et al (2024) Stimulus-responsive nanodelivery and release systems for cancer gene therapy: efficacy improvement strategies. Int J Nanomed 19:7099–7121. https://doi.org/10.2147/IJN.S470637
81. Wang Z, Zhang J, Wang Y et al (2024) Overcoming endosomal escape barriers in gene drug delivery using de novo designed pH-responsive peptides. ACS Nano 18:10324–10340. https://doi.org/10.1021/acsnano.4c02400
82. Chen W, Lai J, Dong S et al (2024) Engineering logic DNA nanoprobes on live cell membranes for simultaneously monitoring extracellular pH and precise drug delivery. Anal Chem 96:3462–3469. https://doi.org/10.1021/acs.analchem.3c05064
83. Wang R, Yang Y, Wang Z et al (2025) Stimuli-responsive peptide nanocarriers for tumor-specific CRISPR/Cas9 delivery and precision genome editing. J Colloid Interface Sci 697:137932. https://doi.org/10.1016/j.jcis.2025.137932
84. Meng X, Shen Y, Zhao H et al (2024) Redox-manipulating nanocarriers for anticancer drug delivery: a systematic review. J Nanobiotechnol 22:587. https://doi.org/10.1186/s12951-024-02859-w
85. Lin S, Cheng X, Chen X et al (2025) Synthetic redox-responsive nanocomplexes facilitate spatially controlled mRNA release and tumor-selective expression. Bioconjug Chem 36:1775–1785. https://doi.org/10.1021/acs.bioconjchem.5c00264
86. Xu L, Cao Y, Xu Y et al (2024) Redox-responsive polymeric nanoparticle for nucleic acid delivery and cancer therapy: progress, opportunities, and challenges. Macromol Biosci 24:2300238. https://doi.org/10.1002/mabi.202300238
87. Xuan J, Wang Z, Huang Y et al (2024) DNA response element-based smart drug delivery systems for precise drug release. Biomater Sci 12:3550–3564. https://doi.org/10.1039/D4BM00138A
88. Bian X, Luo Z, Peng B et al (2025) Engineered bionanomaterials for precision delivery of nucleic acid drugs. Small 21:e02667. https://doi.org/10.1002/smll.202502667
89. Leung HM, Chan HY, Klimezak M et al (2025) Robust photocleavable linkers for DNA synthesis: enabling visible light-triggered antisense oligonucleotide release in 3D DNA nanocages. Biomacromolecule 26:3113–3127. https://doi.org/10.1021/acs.biomac.5c00162
90. Cho HY, Yoo M, Pongkulapa T et al (2024) Magnetic nanoparticle-assisted nonviral CRISPR-Cas9 for enhanced genome editing to treat Rett syndrome. Adv Sci 11:2306432. https://doi.org/10.1002/advs.202306432

91. Xie Y, Zhou Y, Guo J et al (2025) Illuminating gene delivery: insights into the light-induced gene delivery systems with emphasis on mRNA therapeutics. Mater Today 88:959–978. https://doi.org/10.1016/j.mattod.2025.06.034
92. Solak K, Yildiz Arslan S, Acar M et al (2025) Combination of magnetic hyperthermia and gene therapy for breast cancer. Apoptosis 30:99–116. https://doi.org/10.1007/s10495-024-02026-4
93. Li M, Li S, Guo Y et al (2024) Magnetothermal-activated gene editing strategy for enhanced tumor cell apoptosis. J Nanobiotechnol 22:450. https://doi.org/10.1186/s12951-024-02734-8
94. Solak K, Mavi A, Taghizadehghalehjoughi A et al (2025) Enhanced antitumor effects of interferon-stimulating DNA and magnetic hyperthermia in a breast cancer rat model. J Drug Deliv Sci Technol 110:107044. https://doi.org/10.1016/j.jddst.2025.107044
95. Wang H, Liu Y, Li Y et al (2025) Ultrasound-responsive drug delivery system for cancer therapy. Macromol Biosci 25:e00082. https://doi.org/10.1002/mabi.202500082
96. Li M, Li Y, Zheng J et al (2024) Ultrasound-responsive nanocarriers with siRNA and Fe_3O_4 regulate macrophage polarization and phagocytosis for augmented non-small cell lung cancer immunotherapy. J Nanobiotechnol 22:605. https://doi.org/10.1186/s12951-024-02883-w
97. Tang X, Zhao S, Luo J et al (2024) Smart stimuli-responsive spherical nucleic acids: cutting-edge platforms for biosensing, bioimaging, and therapeutics. Small 20:2408898. https://doi.org/10.1002/smll.202310732
98. Chen T, Lv W, Yang M et al (2025) Treatment of atherosclerosis by biomimetic responsive-targeting hybrid nanoparticles enabled by genetically engineering cell membranes and nucleic acid nanostructures for atorvastatin delivery. Chem Eng J 511:162130. https://doi.org/10.1016/j.cej.2025.162130
99. Zeng W, Zheng L, Li Y et al (2024) Engineered extracellular vesicles for delivering functional Cas9/gRNA to eliminate hepatitis B virus cccDNA and integration. Emerg Microbes Infect 13:2284286. https://doi.org/10.1080/22221751.2023.2284286
100. Hu M, Li X, You Z et al (2024) Physiological barriers and strategies of lipid-based nanoparticles for nucleic acid drug delivery. Adv Mater 36:2303266. https://doi.org/10.1002/adma.202303266
101. Teodori L, Omer M, Kjems J (2024) RNA nanostructures for targeted drug delivery and imaging. RNA Biol 21:391–409. https://doi.org/10.1080/15476286.2024.2328440
102. Syed Altaf RR, Simon K, Mohan A et al (2025) Fusing biomimetic engineering with nanotechnology: the rise of platelet membrane-coated nanoparticles as a versatile platform for biomedical innovation. Adv Therap 8:e00262. https://doi.org/10.1002/adtp.202500262
103. Li Y, Sun H, Cao D et al (2025) Overcoming biological barriers in cancer therapy: cell membrane-based nanocarrier strategies for precision delivery. Int J Nanomed 20:3113–3145. https://doi.org/10.2147/IJN.S497510
104. Li Y, Zhang Y, Zhang Z et al (2024) Clathrin-mediated endocytosis of multiple nanoparticles tends to be less cooperative: a computational study. J Phys Chem B 128:9785–9797. https://doi.org/10.1021/acs.jpcb.4c05025
105. Hurst J, Adams F, Schnichels S (2025) The future of nanomaterials tackling the challenge of delivering nucleic acids to the retina. Adv Funct Mater 35:2407173. https://doi.org/10.1002/adfm.202407173
106. Rodríguez DA, Lefebvre GPR, Yang Q et al (2025) Incorporation of cellular membrane protein extracts into lipid nanoparticles enhances their cellular uptake and mRNA delivery efficiency. J Control Release 382:113676. https://doi.org/10.1016/j.jconrel.2025.113676
107. Huang P, Deng H, Wang C et al (2024) Cellular trafficking of nanotechnology-mediated mRNA delivery. Adv Mater 36:2307822. https://doi.org/10.1002/adma.202307822
108. Grau M, Wagner E (2024) Strategies and mechanisms for endosomal escape of therapeutic nucleic acids. Curr Opin Chem Biol 81:102506. https://doi.org/10.1016/j.cbpa.2024.102506
109. Narum S, Deal B, Ogasawara H et al (2024) An endosomal escape Trojan horse platform to improve cytosolic delivery of nucleic acids. ACS Nano 18:6186–6201. https://doi.org/10.1021/acsnano.3c09027

110. Gu J, Liang J, Tian T et al (2025) Current understanding and translational prospects of tetrahedral framework nucleic acids. JACS Au 5:486–520. https://doi.org/10.1021/jacsau.4c01170
111. Chandran A, Bhardwaj A, Kansara K et al (2025) From science to solutions: translating DNA-based nanodevices into clinical applications. Small 21:e08155. https://doi.org/10.1002/smll.202508155
112. Ma X, Tian Y, Yang R et al (2024) Nanotechnology in healthcare, and its safety and environmental risks. J Nanobiotechnol 22:715. https://doi.org/10.1186/s12951-024-02901-x
113. Yousefiasl S, Ghovvati M, Alibakhshi A et al (2025) Smart mesoporous silica nanoparticles in cancer: diagnosis, treatment, immunogenicity, and clinical translation. Small 21:2408898. https://doi.org/10.1002/smll.202408898
114. Tenchov R, Hughes KJ, Ganesan M et al (2025) Transforming medicine: cutting-edge applications of nanoscale materials in drug delivery. ACS Nano 19:4011–4038. https://doi.org/10.1021/acsnano.4c09566
115. Joyce P, Allen CJ, Alonso MJ et al (2024) A translational framework to deliver nanomedicines to the clinic. Nat Nanotechnol 19:1597–1611. https://doi.org/10.1038/s41565-024-01754-7
116. Makkar SK (2025) Advances in RNA-based therapeutics: current breakthroughs, clinical translation, and future perspectives. Front Genet 16:1675209. https://doi.org/10.3389/fgene.2025.1675209
117. Herdiana Y (2025) Bridging the gap: the role of advanced formulation strategies in the clinical translation of nanoparticle-based drug delivery systems. Int J Nanomed 20:13039–13053. https://doi.org/10.2147/IJN.S554821
118. Cunha J, Ventura FV, Charrueau C et al (2024) Alternative routes for parenteral nucleic acid delivery and related hurdles: highlights in RNA delivery. Expert Opin Drug Deliv 21:1415–1439. https://doi.org/10.1080/17425247.2024.2405207
119. Valatabar N, Oroojalian F, Kazemzadeh M et al (2024) Recent advances in gene delivery nanoplatforms based on spherical nucleic acids. J Nanobiotechnol 22:386. https://doi.org/10.1186/s12951-024-02648-5
120. Omidi Y, Pourseif MM, Ansari RA et al (2024) Design and development of mRNA and self-amplifying mRNA vaccine nanoformulations. Nanomedicine 19:2699–2725. https://doi.org/10.1080/17435889.2024.2419815
121. Zhao J, He J, Ding X et al (2024) DENV peptides delivered as spherical nucleic acid constructs enhance antigen presentation and immunogenicity *in vitro* and *in vivo*. Int J Nanomed 19:9757–9770. https://doi.org/10.2147/IJN.S467427
122. Wen F, Wang L, Li X et al (2025) Precision nanomedicine for cancer: innovations, strategies, and translational challenges. Onco Targets Ther 18:1125–1148. https://doi.org/10.2147/OTT.S550104
123. Wang J, Huang S, Wei H et al (2025) A dissolvable microneedle platform for the delivery of tumor-derived total RNA nanovaccines for enhanced tumor immunotherapy. Acta Biomater 199:120–131. https://doi.org/10.1016/j.actbio.2025.04.039
124. Rebolledo LP, Andrade LNS, Bajgelman MC et al (2025) Nucleic acid nanobiosystems for cancer theranostics: an overview of emerging trends and challenges. Nanomedicine 20:1281–1298. https://doi.org/10.1080/17435889.2025.2501919
125. De Souza Cardoso Delfino C, De Paula Pereira MC, Dos Santos Oliveira M et al (2025) Scaling nanopharmaceutical production for personalized medicine: challenges and strategies. J Nanopart Res 27:108. https://doi.org/10.1007/s11051-025-06293-3
126. Devos C, Udepurkar A, Sagmeister P et al (2025) Manufacturing mRNA-loaded lipid nanoparticles with precise size and morphology control. ACS Nano 19:33991–34002. https://doi.org/10.1021/acsnano.5c09800
127. Anchordoquy T, Artzi N, Balyasnikova IV et al (2024) Mechanisms and barriers in nanomedicine: progress in the field and future directions. ACS Nano 18:13983–13999. https://doi.org/10.1021/acsnano.4c00182
128. Yingngam B (2024) AI models for biopharmaceutical property prediction. In: Khanna A, Barachi ME, Jain S, Kumar M, Nayyar A (eds) Artificial intelligence and machine learning

in drug design and development. Wiley-Scrivener, Beverly, MA, pp 407–449. https://doi.org/10.1002/9781394234196.ch13
129. Nandy U, Azad Mandal AK, Pranav (2025) Nanomaterial-mediated nucleic acid delivery for pancreatic cancer therapeutics. ACS Appl Bio Mater 8:6676–6700. https://doi.org/10.1021/acsabm.5c00968
130. Hamouda AEI, Filtjens J, Brabants E et al (2024) Intratumoral delivery of lipid nanoparticle-formulated mRNA encoding IL-21, IL-7, and 4–1BBL induces systemic anti-tumor immunity. Nat Commun 15:10635. https://doi.org/10.1038/s41467-024-54877-9
131. Chen K, Han H, Zhao S et al (2025) Lung and liver editing by lipid nanoparticle delivery of a stable CRISPR–Cas9 ribonucleoprotein. Nat Biotechnol 43:1445–1457. https://doi.org/10.1038/s41587-024-02437-3
132. Rouatbi N, Walters AA, Costa PM et al (2024) RNA lipid nanoparticles as efficient *in vivo* CRISPR-Cas9 gene editing tool for therapeutic target validation in glioblastoma cancer stem cells. J Control Release 375:776–787. https://doi.org/10.1016/j.jconrel.2024.09.019
133. Renzi S, Digiacomo L, Pozzi D et al (2024) Structuring lipid nanoparticles, DNA, and protein corona into stealth bionanoarchitectures for *in vivo* gene delivery. Nat Commun 15:9119. https://doi.org/10.1038/s41467-024-53569-8
134. Singh RR, Kulshreshtha R, Popat A (2025) Ultrasmall nanoparticles for codelivery of antisense oligonucleotides targeting miR-21 and miR-210 to treat glioblastoma. J Nanobiotechnol 23:482. https://doi.org/10.1186/s12951-025-03529-1
135. Guo Z, Huang Q, Cui Z et al (2025) Targeting YTHDF2 with pH-responsive siRNA nanoparticles suppresses MYC m6A modification and restores antitumor immunity in hepatocellular carcinoma. J Nanobiotechnol 23:469. https://doi.org/10.1186/s12951-025-03538-0
136. Oh CY, Kaur H, Tuteja G et al (2024) DNA origami drives gene expression in a human cell culture system. Sci Rep 14:27364. https://doi.org/10.1038/s41598-024-78399-y
137. Opsomer L, Jana S, Mertens I et al (2024) Efficient *in vitro* and *in vivo* transfection of self-amplifying mRNA with linear poly(propylenimine) and poly(ethylenimine-propylenimine) random copolymers as nonviral carriers. J Mater Chem B 12:3927–3946. https://doi.org/10.1039/D3TB03003B
138. Grimaldi MC, Bozzer S, Sjöström DJ et al (2025) DNA-loaded targeted nanoparticles as a safe platform to produce exogenous proteins in tumor B cells. Front Immunol 15:1509322. https://doi.org/10.3389/fimmu.2024.1509322
139. Dabas R, Navaratnam N, Iino H et al (2025) Precise intracellular uptake and endosomal release of diverse functional mRNA payloads via glutathione-responsive nanogels. Mater Today Bio 30:101425. https://doi.org/10.1016/j.mtbio.2024.101425
140. Feja M, Drath I, Weiß S et al (2025) Nose-to-brain siRNA delivery by PEI/PPI-based nanoparticles reduces α-synuclein expression in a Parkinson's disease mouse model. Mol Ther Nucleic Acids 36:102671. https://doi.org/10.1016/j.omtn.2025.102671
141. Zhang J, Zhang YP, Sun Q et al (2025) Cancer cell membrane-camouflaged pH-responsive nanoparticles for enhancing siRNA effect and synergistic anti-tumor therapy. J Nanobiotechnol 23:471. https://doi.org/10.1186/s12951-025-03508-6
142. Zamolo S, Zakharova E, Boursinhac L et al (2024) Peptide dendrimers transfecting CRISPR/Cas9 plasmid DNA: optimization and mechanism. RSC Chem Biol 5:891–900. https://doi.org/10.1039/D4CB00116H
143. Padilla MS, Mrksich K, Wang Y et al (2025) Branched endosomal disruptor (BEND) lipids mediate delivery of mRNA and CRISPR-Cas9 ribonucleoprotein complex for hepatic gene editing and T cell engineering. Nat Commun 16:996. https://doi.org/10.1038/s41467-024-55137-6
144. Bae H, Ji H, Konstantinov K et al (2025) Artificial intelligence-driven nanoarchitectonics for smart targeted drug delivery. Adv Mater 37:e10239. https://doi.org/10.1002/adma.202510239
145. Yingngam B, Navabhatra A, Sillapapibool P (2024) AI-driven decision-making applications in pharmaceutical sciences. In: Nguyen TVT, Vo NTM (eds) Using traditional design methods to enhance AI-driven decision making. IGI Global Scientific Publishing, Hershey, PA, pp 1–63. https://doi.org/10.4018/979-8-3693-0639-0.ch001

146. Song N, Wang L, Zhang L et al (2025) Precision delivery of CRISPR/Cas systems via DNA nanostructures for gene therapy and intracellular detection. ChemBioChem 26:e202500357. https://doi.org/10.1002/cbic.202500357
147. Patil C, Priyanka R, Harshitha BM et al (2025) Advanced nanotheranostic approaches for targeted glioblastoma treatment: a synergistic fusion of CRISPR-Cas gene editing, AI-driven tumor profiling, and BBB-modulation. Med Oncol 42:413. https://doi.org/10.1007/s12032-025-02944-6
148. Li F, Fornasiero P, Xu P et al (2025) Nature-inspired living materials for health, energy, and sustainability. ACS Appl Bio Mater 8:10483–10505. https://doi.org/10.1021/acsabm.5c01099
149. Beyki MS, Mashhadi E, Rashidi J et al (2025) Nature-inspired chemical methods. In: Green biomaterials in tissue engineering. American Chemical Society, Washington, DC, pp 1–53. https://doi.org/10.1021/bk-2025-1497.ch001

Safety and Environmental Risks in Medical Applications of Nanotechnology

Debnath Das, Manish Kumar, Shruti Srivastava, Uma Shankar Maurya, and Rupesh K. Gautam

Abstract Nanotechnology has developed as a revolutionary platform in the current medical practice, which has allowed novel methods to detect a disease, targeted therapy, and customized health care. Through the utilization of physicochemical characteristics on a nanoscale, nanotechnology-based medical products provide higher bioavailability, controlled drug delivery, and high therapeutic efficacy levels than traditional systems. Nevertheless, the same characteristics that make them have those benefits also pose significant safety and environmental issues, which should be tackled with care to provide responsible clinical translation. This chapter gives a critical assessment of the safety and environmental hazards of medical uses of nanotechnology. It discusses the effect of the physicochemical characteristics of nanoparticles on biological interactions, biodistribution, and toxicity, and focuses specifically on organ-specific toxicity, immunological responses, genotoxicity, and long-term health effects. The major nanotoxicological processes, such as oxidative stress, inflammation, mitochondrial dysfunction, and the cell death pathway, are also analyzed. Also, the chapter points out the environmental fate and ecotoxicological consequences of medical nanomaterials released during the manufacturing process, clinical use, disposal, and excretion in patients. Problems of regulation, ethics, and risk evaluation are also discussed with special attention to the shortcomings of traditional models when used to address the nanoscale systems. The chapter goes further to discuss the safe-by-design strategies, life-cycle assessment methods,

D. Das
Department of Pharmaceutics, ISF College of Pharmacy, Moga, Punjab, India

M. Kumar (✉)
KIET School of Pharmacy, Krishna Institute of Engineering & Technology (KIET), Ghaziabad Delhi NCR, Uttar Pradesh, India

S. Srivastava
Directorate of Research, Innovation & Consultancy, Chandigarh University, Uttar Pradesh, India

U. S. Maurya
Goel Institute of Pharmacy and Sciences, Lucknow, India

R. K. Gautam (✉)
Centre for Pharmacology, Amity Institute of Pharmacy, Amity University Uttar Pradesh, Noida, India

J. Chen (ed.), *AI-Driven Nanotechnology in Precision Medicine and Healthcare*, Smart Nanomaterials Technology, https://doi.org/10.1007/978-981-92-1690-1_15

and predictive toxicology tools as viable ways to reduce risks. Overall, this chapter highlights the need to incorporate safety, environmental sustainability, and ethical aspects of nanomedicine development to ensure safe and sustainable development of nanomedicine in the medical sector.

Keywords Nanomedicine · Nanotoxicology · Medical nanotechnology · Safety assessment · Environmental risk · Safe-by-design · Regulatory challenges

1 Introduction

Medical nanotechnology has proven to be a revolutionary field in the current health-care industry, as nanoscale materials can now be designed to interact with biological systems at molecular and cellular scales to be used in diagnosis and treatment [1–5]. Nanomaterials, as unique physicochemical substances, have been utilized to enhance the drug solubility, targeted delivery, and therapeutic efficacy in a broad spectrum of diseases through their small size, high surface area, and ability to tune surfaces [2, 3, 6].

Nevertheless, safety is a critical issue with these same properties. Nanoparticles may also have different pharmacokinetics, like longer circulation times, increased tissue penetration, and penetration through biological barriers, such as the blood–brain and placental barriers [1, 7]. This kind of behavior predisposes the chance of unintended contacts with healthy tissues and could lead to organ accumulation, oxidative stress, and inflammatory reactions [1, 1].

There is increasing evidence that conventional toxicological paradigms with their focus on the mass-based dose parameters and their long-term exposure paradigms are inadequate in the assessment of nanoscale risks [2]. The toxicity of nanomaterials is typically determined by size, surface charge, shape, and the formation of protein corona and is not sufficiently considered in traditional safety testing procedures [1, 9]. Consequently, combined with the growing awareness of the necessity of specific nanotoxicology methods, based on the combination of physicochemical characterization and biological and environmental evaluation [10, 11].

Simultaneously, issues have been raised in terms of the environmental impact of medical nanotechnology. Nanomaterials that are discharged in the manufacturing, clinical use, or excretions by patients might be deposited in wastewater systems and natural ecosystems, where their fate and their overall ecological impact are not well understood [12, 13]. The uncertainties underscore the need to implement life-cycle-based risk assessment measures that focus on human health as well as environmental sustainability [11, 14, 15].

2 Overview of Medical Nanotechnology

Medical nanotechnology, also known as nanomedicine, is the use of nanoparticles and systems in diagnostics, treatment, and prevention of illness [3, 4]. Nanotechnology can be used to interact with cellular and molecular pathways with high precision because it is possible to produce materials on the scale of biological macromolecules [2, 4].

Medical nanomaterials can be mainly categorized into organic, inorganic, and hybrid systems [3, 4]. Lipid nanoparticles, polymeric nanoparticles, dendrimers, and micelles are organic nanomaterials, which are currently used in drug and gene delivery due to their biocompatibility, biodegradability, and structural flexibility [3, 16]. Nanoparticles based on lipids have also become clinically important in the delivery of nucleic acids with controlled release of drugs and functionalization of surfaces of polymeric nanoparticles to target therapies, becoming particularly important [2, 3].

Inorganic nanomaterials, including gold nanoparticles, iron oxide nanoparticles, silica nanoparticles, and quantum dots, have optical, magnetic, and electronic properties that are not available in other materials, thereby rendering them useful in imaging, diagnostics, and theranostic applications [4]. Nanoparticles of iron oxide and gold have been widely used in magnetic resonance imaging contrast agents and photothermal therapy and biosensing, respectively. Nevertheless, safety issues of poor biodegradability, tissue persistence over time, and the possibility of toxicity have limited the overall broad clinical application of most inorganic nanomaterials [7, 9].

Hybrid nanomaterials are also organic and inorganic materials that are used to integrate diagnostic and therapeutic capabilities together as a single platform [4]. These systems are multifunctional and have high potential in customized medicine, but add complexity regarding the consistency of manufacturability, safety testing, and regulatory authorization [17].

The uses of nanomedicine in clinical applications are in the field of diagnosis, therapy, and regeneration [3, 4]. Diagnostic nanotechnology encompasses nanoparticle-based imaging agents and nanosensors in diagnosing diseases early; the therapeutic application in nanotechnology is predominantly in oncology and gene therapy [2, 3]. Although the regulatory approvals continue to rise, inadequate knowledge about nanoparticle-biological interactions, variability in formulations, and unresolved safety issues continue to limit the clinical translation of nanomedicines [2].

Moreover, the life cycle of the medical nanomaterials, such as manufacturing, clinical applications, excretion, and disposal, presents several chances of human and environmental contact [11, 13]. These aspects underscore the importance of considering safety and environmental issues in the formulation and introduction of medical nanotechnology [11].

3 Physicochemical Properties Influencing Nanotoxicity

The physicochemical properties of nanomaterials employed in the medicine field dictate the toxicological behavior of these nanomaterials. Contrary to traditional pharmaceuticals, nanoparticles are attracted to biological systems based on the highly dependent parameters of the particle size, shape, surface charge, chemical composition, and aggregation state [8, 9]. The sum of these properties establishes biodistribution, cellular uptake, clearance, and possible adverse effects of nanoparticles.

3.1 Particle Size and Size Distribution

One of the most important factors that can define nanotoxicity is particle size. The nanoparticles in the nanoscale size range are more likely to be internalized by cells because of their capacity to penetrate the cells through endocytic mechanisms [8, 18]. The smaller-sized nanoparticles are usually more reactive and have a higher surface area that can increase oxidative stress and intracellular injury [1, 9]. In addition, nanoparticles that are smaller than specific size limits may penetrate biological barriers, including the blood–brain barrier and the placental barrier, which creates the issue of neurotoxicity and developmental toxicity [19].

The size distribution also plays an important role in the calculation of safety. Polydisperse nanoparticle systems can display heterogeneous biological behavior (the smaller portions of the system can play a disproportionately larger role than the larger particles) [8].

3.2 Shape and Morphology

The shape of nanoparticles has a great impact on cellular response and interaction. The rabbits of nanomaterials may be internalized more easily compared to rod-shaped or fibrous ones (spherical nanoparticles) [9]. Conversely, the nanostructure having a high aspect ratio, including nanotubes and nanofibers, has been linked to greater inflammatory reaction and decreased clearance because of frustrated phagocytosis [9, 20].

3.3 Surface Charge and Surface Chemistry

Surface charge is also a major factor in the interactions of nanoparticles with cell membranes. Nanoparticles with positive charges are more likely to have better electrostatic interactions with negatively charged cellular membranes, which results in more uptake, more cytotoxicity, and membrane disruption [1, 18]. Nonspecific interactions and biocompatibility are usually reduced with neutral or negatively charged nanoparticles [1, 9].

Surface chemistry (functional groups and coating) also has an impact on nanoparticle stability and biological identity. To evade immune recognition and increase circulation time, the modification of the surface using polymers, i.e., polyethylene glycol, is typically used [1, 16]. However, under physiological conditions, reactive cores may be exposed by the labile or degradation of surface coatings and alter the toxicity profiles [9].

3.4 Protein Corona Formation

When nanoparticles are exposed to biofluids, they quickly adsorb proteins and other biomolecules and form a dynamic protein corona that determines their biological identity [1, 9]. Each protein corona composition determines cellular uptake, immune activation, and biodistribution. Whereas some coronas can work to neutralize poisoning by covering active sites, some can increase the immune cell recognition and promote inflammation [1].

3.5 Chemical Composition, Solubility, and Aggregation

Toxicity of nanomaterials is significantly contributed to by the intrinsic chemical composition. The biodegradable nanomaterials are usually linked with reduced long-term risk because of their capability to break down into non-toxic degradable products [3]. Conversely, the inorganic nanoparticles can be retained in tissues and lead to reactive oxygen species caused by oxidative stress or toxic ions release, which cause chronic toxicity [9, 18, 21].

Aggregation behavior also controls toxicity by modification of biodistribution and effective size of particles. Aggregated nanoparticles are also likely to be deposited in the liver and spleen, whereas unstable nanoparticles disaggregated in vivo would release more reactive species [8, 12].

4 Safety Risks in Medical Applications of Nanotechnology

Though nanotechnology has allowed for major breakthroughs in the area of medical diagnostics and therapeutics, it is also accompanied by a variety of safety risks that should be thoroughly considered. The main feature of these risks is due to the biological peculiarities of nanomaterials, which are fundamentally different from conventional pharmaceuticals and medical devices [1, 2]. Small size, high surface area, and increased reactivity of nanoparticles enable their easy uptake and targeting to cells, but also enhance the risk of unwanted biological responses [1, 8].

4.1 Organ-Specific Toxicity

After systemic injection, nanoparticles have a tendency to be trapped in the reticuloendothelial system and therefore accumulate mainly in the liver and the spleen [8, 18]. Hepatic accumulation can cause oxidative stress, inflammatory signaling, and impairment of the normal metabolic pathways to cause hepatotoxicity [18]. The further causes of liver damage during repeated or chronic exposure are the Kupffer cell activation and the retention of the nanoparticles [8].

Another major concern is renal toxicity, especially with nanoparticles that have been developed to be cleared from the body by the renal system. Even though small nanoparticles can be excreted effectively, repeated exposure or defective kidney functioning may cause the build-up in renal tissues, which may cause nephrotoxicity. The contacts with renal epithelial cells can impair the filtration process and cause stress in the cells [18].

Pulmonary toxicity can be caused when the nanoparticles are inhaled or reach the lungs through the systemic circulation. Nanoparticles can also cause inflammation, oxidative, and fibrotic reactions in pulmonary tissues, particularly when they become permanent residents in the alveolar macrophages [9, 20]. The fact that some nanoparticles can pass through the blood–brain barrier also brings additional worries about neurotoxicity because their accumulation in the neural tissues can impair neuronal communication and facilitate neuroinflammation [19].

4.2 Immunotoxicity and Hypersensitivity Reactions

Nanoparticles have the potential to significantly regulate the functionality of the immune system, causing immunostimulatory and immunosuppressive effects [1, 20]. Immune cell recognition of nanoparticles can lead to macrophage, dendritic cell, and complement pathway activation, which leads to cytokine release and inflammatory reactions [1, 8]. It has been reported that acute immune responses, including pseudo-allergy due to the complement activation, occur after intravenous injection of some nanomaterials [20].

On the other hand, certain nanomaterials can hide immune responses through altering antigen presentation or immune cell signaling, enhancing vulnerability to infection or weakening immune response against malignant cells [1]. Repeated dosage is also known to cause immune sensitization, thereby risking the chance of exaggerated responses at the next exposure [20].

4.3 Genotoxicity and Carcinogenic Concerns

A significant safety issue of medical nanotechnology in the long term is genotoxicity. Nanoparticles can cause DNA damage either by direct contact with genetic material or indirectly by oxidative stress and inflammatory mediators [9, 18]. Some classes of nanomaterials have been reported to cause DNA strand breaks, chromosomal aberrations, and changes in the pattern of gene expression in experimental models [18].

Incessant oxidative stress and chronic inflammation can provide the environment that facilitates carcinogenesis, especially when dealing with non-biodegradable and high-aspect-ratio nanomaterials [9, 17]. Even though there is still no conclusive data that nanoparticles cause cancer in humans, these results suggest that long-term safety assessment and post-marketing surveillance are crucial [17].

4.4 Dose, Exposure Duration, and Cumulative Effects

Nanoparticle dose toxicity relationship is nonlinear, and chronic exposure at low doses may be of more concern than high-dose exposure, acute exposure in the clinical risk assessment [8, 9]. Nanoparticles can have long-term retention in tissues and therefore cumulative effects in repeated dose regimens that are common in cancer treatment and chronic diseases. The absence of standardized dose measures also makes it difficult to compare risks and compare studies across studies [17].

4.5 Clinical Translation and Vulnerable Populations

Not all preclinical models are good predictors of human behavior with respect to nanomaterials, hence the uncertainties during clinical translation [2]. Prior dysfunction of age, immune condition, genetic background, and patient-specific factors can be a major determinant of nanoparticle safety [19]. There is a risk of increased vulnerability in vulnerable populations such as pregnant women, children, and frail patients, since some nanoparticles can penetrate placental barriers [19].

5 Nanotoxicological Mechanisms

Nanotoxicological mechanisms refer to the molecular and cellular events by which nanomaterials cause adverse biological impacts. Compared to traditional toxicants, nanoparticles can cause a combination of interrelated, simultaneous pathways because of the nanoscale size, high surface reactivity, and dynamic reactions between the biological systems and nanoparticles [1, 8]. Such mechanisms should be known to predict safer outcomes of nanomedicine and design safer nanomedicine platforms.

5.1 Oxidative Stress and Reactive Oxygen Species Generation

The role of oxidative stress is generally accepted as one of the key pathways of nanoparticle toxicity. Nanomaterials can be used to produce reactive oxygen species either directly by surface redox reactions or indirectly by destabilizing mitochondrial processes and cellular antioxidant protection systems [8, 9, 18]. The extreme production of reactive oxygen species may overload the intrinsic production of antioxidant mechanisms, resulting in lipid peroxidation, oxidation of proteins, and damage to DNA [18].

Oxidative stress is especially attributed to metal and metal oxide nanoparticles because of their catalytic properties and possible ion release [9]. Continuous oxidative stress not only causes injury to the cellular components but also triggers downstream signaling pathways, which induce inflammation, apoptosis, and fibrosis [8, 9].

5.2 Inflammation and Immune Signaling Pathways

Nanoparticles can initiate inflammatory responses by interacting with other immune cells, such as the macrophages and dendritic cells. Nanoparticle recognition as foreigners would lead to pattern recognition receptor activation and subsequent release of pro-inflammatory cytokines [1, 8]. Acute inflammation can be considered a defense mechanism, but chronic inflammation and tissue damage can be caused by repeated or prolonged exposure to nanoparticles [9].

Nanoparticles have been connected to inflammatory signaling that causes organ-specific toxicity and can worsen underlying inflammatory or autoimmune disease [20]. The disease progression is also linked with a chronic inflammatory microenvironment and can also cause a long-term pathological outcome [9].

5.3 Mitochondrial Dysfunction

The vulnerability of mitochondria to the effects of nanoparticles is especially high because they are the ones that carry out energy metabolism in the cell and apoptosis regulation. Localization to mitochondria nanoparticles could disrupt electron transport chains, inhibit the production of adenosine triphosphate, and increase the production of mitochondrial reactive oxygen species [18]. Mitochondrial membrane potential loss is also a typical sign of cytotoxicity caused by nanoparticles and can normally precede cell death [8].

When mitochondrial dysfunction occurs, intrinsic apoptotic pathways may be activated, resulting in programmed cell death. Apoptosis can be useful in the treatment of cancer, but it has serious safety consequences when accidentally activated in normal tissues [9, 18].

5.4 Cell Death Pathways

Depending on the characteristics of nanoparticles, concentration, and time of exposure, nanoparticle exposure can cause various types of cell death. Apoptosis has been commonly found after the oxidative stress and mitochondrial damage induced by nanoparticles and is marked by a regulated cellular degradation [18]. Conversely, extreme concentrations or acute exposure can cause necrosis, which involves rupture of the membrane and intense inflammatory reactions [9].

Another cellular reaction to stress caused by nanoparticles is autophagy. In early studies, autophagy was viewed primarily as a pro-survival process, the clearance of damaged organelles, autophagy may lead to cell death and tissue malfunction under conditions of dysregulation or over-expression [8, 9].

5.5 Genotoxic and Epigenetic Effects

Nanoparticles can have a genotoxic effect by direct interaction with DNA or indirectly by oxidative stress and inflammation [9, 18]. Experimental research on different nanomaterials has reported DNA strand breaks, chromosomal aberrations, and modified gene expression patterns [18]. This would lead to a reduction in genomic stability and predisposition to diseases.

Besides having direct genotoxicity, nanoparticles can have epigenetic effects, such as DNA methylation and histone modification patterns [9]. The disturbing aspect of these epigenetic effects is that they might be long-lasting and might affect long-term cellular behavior even though the DNA sequence does not change.

5.6 Disruption of Cellular Barriers and Homeostasis

Nanoparticles can damage cellular and tissue barriers such as epithelial and endothelial layers and tight junction integrity [1]. Increased barrier permeability may add to undesired translocation and systemic translocation of nanoparticles. Further, the changes that occur in the calcium signaling, membrane fluidity, and ion homeostasis lead to the dysfunction and toxicity of the cell [8].

6 Environmental Risks and Ecotoxicology of Medical Nanotechnology

Besides patient issues of safety, the growing application of nanotechnology in medicine brings up some serious questions of environmental risk and ecological sustainability. Medical nanomaterials may be released into the environment on a variety of different life cycles, such as during their manufacture, during clinical use, when excreted into patients, and during the disposition of nanotechnology-containing medical waste [12, 13]. When set free, they can interact with the environmental systems in unpredictable and unconventional manners, which are likely to affect the ecosystem and indirectly impact the well-being of the human population.

6.1 Environmental Release Pathways

Nanomaterials that are used in medicine can be accidentally discharged into the environment during their manufacturing, which involves insufficient containment or waste management systems that release the nanoparticles into the air, water, or soil [13, 14]. When using nanoparticles in a clinical setting, whether as a diagnostic or treatment agent, the nanoparticles released into the environment (through hospital effluents or released by the patient after treatment) may be discharged [12, 13]. Traditional wastewater plants are not specifically designed to eliminate nanoscale particles, and a portion of nanomaterials will remain in effluents or sewage sludge [13].

Nanomedical medical equipment, packaging, and formulations that are not in use are also disposed of, which leads to environmental exposure. The release pathways emphasize the significance of life-cycle assessment of medical nanomaterials when deciding on the environmental risk [11, 14].

6.2 Persistence, Transformation, and Environmental Fate

On entry into environmental compartments, nanomaterials can be subjected to physical, chemical, and biological changes that affect their mobility, bioavailability, and toxicity [12, 14]. The environment (e.g., pH, ionic strength, temperature, natural organic matter) can be conducive to nanoparticle aggregation, dissolution, or surface modification [12]. The transformations can neutralize toxicity through lessening the reactivity or boosting the toxicity through creating increased bioavailable or reactive forms [14].

Some of the inorganic nanomaterials being used in medicine are very persistent in the environment as they are not easily degraded [9]. The chronic nanoparticles can be deposited in the sediment and soil, where chances of sustained exposure of the organisms in the environment are likely to occur. Biodegradable nanomaterials are also able to release smaller units that can be biologically active or toxic, which outlines the importance of studying the environmental fate of nanomaterials thoroughly [12].

6.3 Aquatic Ecotoxicity

Contamination of nanomaterials is also consistent in aquatic ecosystems, in which surface waters tend to provide the ultimate location of wastewater discharges [13]. Nanoparticles found in water bodies may react with algae, plankton, invertebrates, and fish. It has been experimentally proven that some nanomaterials may prevent the growth of algae, interfere with photosynthesis, and cause oxidative stress in aquatic life [12, 14].

The invertebrates can absorb nanoparticles either by filtration or feeding, which contributes to impaired reproduction and altered behavior [14]. Exposure to nanoparticles in fish has been linked with gill damage, oxidative stress in important organ systems, and metabolic disruptions [12]. Although acute toxicity is usually apparent in higher levels, low environmental levels of chronic exposure can lead to non-lethal effects in the long term with ecological impacts[14].

6.4 Terrestrial Ecosystems and Soil Health

The nanomaterials released to the terrestrial environment by biosolids, landfill leachate, or agricultural runoff can influence the soil organisms and plant health [12]. The soil microorganisms are the most vulnerable to the exposure of nanoparticles, and any change in the microorganism community structure or enzyme activity may interfere with nutrient cycling and soil fertility [14].

Nanoparticles can be absorbed by plants via the root system, whereby internal translocation and accumulation can occur in the edible tissues [12]. Whereas other nanomaterials have been reported to promote plant growth at low doses, others have induced phytotoxicity in plants, including oxidative stress, biomass reduction, and gene expression changes [14]. These results are of concern as they suggest the entry of nanoparticles into the food chain and indirect exposure to humans.

6.5 Bioaccumulation and Trophic Transfer

One of the ecological issues that has been linked to medical nanotechnology is the possibility of bioaccumulation and trophic transfer of the nanomaterials [13, 14]. Lower troicola organisms that accumulate nanoparticles could further be passed to the higher troicola levels by predation, resulting in biomagnification. There is still no clear evidence on the existence of extensive biomagnification, but it still cannot be ruled out, especially in the case of persistent and poorly degradable nanomaterials [9].

6.6 Environmental Risk Assessment Challenges

Evaluation of the environmental hazards of medical nanotechnology is a complex issue due to restrictions brought about by the conventional ecotoxicological testing methodology, which had mainly been designed to assist in the evaluation of dissolved chemicals [11, 17]. The analytical measures exclusively involving the mass concentrations might not provide enough information about nanoparticles, whereas measures, like the size of particles, their count, and their surface characteristics, can prove to be more important agents of ecological effect [14]. All these challenges precondition the importance of a nanoparticle-specific environmental risk assessment paradigm and standardized testing methods [11, 17].

7 Regulatory, Ethical, and Risk Assessment Challenges

The fast pace at which nanotechnology has evolved in the medical field has exceeded the establishment of proper regulatory frameworks that would be able to tackle the special safety and environmental issues that come hand in hand with nanoscale materials. The existing pharmaceutical and medical device regulatory frameworks were created mostly around bulk substances and small molecules and, in general, do not take into consideration the specific physicochemical characteristics of nanomaterials, including size-dependent behavior, surface reactivity, or altered biodistribution [11, 22]. This has often seen nanomedicines being reviewed within existing regulations without due regard to the property to nano-specific risks, resulting in a high level of regulatory uncertainty and unequal safety evaluations across the regions [10, 17].

A major regulatory issue that exists is that there is no standardized definition and classification of nanomaterials. Differences in the size limit, surface changes, and functional features make regulatory approval processes difficult and impede international harmonization [11, 22]. Also, the conventional measures of dose in terms of mass concentration can be unsuitable to measure nanotoxicity because nanoparticles can alter toxicity with surface area, number of particles, or reactivity [9, 11]. This is a weakness that can undermine the validity of preclinical safety data and complicate the association of laboratory results with clinical and environmental situations.

Nanomedicines are also difficult to evaluate with regard to risk due to inadequate data on validated long-term toxicity. The majority of the preclinical studies are put into the realm of acute or short exposure, but the chronic exposure situations, such as repeated dosing, longer-term implantation, or environmental persistence, are not much studied [2, 17]. Lack of standard testing procedures regarding immunotoxicity, genotoxicity, and reproductive toxicity complicates the uncertainty concerning cumulative and delayed effects [9, 18]. Besides, the active change in nanomaterials in biology and in the environment, such as aggregation, dissolution, or the formation of protein coronas, further complicates the models of risk prediction [1, 12].

The encroachment and implementation of medical nanotechnology in ensuring the ideal development and utilization of nanotechnology are fundamentally ethical. The concept of informed consent is especially demanding in the case of long-term risks that are either uncertain or poorly understood, and the matter of patient autonomy and transparency is left to be questioned [10, 20]. Ethical concerns do not only limit themselves to individual patients, but also to larger societal and environmental concerns. The possibility of nanomaterials escaping into the ecology due to manufacturing processes, medical waste, and excretion by patients would pose the question of intergenerational equity and environmental justice [12, 14]. The consideration that the benefits of nanomedicine should not be at the cost of environmental health is one of the ethical imperatives.

Also, fair access to nanotechnology-based medical technologies is yet another ethical and regulatory issue. The development costs are very high, manufacturing processes are complicated, and the regulatory requirements are very strict, which may restrict its accessibility, especially in low- and middle-income countries [10, 20]. To handle such disparities, it is necessary to have concerted policy action, align the regulatory framework on a global scale, and communicate risk benefits clearly.

All in all, the regulatory, ethical, and risk evaluation issues revolving around medical nanotechnology demonstrate the critical importance of having nano-specific regulatory policies, standardized testing systems, and combined life-cycle risk assessment strategies. A greater cooperation between the scientists, regulators, ethicists, and policymakers will be necessary in ensuring that nanomedicine can grow in a safe, ethical, and sustainable manner that protects the environment and maintains the trust of the population [11, 16, 22].

8 Risk Mitigation Strategies and Safe-by-Design Approaches

The growing clinical translation of nanotechnology-based medical products has manifested the need for proactive risk mitigation approaches that deal with safety and environmental issues across the product life cycle. One of such strategies is the safe-by-design (SbD) notion, with its focus on the incorporation of safety concerns into the very initial phases of nanomaterial development as opposed to the application of toxicological tests only after a certain material is created [1, 9]. This approach is to create a balance between therapeutic efficacy, fewer biological and ecological risks, and quick regulatory acceptance.

Safe-by-design approaches are initiated by the rational choice of the nanomaterial composition, with the preference of biodegradable, biocompatible, and metabolizable materials over the persistent inorganic systems where possible [11, 16]. Polymer-based nanoparticles, lipid-based carriers, and naturally found nanomaterials have become the preferred nanostructures due to their lower accumulation and better clearance profiles over metallic or carbon-based nanostructures [8, 9]. Surface modification is a very important aspect of risk mitigation since any functionalization with polyethylene glycol (PEG), zwitterionic surfaces, or biomimetic ligands can decrease nonspecific adsorption of proteins, immune activation, and clearance by the reticuloendothelial system[1, 17].

The other fundamental feature of the SbD paradigm is the optimization of the physicochemical parameters, including particle size, shape, surface charge, and aggregation behavior. Optimizing these properties can have important cellular uptake, biodistribution, and toxicity consequences [8]. As an illustration, nanoparticles in the size range of 50–150 nm may exhibit desirable pharmacokinetics and evade overfiltration in the kidney or unwanted entry into the body via sensitive types of biological barriers such as the blood–brain barrier [9]. Likewise, the neutral or slightly negative surface charges are typically linked to reduced cytotoxicity as well as reduced hemolytic activity [8, 17].

There are growing uses of advanced in vitro and in silico models in early-stage screening as part of the safety of new products to minimize animal usage and more effectively determine potential risks [1, 11]. Mechanistic data of nanoparticle-induced oxidative stress, inflammation, and genotoxicity under physiologically relevant conditions are provided by the use of three-dimensional cell cultures, organ-on-chip models, and computational toxicology models and tools [9, 17]. These predictive tools can permit modifications in the design iteratively, and therefore, unsafe formulations can be removed before large-scale manufacturing or clinical testing.

The mitigation of risks should also not limit itself to the safety of patients, but also to their environmental sustainability. The methods used to measure the environmental exposure risks during the synthesis, clinical, disposal, and post-excretion routes of nanoparticles require life-cycle assessment (LCA) [11, 16]. The principles of green nanotechnology, such as reduction of solvents, efficient energy utilization in the synthesis, and reduction of waste, can significantly reduce the ecological footprint of nanomedicine production [12, 16]. Also, it is important to develop standardized disposal and wastewater treatment procedures to avoid uncontrolled dumping of medical nanomaterials into aquatic and terrestrial environments [13, 14].

The combination of safe-by-design principles and sound regulatory science, predictive toxicology, and environmental stewardship can be seen as a holistic approach toward the reduction of risks related to medical nanotechnology. Nanomedicine can be developed to reach the goal of clinically effective, ethically, and environmentally responsible health care by implementing safety considerations across the innovation pipeline [1, 9, 11, 16].

9 Future Perspectives and Research Gaps

Although the sphere of medical nanotechnology advances at a very fast pace, there are still some research gaps that should be closed to make it safer, clinically reliable, and environmentally sustainable in the future. The prevailing evidence is predominantly based on experimental and preliminary clinical trials at the short-term level, which might not be sufficient to ascertain the chronic, cumulative, and delayed impacts related to exposure to nanomaterials [1, 9, 11]. To solve these shortcomings, it is vital to respond to the responsible development of nanomedicine.

Among the most urgent gaps, the absence of long-term toxicity data should be mentioned. Nanomedicines targeting repeated dosage or chronic therapy are often meant to be used over a long period, but the systematic research assessing the long-term biodistribution, persistence, and clearance of such drugs is not frequently done [9, 17]. Specifically, the possibility of bioaccumulation in critical tissues like the liver, the spleen, the brain, and reproductive tissues is undercharacterized [18, 19]. There is thus an urgent need to have longitudinal in vivo studies that can be used to simulate clinically relevant dosing regimens.

The other significant research gap is that of vulnerable and special populations. The majority of safety tests are performed with models of healthy adults, which are unlikely to be representative of children, the elderly, pregnant, and those with compromised immune system and organ activities [19]. Since it has been demonstrated that there is a possibility that certain nanoparticles may penetrate the placental and blood–brain barriers, more research should be placed on reproductive, developmental, and neurotoxicity [19].

As a methodology, predictive and mechanism-based safety models are required. Conventional toxicology measures cannot typically explain nanoscale-related phenomena, which include protein corona formation, dynamic changes of surfaces, and nonlinear dose–response dynamics [1, 8]. More promising systems are advanced in vitro systems, including organ-on-a-chip systems and multi-organ systems, which provide more physiologically relevant conditions in which to test toxicity [1, 8]. Nevertheless, these models would need additional validation and standardization before they are accepted by the general regulatory policy.

Lapses in environmental research are also important. In spite of laboratory research that has proven the possible ecotoxicological risks of nanomaterials, information on exposure concentrations and long-term environmental ecosystem effects at realistic levels is limited [12, 14]. In future research, it may be advisable to combine both the life-cycle assessment methods with the environmental monitoring so that the actual exposure and the cumulative ecological risk can be well quantified [11, 14]. It is especially important to understand the transformation, persistence, and trophic transfer of nanoparticles in the natural environment to understand the long-term environmental consequences.

Data integration and standardization are another major challenge. The different ways nanomaterials are characterized, the conditions of testing, and reporting practices prevent reproducibility and cross-study comparison [17]. Standardized testing procedures, reference standards, and databases would play a great role in improving the consistency of the safety tests and in the regulation decision-making [11, 22].

Lastly, interdisciplinary and international collaboration is necessary for future developments in the field of medical nanotechnology. The foundation of the process of scientific research alignment with regulatory, ethical, and societal issues will be crucial in addressing the translation of nanomedicine breakthroughs into safe and universal clinical solutions [10, 20]. Further funding in nanosafety studies must be regarded as a strategic necessity that would facilitate, but not limit, the future development of nanotechnology in medicine.

10 Conclusion

Nanotechnology in medicine has been one of the important inventions that has been able to transform the current health care into a new way of healthcare delivery. Medical products that use nanotechnology have proven to exhibit distinct benefits in comparison to their traditional counterparts, especially in the delivery of drugs to specific destinations, imaging, and personalized medicine. Nevertheless, the special physicochemical properties underlying these advantages result in challenging safety and environmental issues, as well.

Nanomaterials may possess increased biodistribution, long tissue residence, and cross into crucial biological barriers, which cause possible organ-selective toxicity, immune regulation, and genotoxicity. The lack of linear dose–response relationships and cumulative exposure also makes it more difficult to assess the safety of nanomedicines, particularly when they are to be used repeatedly or over an extended period of time. These outcomes show how ineffective the traditional models of toxicology can be under the conditions of nanoscale systems.

Another aspect of nanomedicine risk is environmental considerations. The nanomaterials that are discharged during manufacturing, clinical practice, disposal, or excretion by the patient into the environment might remain in the environment, be transformed, and interact with biological systems at different trophic levels. Even when the exposure amounts are small, chronic accumulation and long-term ecological impacts are issues that should not be overlooked, and thus, methodologies of life-cycle-based assessment should be applied.

Regulatory and ethical issues continue because there are no nano-specific regulations, dose measurements, and international standard harmonies. To resolve these gaps, it is necessary to have good governance, sound risk assessment plans, and stakeholder participation.

To sum up, safe-by-design, predictive toxicology, and environmental sustainability represent the key to responsible development of medical nanotechnology on all levels of development. Nanomedicine is capable of developing further as a transformative and sustainable aspect of contemporary health care due to the balance of innovation and safety, as well as ethical responsibility.

References

1. Fadeel B et al (2018) Advanced tools for the safety assessment of nanomaterials. Nat Nanotechnol 13(7):537–543. https://doi.org/10.1038/s41565-018-0185-0
2. Shi J, Kantoff PW, Wooster R, Farokhzad OC (2017) Cancer nanomedicine: progress, challenges and opportunities. Nat Rev Cancer 17(1):20–37. https://doi.org/10.1038/nrc.2016.108
3. Pelaz B et al (2017) Diverse applications of nanomedicine. ACS Nano 11(3):2313–2381. https://doi.org/10.1021/acsnano.6b06040
4. Se M (2005) Nanotechnology for the biologist. J Leukoc Biol 78(3):585–594
5. Ma X et al (2024) Nanotechnology in healthcare, and its safety and environmental risks. J Nanobiotechnology 22(1):715. https://doi.org/10.1186/s12951-024-02901-x
6. Fortune A, Aime A, Raymond D, Kumar S (2025) Nanotechnology in medicine: a double-edged sword for health outcomes. Health Nanotechnol 1(1):9. https://doi.org/10.1186/s44301-025-00008-2
7. Gupta R, Xie H (2018) Nanoparticles in daily life: applications, toxicity and regulations. J Environ Pathol Toxicol Oncol 37(3):209–230. https://doi.org/10.1615/JEnvironPatholToxicolOncol.2018026009
8. Nel A et al (2013) Nanomaterial toxicity testing in the 21st century: use of a predictive toxicological approach and high-throughput screening. Acc Chem Res 46(3):607–621. https://doi.org/10.1021/ar300022h
9. Krug HF (2014) Nanosafety research–are we on the right track? Angew Chem Int Ed Engl 53(46):12304–12319. https://doi.org/10.1002/anie.201403367
10. Clogston JD et al (2024) Current state of nanomedicine drug products: an industry perspective. J Pharm Sci 113(12):3395–3405. https://doi.org/10.1016/j.xphs.2024.09.005
11. ki te Pirimia MPM (2021) Nanotechnology regulation in Aotearoa New Zealand: current developments and comparison with overseas regulations, pp 1–46
12. Maurer-Jones MA, Gunsolus IL, Murphy CJ, Haynes CL (2013) Toxicity of engineered nanoparticles in the environment. Anal Chem 85(6):3036–3049. https://doi.org/10.1021/ac303636s
13. Gottschalk F, Nowack B (2011) The release of engineered nanomaterials to the environment. J Environ Monit 13(5):1145–1155. https://doi.org/10.1039/c0em00547a
14. Bundschuh M et al (2018) Nanoparticles in the environment: Where do we come from, where do we go to? Environ Sci Eur 30(1):6

15. Muazu RI, Rothman R, Maltby L (2021) Integrating life cycle assessment and environmental risk assessment: a critical review. J Clean Prod 293:126120. https://doi.org/10.1016/j.jclepro.2021.126120
16. Fadeel B, Pietroiusti A, Shvedova AA (2017) Adverse effects of engineered nanomaterials: exposure, toxicology, and impact on human health. Academic Press
17. Hristozov DR, Gottardo S, Critto A, Marcomini A (2012) Risk assessment of engineered nanomaterials: a review of available data and approaches from a regulatory perspective. Nanotoxicology 6(8):880–898. https://doi.org/10.3109/17435390.2011.626534
18. George S et al (2011) Use of a high-throughput screening approach coupled with in vivo zebrafish embryo screening to develop hazard ranking for engineered nanomaterials. ACS Nano 5(3):1805–1817. https://doi.org/10.1021/nn102734s
19. Wick P et al (2010) Barrier capacity of human placenta for nanosized materials. Environ Health Perspect 118(3):432–436. https://doi.org/10.1289/ehp.0901200
20. Schulte PA et al (2014) Occupational safety and health criteria for responsible development of nanotechnology. J Nanopart Res 16(1):2153. https://doi.org/10.1007/s11051-013-2153-9
21. Pietroiusti A (2012) Health implications of engineered nanomaterials. Nanoscale 4(4):1231–1247. https://doi.org/10.1039/c2nr11688j
22. 欧州医薬品庁. Reflection paper on nanotechnology-based medicinal products for human use. http://www.ema.europa.eu/docs/en_GB/document_library/Regulatory_and_procedural_guideline/2010/01/WC500069728.pdf

www.ingramcontent.com/pod-product-compliance
Ingram Content Group UK Ltd.
Pitfield, Milton Keynes, MK11 3LW, UK
UKHW020435270926
475829UK00013B/22
* 9 7 8 9 8 1 9 2 1 6 8 9 5 *